W9-AXU-296

ELECTRIC MOTOR CONTROL FUNDAMENTALS

ELECTRIC MOTOR CONTROL FUNDAMENTALS

Third Edition

Paul Pratt

R. L. McINTYRE

Electrical Instructor
20 years
Local Electrical
Training Director
2 years

Former Assistant Director
National Joint Apprenticeship
and Training Committee
for the Electrical Industry
9 years

McGRAW-HILL BOOK COMPANY

New York
St. Louis
San Francisco
Düsseldorf
Johannesburg
Kuala Lumpur
London
Mexico
Montreal
New Delhi
Panama
Rio de Janeiro
Singapore
Sydney
Toronto

Library of Congress Cataloging in Publication Data
McIntyre, R. L.
Electric motor control fundamentals.

First published in 1960 under title: A-c motor-control fundamentals.
1. Electric controllers. 2. Electric motors, Alternating current. 3. Automatic control.
I. Title.
TK2851.M25 1974 621.46′2 73-12983
ISBN 0-07-045103-6

ELECTRIC MOTOR CONTROL FUNDAMENTALS

9 0 DODO 8 3 2 1 0

The editors for this book were Alan W. Lowe and Cynthia Newby, the designer was Marsha Cohen, and its production was supervised by Patricia Ollague.
It was set in Times Roman by Monotype Composition Company, Inc.

CONTENTS

PREFACE

The purpose of this book is to present the technical subject of electric motor controls and their application in language as nontechnical as possible. The book is designed for use as a text in apprentice and industrial training programs and courses in technical, vocational, and trade schools; it should also be useful to industrial technicians, servicemen, and others.

The third edition of this book is the result of suggestions from many of the longtime users of the previous editions.

The chapters dealing with static control have been updated to incorporate changes in this type of control. Chapters 11 to 13 represent the latest information on digital static control. Most of the information in these chapters has been adapted from manufacturers' training programs. Chapter 15, dealing with shift registers and counters, has been expanded.

Every effort has been made to preserve the original and rather successful treatment of magnetic control while bringing the total content of the book up to date in every respect.

To make use of this text, the reader needs a working knowledge of basic electrical theory; but the mathematics requirement has been kept to a minimum. Where new terms are used for the first time, they are fully explained. Most texts on this subject approach control circuits from the standpoint of memorizing basic circuits, which leaves the student handicapped because of the endless variations in circuits in actual practice. This text shows how to develop circuits step by step to perform any desired function. In order to build a background for circuit development, the text presents a discussion of the possible functions of control and the components used in control circuits. The sections on components discuss the construction and operation of each general type rather than any one manufacturer's product.

The subject of motors and their characteristics has been well presented in several other books; therefore only a limited amount of material on motors as such has been included. (Students who are interested in other than digital electric control of motors, are referred to "Electronics in Industry," by Chute, published by McGraw-Hill, chapters 15 and 21.) There are sections devoted to the maintenance and troubleshooting of controls and an introduction to static control. A detailed study of three systems of static control is presented along with the appropriate symbols and circuit diagrams.

The beginning student should study the material in the order in which it is presented. The sections preceding circuit development are necessary if one is to apply the technique used. It is recommended that the student supplement his study of components with catalogs, published by the various component manufacturers. This will give him a broad understanding of the different applications of the principles. While studying the section on circuit analysis, the student should obtain additional circuits, preferably for equipment with which he is familiar, and practice the principles set forth in the text. Each chapter is followed by a brief summary and a set of review questions as instructional aids.

The book presents a system of study, application, and analysis of control circuits which has been applied by the author with gratifying results, both on the job and in class for the past 30 years.

R. L. McIntyre

1
FUNDAMENTALS OF CONTROL

Since the advent of mass production, the machine has become a vital part of our economy. In the beginning machines were operated chiefly by hand and powered from a common line shaft. This line shaft was driven by a large motor that ran continuously and was connected to the individual machine by a belt when needed. It should not be hard to see that this type of power did not lend itself to quantity production.

With the demand for more and more production, the machine took on a new look. Down came the line shaft, and the electric motor went into the individual machine. This change allowed more frequent and more rapid starts, stops, and reversals of the machine. A small machine could have a small high-speed motor, while next to it a large machine could have a large constant- or variable-speed motor. In other words, the machine shop or factory became more flexible. Once the drive motor was put on the machine, with the sole function of operating one piece of equipment, it was possible to introduce some automatic operations.

Today in our industrial plants, more and more machines are being made fully automatic. The operator merely sets up the original process, and most or all operations are carried out automatically. The automatic operation of a machine is wholly dependent upon motor and machine control. Sometimes this control is entirely electrical and sometimes a combination of electrical and mechanical control is used. The same basic principles apply, however.

A modern machine consists of three separate divisions which need to be considered. First is the machine itself, which is designed to do a specific job or type of job. Second is the motor, which is selected according to the requirements of the machine as to load, duty cycle, and type of operation. Third, and of chief concern in this book, is the control system. The control-system design is dictated by the operating requirements of the motor and the machine. If the machine needs only to start, run for some time, and stop, then the only control needed would be a simple toggle switch. If, however, the machine needs to start, perform several automatic operations, stop for a few seconds, and then repeat the cycle, it will require several integrated units of control.

It is the intent of this book to present the basic principles and components of control and then show how they are put together to make a control system.

1·1 MEANING OF CONTROL

"What is motor control?" This is a question that has no simple answer. It is not, however, the mysterious, complicated subject that some people believe it to be.

The word *control* means to govern or regulate, so it must follow that when we speak of motor or machine control, we are talking about governing or regulating the functions of a motor or a machine. Applied to motors, controls perform several functions, such as starting, acceleration, speed, power, protection, reversing, and stopping.

Any piece of equipment used to regulate or govern the functions of a machine or motor is called a *control component.* Each component will be taken up in a separate section of this book.

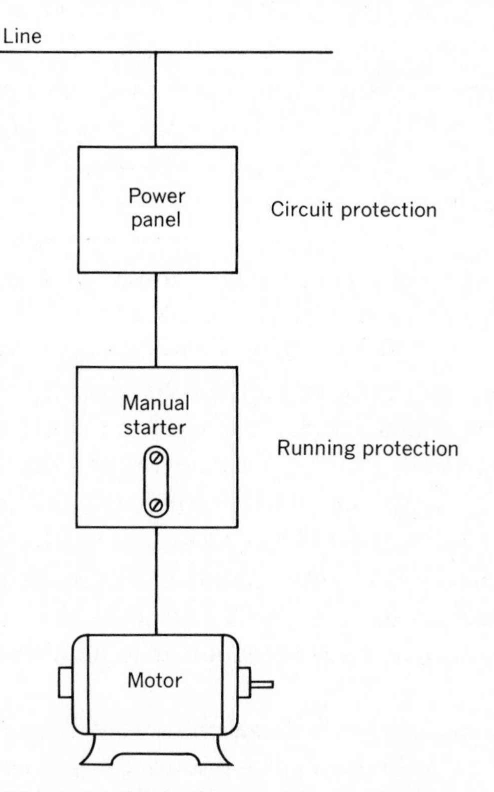

Fig. 1·1 Manual control for a motor.

An *electric controller* is a device or group of devices that controls or regulates the functions of a motor or machine in a predetermined manner or sequence.

1·2 MANUAL CONTROL

A *manual controller* is one having its operations controlled or performed by hand at the location of the controller (Fig. 1·1). Perhaps the most popular single type in this category is the manual full-voltage motor starter in the smaller sizes. This starter is used frequently where the only control function needed is to start and stop the motor. Probably the chief reason for the popularity of this unit is the fact that its cost is only about one-half that of an equivalent magnetic starter. The manual starter generally gives overload protection (Sec. 2·7) and low-voltage release (Sec. 2·12), but does not give low-voltage protection.

Manual control which provides the same functions as those achieved by the manual full-voltage motor starter can be had by the use of a switch with fusing of the delayed-action type, which will provide overload protection for the motor.

Examples of this type of control are very common in small metalworking and woodworking shops, which use small drill presses and lathes and pipe-threading machines. Another good example is the exhaust fan generally found in machine shops and other industrial operations. In this installation the operator or maintenance man generally pushes the START button for the fan in the morning when the plant opens, and it continues to run throughout the day. In the evening, or when the plant is shut down, the operator then pushes the STOP button, and the fan shuts down until needed again. The welding machines of the motor-generator type are a very common example of this kind of control and should be familiar to most students of motor control.

The compensator, or manual reduced-voltage starter, is used extensively to control polyphase squirrel-cage motors where reduced-voltage starting is required and the only control functions required are start and stop. The compensator gives overload protection, low-voltage release, and low-voltage protection. The compensator type of starter is quite frequently used in conjunction with a drum controller on wound-rotor motors (Fig. 1·2). This combination gives full manual control of start, stop, speed, and direction of rotation.

The compensator, being a reduced-voltage starter, is generally found only on the larger horsepower motors. A very common use for the compensator with the addition of a drum controller is found in the operation of many centrifugal-type air-conditioning compressors. The reduced-voltage feature is used to enable the motor to overcome the inertia of the compressor during starting without undue current loads on the system. The drum controller, through its ability to regulate speed of a wound-rotor motor, provides a means of varying the capacity of the air-conditioning system, thus giving a flexibility that would not be possible with a constant-speed full-voltage installation.

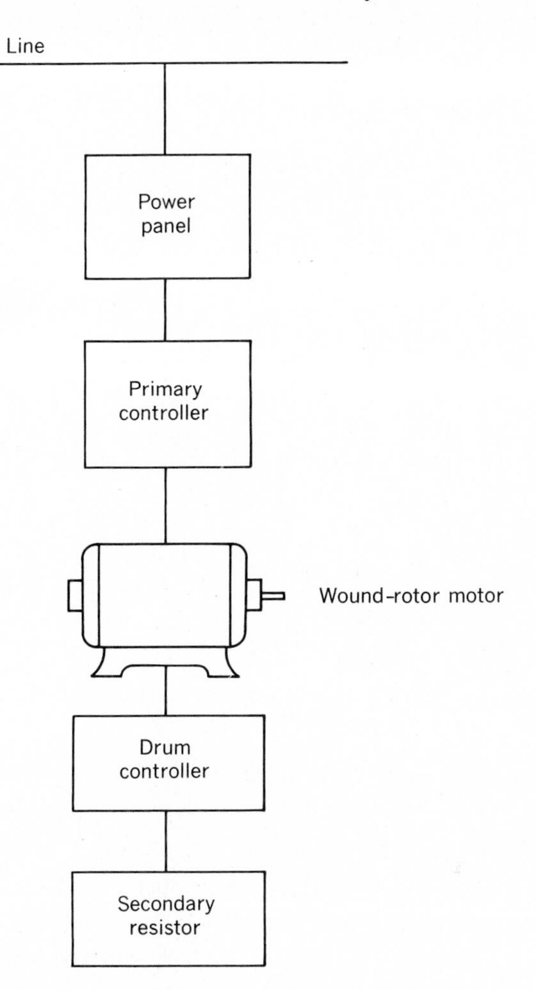

Fig. 1·2 Control for wound-rotor motor.

These are just a few of the manual controllers, but you should have little trouble classifying any unit of this type because it has no automatic functions of control. Manual control is characterized by the fact that the operator must move a switch or push a button to initiate any change in the condition of operation of the machine or equipment in question.

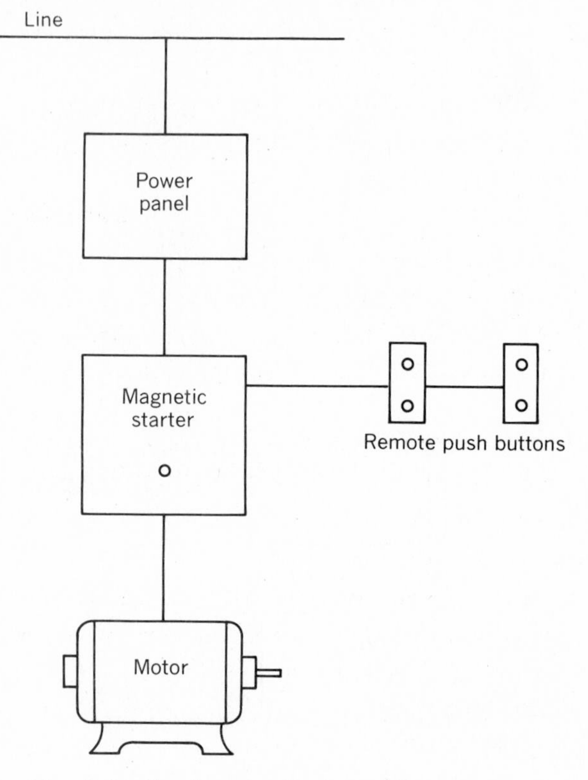

Fig. 1·3 Semiautomatic control for a motor.

1·3 SEMIAUTOMATIC CONTROL

Controllers that fall in this classification use a magnetic starter and one or more manual pilot devices such as push buttons, toggle switches, drum switches, and similar equipment (Fig. 1·3). Probably the most used of these is the push-button station because it is a compact and relatively inexpensive unit. Semiautomatic control is used mainly to give flexibility of control position to installations where manual control would otherwise be impractical.

The key to classification as a semiautomatic control system lies in the fact that all the pilot devices are manually operated and that the motor starter is the magnetic type. There are probably more machines operated by semiautomatic control than by either manual or automatic. This type of control requires the operator to initiate any change in the attitude or operating condition of the machine. Through the use of the magnetic starter, however, this change may be initiated from any convenient location, as contrasted to the manual-control requirement that the control point be at the starter.

1·4 AUTOMATIC CONTROL

An automatic controller is a magnetic starter or contactor whose functions are controlled by one or more automatic pilot devices (Fig. 1·4). The initial start may be automatic, but usually it is a manual operation, activated by a push-button station or switch.

In some cases there may be a combination of manual and automatic pilot devices in a control circuit. If the circuit contains one or more automatic devices, it would be classed as an automatic control. Consider, for example, a tank that must be kept filled with water between definite limits and a pump to replace the water as it is needed. If we equip the pump motor with a manual starter, and station a man at the pump to turn it on and off as needed, we have manual control. Now let us replace the manual starter with a magnetic starter and put a push-button station at the foreman's desk. If we ring a bell to let him know when the water is low and again when it is high, he can do other work and just push the proper button when the bell rings. This would be semiautomatic control. Now suppose we install a float switch that will close the circuit when the water reaches a predetermined low level and open it when it reaches a predetermined high level. When the water gets low, the float switch will close the circuit and start the motor. The motor will now run until the water reaches the high level, at which time the float switch will open the circuit and stop the motor. This would be automatic control.

Even though the automatic system would cost more to install than the other two, it requires no attention by an operator, thus saving the cost of his labor. This may well make it the cheapest method to use. The automatic control would be more accurate

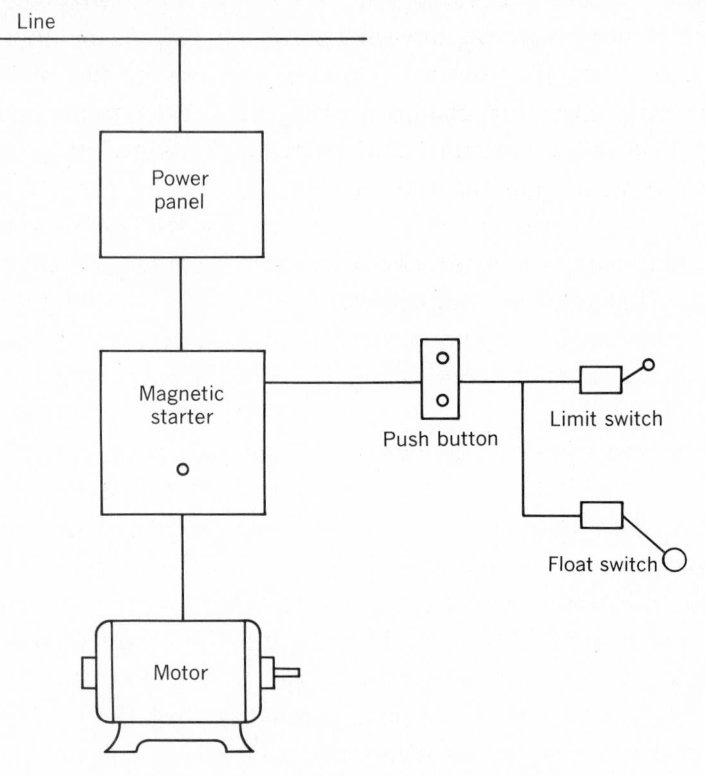

Fig. 1·4 Automatic control for a motor.

because there is no delay between the water reaching the desired level and the closing or opening of the control circuit.

Automatic control systems are to be found in almost all machine-tool installations. Tools such as drill presses, milling machines, shapers, turret lathes, precision grinders, and almost any other type of machine in common use may, through the use of limit switches and other automatic pilot devices, perform their operations more efficiently and rapidly by the use of automatic control systems.

Summary

The basic difference in manual, semiautomatic, and automatic control lies in the flexibility of the system. With manual control the operator must go to the physical location of the starter to make any change in the operation of the machine. With semiautomatic control the operator may have his control point at any convenient location so that he might initiate changes in the operation from the most desirable position. With automatic control the necessity of the operator initiating the necessary changes has been eliminated for each automatic operation that is included in the control system.

2
CONTROL OF MOTOR STARTING

Control circuits and equipment can perform varied functions. These can be grouped into 11 general types according to the effect they have on the motor to be controlled. Each of the general types can be broken down into endless variations, but they each stem from a few basic principles which, if understood, are the key to control work. It is the aim of this chapter to present these principles in as nontechnical a language as possible.

2·1 MOTOR STARTING

There are several general factors to be considered in the selection of motor-starting equipment. The most obvious of these are the current, voltage, and frequency of the motor and control circuits. Motors require protection according to the type of service, the type of motor, and the control functions that will be needed.

Whether to use a full-voltage starter or a reduced-voltage

starter may depend on the current-carrying capacity of the plant wiring and the power company lines and rates. Other factors, such as the need for jogging or inching, acceleration control, or the type of motor to be used, will also affect this selection.

Full-voltage Starting. The requirement of this type of starting is simply that the motor leads and line leads be connected (Fig. 2·1). This could be accomplished merely by using a knifeblade

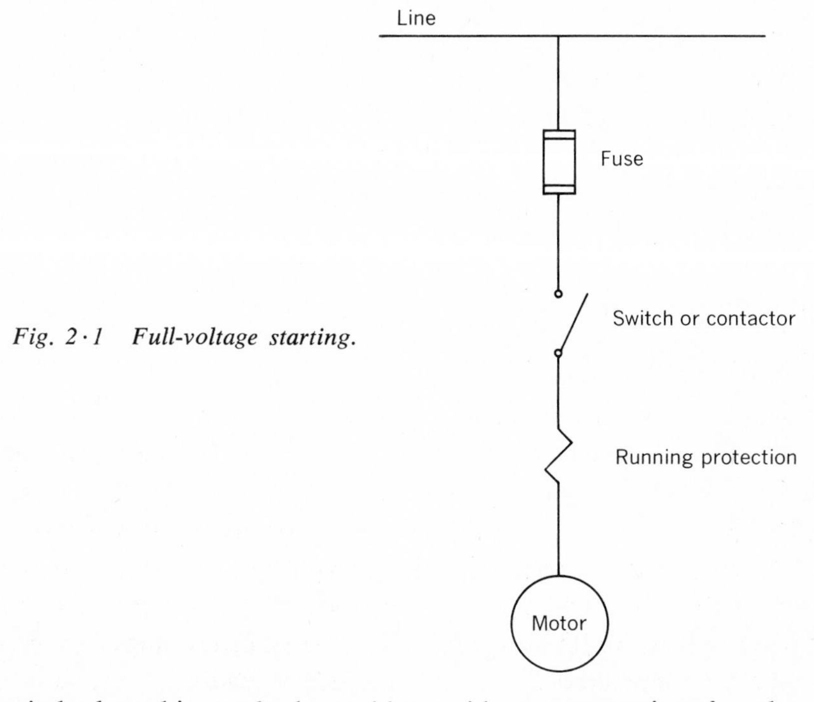

Fig. 2·1 Full-voltage starting.

switch, but this method would provide no protection for the motor except the circuit fusing.

For small fractional-horsepower motors on low-amperage circuits, the simple switch may be satisfactory and is used frequently. Many appliances use nothing more than the cord and plug as the disconnecting means, with a small toggle switch to start and stop the motor. Because the motor is not disconnected from the line on power failure, this type of starting control can be an advantage for fans and other devices that otherwise would have to be restarted.

For motors up to 10 horsepower (hp) and not over 600 volts, the manual across-the-line motor starter may be used to give manual control. Most of these units give overload protection and undervoltage release.

The most popular starter for motors up to 600 hp and 600 volts or less is the magnetic across-the-line starter. This starter, combined with pilot devices, can give full protection to the motor and fully automatic operation.

The vast majority of motors today are built to withstand the surge of current that occurs when they are suddenly thrown across the line. Not all our plant circuits can stand this surge, however, nor can all the power company equipment. When a large motor starts at full voltage, it may cause a voltage drop large enough to drop out other control equipment. Should the voltage drop be serious enough, it might even cause a dimming of lights in other buildings.

In most industrial installations the utility company penalizes, in the form of higher rates, for surges of excessive current on the line by the use of a demand meter. The demand meter registers the maximum average demand of power on the line for a given period of time, generally a 15-minute period. This factor should always be taken into consideration when the method of starting large motors is being decided. The extra cost of power because of the excessive demand for starting large motors across the line may well exceed the cost of reduced voltage starting, which would materially reduce the demand reading.

When considering using full-voltage starting, always check the building wiring and distribution system capacity. Should the wiring be inadequate, either it must be increased in capacity or reduced-voltage starting must be used.

Reduced-voltage Starting. Whenever the starting of a motor at full voltage would cause serious voltage dips on the power company lines or the plant wiring, reduced-voltage starting becomes almost a necessity (Fig. 2·2). There are, however, other reasons for using this type of control. The effect on the equipment must be taken into account in the selection of motor

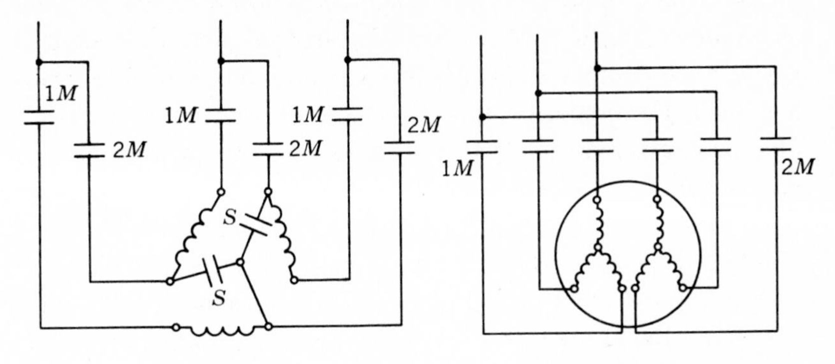

Wye - delta

Contactor sequence			
Contactor	Start	Transition	Run
1*M*	X	X	X
2*M*			X
S	X		

Part winding

Contactor sequence		
Contactor	Start	Run
1*M*	X	X
2*M*		X

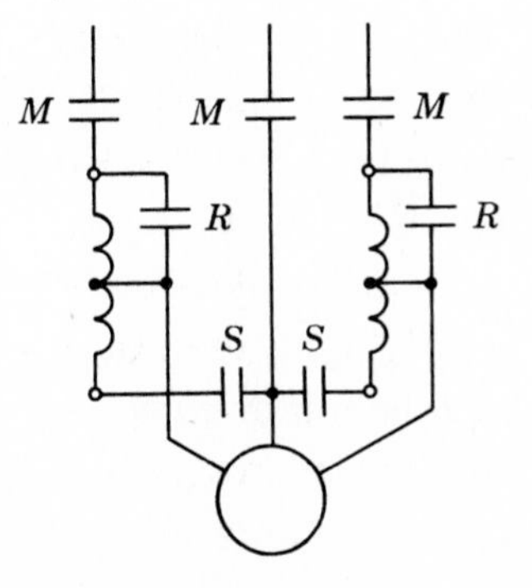

2 coil auto transformer

Contactor sequence			
Contactor	Start	Transition	Run
M	X	X	X
S	X		
R			X

Fig. 2·2 Reduced-voltage starting.

starters. When a large motor is started across the line, it puts a tremendous strain or shock on such things as gears, fan blades, pulleys, and couplings. Where the load is heavy and it is hard to bring it up to speed, reduced-voltage starting may be necessary. Belt drives on heavy loads are apt to have excessive slippage unless the torque is applied slowly and evenly until full speed is reached.

Reduced-voltage starting is accomplished by the use of resistors, autotransformers, or reactors in order to reduce the line voltage to the desired value during starting. Many electronic devices are also used to reduce starting current and starting torque. Regardless of the means of reducing the voltage, it must be designed to fit the particular motor to be started. It is not within the scope of this book to go into the design of reduced-voltage starters, but rather to point out the need for proper selection according to the specifications furnished by the motor manufacturer.

Another method of achieving reduced-voltage starting is to use a wound-rotor motor with secondary control (Fig. 1·2). With this system, a full-voltage starter is used on the primary or stator winding, and resistance grids are put in series with the secondary or rotor winding to reduce torque and starting current. The secondary control device shorts out the resistance grids as needed for acceleration, until at full speed all the resistance is shorted out and the motor runs as a squirrel-cage motor. The advantage of secondary control is that it gives speed control as well as reducing the starting current.

Regardless of the method used to provide reduced-voltage starting, it must be kept in mind that the starting torque of the motor is also reduced. If a motor is not capable of starting its load under across-the-line conditions, the application of reduced-voltage starting will only aggravate the situation because of the reduced starting torque. The torque of an induction motor is a function of the square of the rotor current, or approximately the square of the line current. If the starting voltage is reduced by 50 percent, the motor current will be reduced to 50 percent of normal, but the torque will be reduced to 25 percent of normal.

Some of the methods of obtaining reduced-voltage starting will result in very little or no acceleration under starting conditions. This requires that the total acceleration occur after full voltage has been applied. The starting current during reduced-voltage conditions will be somewhat less than across-the-line starting current. When full voltage is applied, however, the starting current will be approximately the same as it would have been if the motor had been placed across the line in the beginning. This type of starting is generally referred to as *increment starting* and is used generally to spread the rate of change of current demanded from the line over a longer period of time. Part-winding squirrel-cage motor starting and wye-delta starting of squirrel-cage motors generally fall in the increment-start category (Fig. 2·2).

Jogging and Inching. Printing presses, cranes, hoists, and similar equipment require that the motor be started repeatedly for short periods of time in order to bring some part of the machine into a given position. This process is known as *jogging* or *inching*. Even though these terms are often used interchangeably, there is a slight difference in their meaning. If the motor is started with full power in short jabs, it is jogging. If the motor is started at reduced speed so as to let the machine creep to the desired spot, then it is inching.

When jogging service is required, the starter must be derated. For instance, a size 3 starter rated at 30-hp 230-volt three-phase normal duty should be derated to 20 hp for jogging duty. The manufacturer's literature should be consulted for ratings of starters in jogging service. The standards shown in Tables 2·1 and 2·2 have been set up by NEMA (National Electrical Manufacturers Association).

2·2 ACCELERATION CONTROL

Squirrel-cage motors do not generally lend themselves very well to speed or acceleration control. There are special types of squirrel-cage motors designed for two-, three-, or four-speed applica-

*Table 2·1 Ratings for polyphase single-speed full-voltage magnetic controllers for nonplugging and nonjogging duty**

Size of controller	Continuous current rating, amp	Three-phase horsepower at 200 volts	230 volts	460/475 volts	Service-limit current rating, amp
00	9	1½	1½	2	11
0	18	3	3	5	21
1	27	7½	7½	10	32
2	45	10	15	25	52
3	90	25	30	50	104
4	135	40	50	100	156
5	270	75	100	200	311
6	540	150	200	400	621
7	810	......	300	600	932
8	1,215	......	450	900	1,400
9	2,250	......	800	1,600	2,590

* Standard ICS 2-321B.20 (1971).

*Table 2·2 Ratings for polyphase single-speed full-voltage magnetic controllers for plug-stop, plug-reverse, or jogging duty**

Size of controller	Continuous current rating, amp	Three-phase horsepower at 200 volts	230 volts	460/475 volts	Service-limit current rating, amp
0	18	1½	1½	2	21
1	27	3	3	5	32
2	45	7½	10	15	52
3	90	15	20	30	104
4	135	25	30	60	156
5	270	60	75	150	311
6	540	125	150	300	621

* Standard ICS 2-321B.21 (1971).

tions. This type of multispeed squirrel-cage motor does not have a true variable speed but rather has several definite speeds which may be used as desired or in steps. When adjustable speed is required, the wound-rotor motor with secondary control or the adjustable-speed a-c commutator motor is probably the most practical.

Manual control of acceleration or speed may be accomplished

with multispeed squirrel-cage motors by having the operator close the proper contactor as determined by the load or speed requirements. With wound-rotor motors, the secondary or drum controller is moved as needed to give the desired speed.

Automatic control of acceleration may be accomplished by several methods. Probably the simplest is *definite-time control.* With this method, a time-delay relay is used for each step or speed. When the motor is started in its lowest speed, the first time-delay relay is energized. When this relay times out, it energizes the second contactor, increasing the speed to its second step. This process may be carried through as many steps as necessary to give the speed and acceleration desired. The chief disadvantage of this method is that it is not affected by the conditions of the machine, its load, or the motor current.

For machines or equipment that cannot stand full torque until the load has reached a given speed, the system called *current-limit control* should be used to accelerate the motor. In this system each step is brought in by a current relay which will not close the circuit for its speed until the current has dropped to a safe value. The current, of course, will not drop until the motor and the load are running at nearly the same speed. This system is very well suited for belt or gear drives with heavy high-inertia loads. This method of control must be designed for the particular application, and the relays must be set for the particular machine and its requirements. For this reason, they are not available as standard stock controllers.

Another system of acceleration control is *slip-frequency control.* This system is used on wound-rotor motors and is also used to energize the field of synchronous motors.

Because the secondary voltage and frequency of wound-rotor motors is inversely proportional to the speed, a frequency-sensitive relay may be used to energize each progressive step or speed. One disadvantage to slip-frequency control is that it must be started on the first, or lowest, speed.

Several features must be kept in mind when selecting the method of acceleration control. Manual control is sensitive only to the operator's reaction. Definite-time control is sensitive only to the lapse of time. Current-limit control is sensitive to the load on the motor. Slip-frequency control is sensitive to the speed of the motor.

It is quite possible that under certain specific applications a combination of any two or more of the above-mentioned control systems might be used to enable the motor to be sensitive to more than one of these factors. Such a controller would of necessity be custom-built to meet the specific requirements of a given installation.

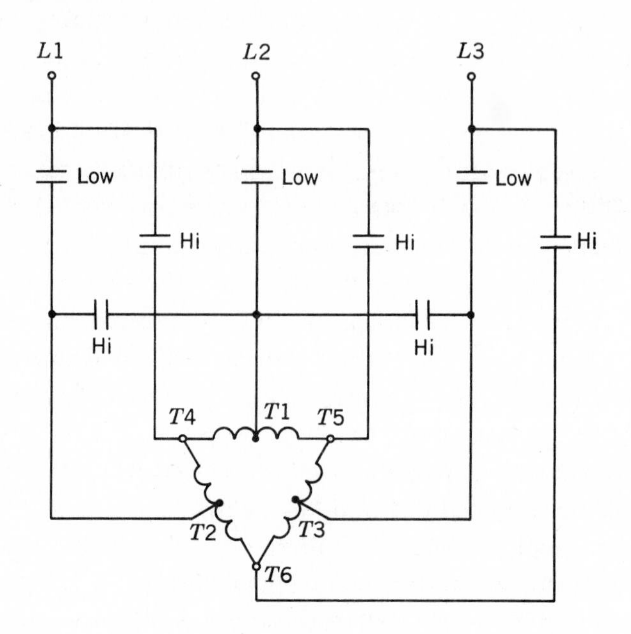

Fig. 2·3 Two-speed squirrel-cage motor and controller.

2·3 STARTING MULTISPEED SQUIRREL–CAGE MOTORS

Single-speed squirrel-cage motors are generally started by the use of across-the-line magnetic starters. A multispeed squirrel-cage motor, however, requires a controller that is built for its particular windings.

Two-speed motors may either have two separate stator windings or be of the consequent-pole type (Fig. 2·3) which have only one stator winding. The two speeds are obtained with the consequent-pole motor by regrouping the coils to give a different number of effective poles in the stator. It is characteristic of this type of motor that it gives a 2 to 1 speed ratio. Three-speed motors usually have one winding for one speed and a second winding that is regrouped to give the other two speeds. Four-speed motors usually are wound with two windings which are regrouped to give two speeds each. The number of contactors, the order in which they close, and the number and types of overload units required depend on the method of obtaining the various speeds.

2·4 STARTING WOUND–ROTOR MOTORS

The wound-rotor motor is essentially the same as a squirrel-cage motor except that it has definite windings instead of short-circuited bars on the rotor. By the introduction of resistance in series with the rotor windings, with a drum controller or with contactors, the speed may be controlled in any number of steps. The most common method of starting wound-rotor motors is by the use of a manual or magnetic full-voltage starter on the primary, interlocked to the secondary controller (Fig. 2·4). The secondary controller may be a manual drum controller, a motor-driven drum controller, a liquid rheostat, or a magnetic contractor designed for secondary control. The secondary controller may be for starting service only and have only two or three steps, or it may be for speed control as well and have any number of steps.

It is necessary that there be an interlock between the primary and secondary controllers that will prevent the motor from starting without all the resistance in the secondary circuit unless the motor and the machine are designed to start at any speed.

2·5 STARTING SYNCHRONOUS MOTORS

The synchronous motor starts as a squirrel-cage motor with a resistor connected across the field winding to dissipate the power generated in this winding (Fig. 2·5) Usually the stator controller is a reduced-voltage starter with the addition of a slip-frequency or field-application relay to apply d-c voltage to the rotor at about 95 percent of synchronous speed. The slip-frequency relay must also remove field excitation and connect the field resistor if the motor should pull out of step. If the excitation is not removed, the stator winding will be subject to damaging

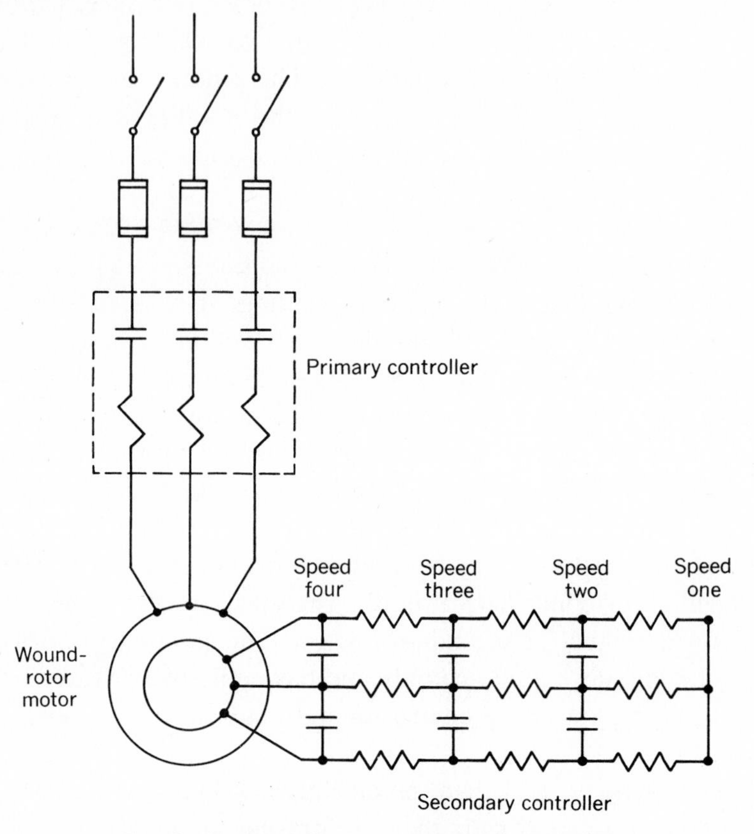

Fig. 2 · 4 Wound-rotor motor and controller.

current. The synchronous motor should be provided with an incomplete-sequence relay to protect the starting winding if the starting sequence should not be completed. Provision must be also made to adjust the field excitation.

While the above description of the starting of a synchronous motor may seem to be oversimplified, it is intended to be general in nature and to apply to all synchronous motors. For a specific application of a definite type of synchronous motor, the manufacturer's literature on the individual motor should be consulted. Many synchronous motors are designed for specific applications and vary somewhat from this general outline for starting in that they require additional steps or equipment.

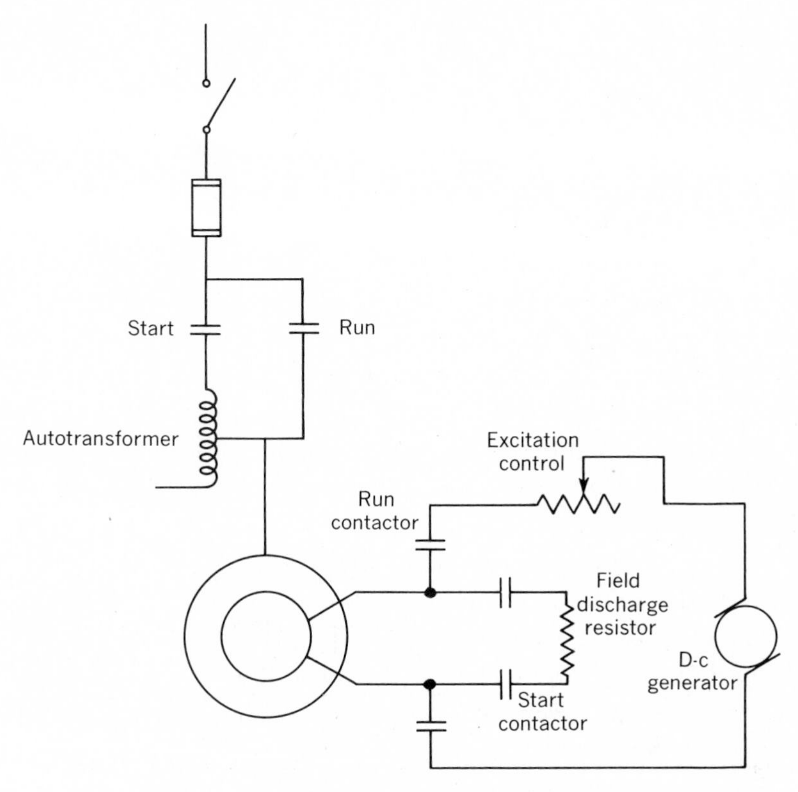

Fig. 2 · 5 Synchronous motor and controller.

2·6 SELECTION OF STARTING CONTROLLERS

There are several points that must be considered when selecting starting controllers. Listed below are some questions that should be asked whenever selection of a controller is necessary:

1. Is it designed for the type of motor to be used?

2. Does the motor require reduced-voltage starting?
3. Is speed control needed?
4. Does the controller offer all the types of protection that will be needed?
5. Are the line and control voltages and frequency correct?

Analyze the needs of the machine and the motor before selecting any controller, and avoid costly mistakes.

2·7 OVERLOAD PROTECTION

Overload of a motor may be mechanical or electrical in origin; therefore, the overload protection must be sensitive to either. The current that a motor draws from the line is directly proportional to the load on the motor, so if this current is used to activate the overload protection device, the machine as well as the motor will be protected.

Overload protection is achieved in almost all controllers by placing heating elements in series with the motor leads on multiphase motors (Fig. 2·6). These heater elements activate electrical contacts, which open the coil circuit when used on magnetic controllers. When used on manual starters or controllers, the heating elements release a mechanical trip to drop out the line contacts. Older controllers use two overloads, while newer units are required to have three units.

The overload relay is sensitive to the percentage of overload; therefore a small overload will take some time to trip the relay, whereas a heavy overload will cause an almost instantaneous opening of the circuit. The overload relay does not give short-circuit protection, however. It is quite possible that under short-circuit conditions the relay might hold long enough to allow considerable damage to the motor and other equipment.

It would be impossible to overstress the necessity of proper selection of overload protective equipment. The manufacturer's rating of running current for a specific motor should be adhered to in the selection of the heating elements for overload relays. The all too frequent practice of increasing the size or rating of the heater element beyond the value called for is probably the greatest single cause of motor failure in industrial plants today. When a motor is tripping its overload units, a careful check of the actual current drawn should be made in order to determine whether the fault lies in the overload protective device or in the motor itself actually drawing excessive current. Should the motor be found to be drawing excessive current, then it must be determined whether this is caused by mechanical overload or by defective windings within the motor itself. Many times, today's heavy production schedules require that the opera-

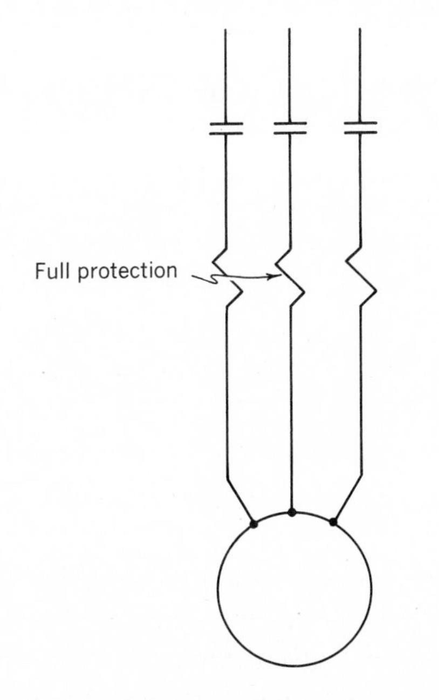

Fig. 2 · 6 Overload heating elements for multiphase motors.

tor demand more from his machine than its motor is capable of producing. The practice of increasing the allowable current through the overload units will only hasten the time when a shutdown of the equipment is necessary in order to rewind or replace the motor.

When one phase of a motor circuit fails, the motor is subject to what is commonly called *single phasing*. This condition causes an excessive current to flow in the remaining motor windings and leads. In most cases, this excessive current will cause the overload units to trip, thus disconnecting the motor from the line and preventing a burnout of its windings. Under certain particular conditions of load characteristic of the individual motor, it is possible for the motor to run single phase and burn out its windings, even though there are two overload units in the control device. Modern devices are required to have three overload units.

2 · 8 SHORT–CIRCUIT PROTECTION

Squirrel-cage motors draw up to 600 percent of full-load current when starting, and some designs draw up to 800 percent or more. The overload relay is designed to pass these large currents for a short period of time. The circuit feeding a motor must have either a fused disconnect or a circuit breaker ahead of the motor to give

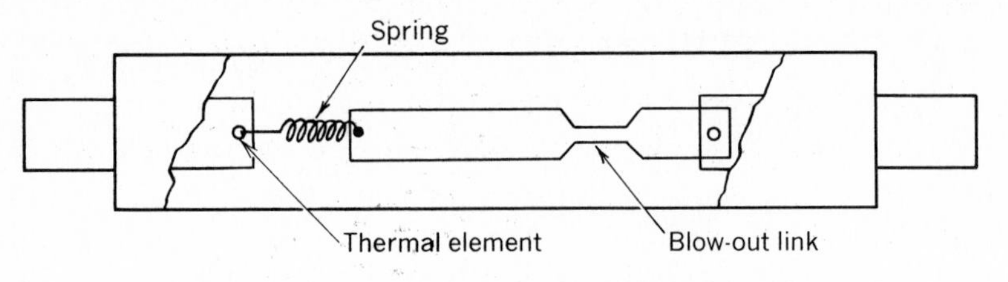

Fig. 2 · 7 Dual-element cartridge-type fuse.

short-circuit protection since the motor controller is not designed to interrupt short-circuit currents greater than 10 times its rating. Class J fuses (or current limiters of the same characteristics) provide the best limiting of peak short-circuit current.

The dual-element fuse (Fig. 2·7) is so constructed that one of its elements consists of a fuse link. This link will open very rapidly under short-circuit conditions. The second element in this type of fuse consists of a thermal element which has considerable time lag in the process of breaking the circuit. The net result of using the dual-element fuse on motor circuits is to give short-circuit protection through the fuse link and yet a degree of overload protection in the thermal element. This type of fuse is used extensively with small fractional-horsepower motors which have built-in thermal protectors.

The use of thermal-magnetic circuit breakers for short-circuit protection offers a degree of time lag for starting loads in that such breakers have a thermal device which requires some time lag to open on currents less than 20 times rating. This time lag is inversely proportional to the amount of current. The larger the overload, the shorter the time required to open the circuit. Such breakers also have a magnetic trip element for instantaneous trip on currents greater than about 25 times rating.

2·9 LIMIT PROTECTION

Limit protection, as its name implies, must limit some function of the machine or its driving motor. The most common type of limit control is used to limit the travel of a cutting tool or table or other part of a machine tool. When the cutter reaches a predetermined setting, it will activate a limit switch, causing the motor to reverse and the machine to return to the other extreme of travel. There are other types of limit protection, such as limits for over- or underspeed of the driven machine. There are also limit controls which do not reactivate the machinery at all but merely stop the motor until corrections have been made by the operator.

This type of protection is accomplished by the use of limit switches, which will be discussed fully later. Basically, a limit switch is merely a switch with a mechanical bumper or arm that will allow some action of the machine to throw the switch mechanism. Limit switches are one of the most frequently used control devices on machines today.

2·10 INTERLOCKING PROTECTION

Interlocking is preventing one motor from running until some other motor or sequence has been activated. A good example of this is found in air conditioning. If the compressor were to run without the cooling-tower pump running, the compressor-head pressure would rise dangerously, and either the high-limit pressure switch would shut the machine down or the compressor would be damaged. To prevent this from occurring, the compressor should be interlocked to the pump so that it cannot start until the pump is running.

Interlocking may be electrical or mechanical or a combination of both. Reversing starters built with both starters in the same box generally have a mechanical interlock and sometimes have an electrical interlock. When the two units to be interlocked are in two separate boxes (Fig. 2·8), electrical interlocking is a necessity. Electrical interlocking is accomplished by connecting an auxiliary contact on one starter in series with the coil circuit of the second starter.

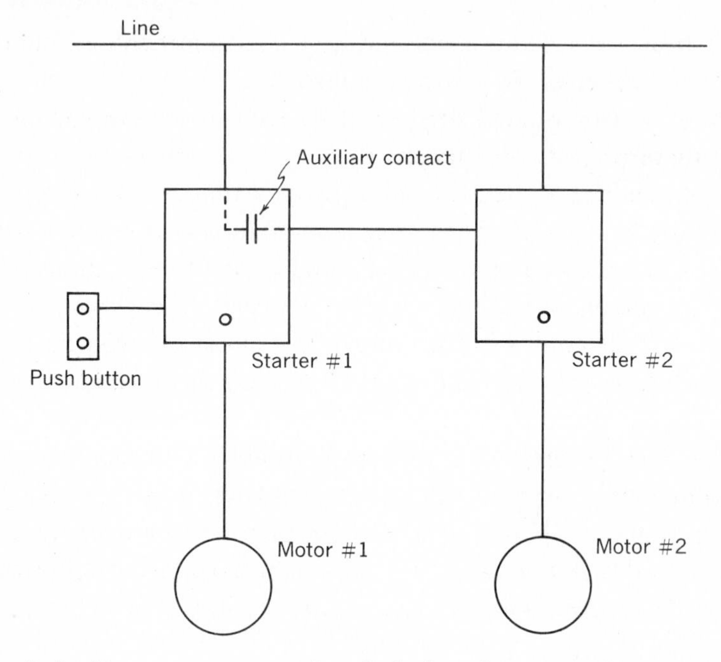

Fig. 2·8 Two motor starters interlocked so that motor 1 must run first and automatically start motor 2.

While we have discussed interlocking only in relation to motors, it is well to realize that interlocking is used in all phases of control wiring whether it is involved in the starting of a motor or in the closing of valves in a process-control installation. The use of interlocking control assures the proper sequence of operation for the entire control system.

Squirrel-cage motors do not lend themselves to any system of continually variable speed control, but rather may be had in two, three, or four speeds according to the design of the motor. These were discussed under acceleration control. The most versatile a-c motor as far as speed control is concerned is the wound-rotor motor, sometimes referred to as a *slip-ring* motor. By the use of secondary control, this type of motor can have as many steps of speed as are desirable. Methods of controlling wound-rotor motors will be discussed more fully when we take up their controllers.

Another type of a-c motor which gives excellent speed control is the a-c commutator type of motor.

There are four general types of speed control, depending on the requirements of the machine:

Constant Speed Control. Many machines require only a reduced speed for starting and then a constant speed for operation. This type of speed control may be accomplished by using a reduced-voltage starter on either a squirrel-cage, wound-rotor, or synchronous motor. It must be kept in mind, however, that reduced-voltage starting also invariably gives reduced starting torque.

Variable Speed Control. Variable speed is a requirement that a motor must operate at several different speeds at the selection of the operator. This type of control may best be accomplished by the use of a wound-rotor motor with a secondary controller or a commutator-type a-c motor. This type of control requires that a speed change be made under load.

Multispeed Control. This type of control differs only slightly from variable-speed control in that it usually does not require speed changes under load. The multispeed squirrel-cage motor is well suited for this type of service.

Predetermined Speed Control. With this type of control, the machine is accelerated through the necessary steps of speed to a preset operating speed. Both multispeed squirrel-cage and wound-rotor motors are suitable for this type of service.

Any of the above types of speed control may be wired so that the operator may vary the sequence of operation. Quite frequently, however, the control system compels the operator to start at a particular place in the sequence and follow it through without variation. When the control is of this type, it is known as *compelling sequence control.* This term also applies to control systems other than speed control and is dependent only on the requirement that it compels the operator to follow a set sequence of operation.

2·12 UNDERVOLTAGE PROTECTION AND RELEASE

The line voltage supplying motor circuits may drop to dangerously low values or may be shut off at almost any time. When the voltage is too low, severe damage may be done to the motor windings if they are allowed to remain on the line. While some large motors employ a special voltage relay to disconnect the motor under low-voltage conditions, most smaller motors depend on the overload units to open the starter contacts.

If the control circuit is such that the motor will restart when the power is restored to its proper value, the protection is referred to as *undervoltage release.* The use of maintained-contact pilot devices on magnetic starters gives this type of protection.

If the protection used requires that the motor be restarted manually, then the protection is referred to as *undervoltage protection.* The use of momentary-contact pilot devices on magnetic starters gives this type of protection.

Whether to use undervoltage protection or undervoltage release depends upon the requirements of the machine. Ventilating fans, unit heaters, and many other small units in a plant may operate more effectively with undervoltage release. This saves the necessity of having to restart them. In any machine where there is the slightest risk that the machine or operator might be injured by an unexpected start, undervoltage protection should by all means be used.

2·13 PHASE–FAILURE PROTECTION

When a three-phase motor has the current interrupted on one

phase, the condition is referred to as *single phasing*. Ordinarily the overload units will trip the starter and remove the motor from the line. There is, however, a condition of loading for each motor where it may quite possibly burn up without causing excessive current to flow through the overload units. This is generally about 65 percent of load for most squirrel-cage motors. For small motors the risk is generally considered too slight to warrant the cost of additional protection. For large motors a voltage relay is placed across each phase, and its contacts are connected in series with the holding coil of the starter. Failure of one phase will drop the starter out at once.

The use of three overload relay units on the starter gives what is generally considered adequate phase-failure protection for most motor installations up to 200 hp. Three overload relay units are now required by the national electrical code.

2·14 REVERSE–PHASE PROTECTION

Some machines could be severely damaged when the motor runs in reverse, as would occur with a reversal of phasing. While this is not a common type of protection, when it is needed it can prevent costly damage.

Reverse-phase protection can be accomplished by the use of a phase-sensitive relay with its contacts in series with the holding coil of the starter.

2·15 INCOMPLETE SEQUENCE PROTECTION

When reduced-voltage starting is used on a motor, there is a danger that the motor windings or the autotransformer or both might be damaged through prolonged operation at reduced voltage. To prevent this condition and to assure the completion of the starting cycle, a thermal relay is placed across the line during starting. This relay is so designed and connected that prolonged starting will cause the thermal unit on the relay to open its contacts and drop out the starter. This type of protection is also necessary on synchronous motor controllers.

Another method of obtaining incomplete sequence protection for starting of motors is by the use of the timing relay which will disconnect the motor if it has not completed its starting sequence in the predetermined length of time.

2·16 STOPPING THE MOTOR

There are several factors that must be considered in stopping a motor. On some machines all that is necessary is to break the motor leads and let the motor coast to rest. Not all machines can be allowed to coast, however. For instance, a crane or hoist not only must stop quickly, but also must hold heavy loads. Other machines, such as thread grinders, must stop very abruptly, but need not hold a load.

The method of stopping may be either manual or automatic. Automatic stopping is accomplished by the use of limit switches, float switches, or other automatic pilot devices. Manual stopping is controlled by push buttons, switches, or other manually operated pilot devices.

The most common method of stopping is merely to remove the motor from the line by breaking the circuit to the starter coil, if it is a magnetic starter, or by tripping the contacts of a manual starter with the STOP button.

For motors that must be stopped very quickly and accurately but which do not need to hold a load, probably the most widely used method is known as the *plugging stop*. This is accomplished by the use of an automatic plugging switch or a plugging push button in conjunction with a reversing starter.

With either of these units, the motor starter is dropped out and then momentarily energized in the reverse direction. The momentary reversal plugs the motor to an abrupt stop. This type of stopping will not do for cranes or hoists because it will not hold a load.

When dealing with equipment such as cranes and hoists, we must consider that the load has a tendency to turn the motor. This is known as an *overhauling* load. When a-c motors are used, they frequently are of the wound-rotor type and stopping is preceded by a slowing of the motor through one or more steps. This slowing helps to nullify the overhauling load. As the motor is dropped from the line, a mechanical brake (Fig. 2·9) is automatically applied which locks the motor shaft connected to the load.

When d-c motors are used, the overhauling load is slowed down by the use of dynamic (regenerative) braking; then a friction brake is applied.

Synchronous motors are sometimes stopped by the use of dynamic braking. This is accomplished by removing the line from the motor while placing a resistance across the motor leads, thus making an a-c generator out of the motor. The resistor presents a heavy load to the generator, causing it to come to a rapid stop. Care must be taken to use a resistor unit capable of dissipating the power generated while stopping the motor. It must also be noted that this type of stop cannot be used for frequent stopping because the resistor units must have time to cool between operations.

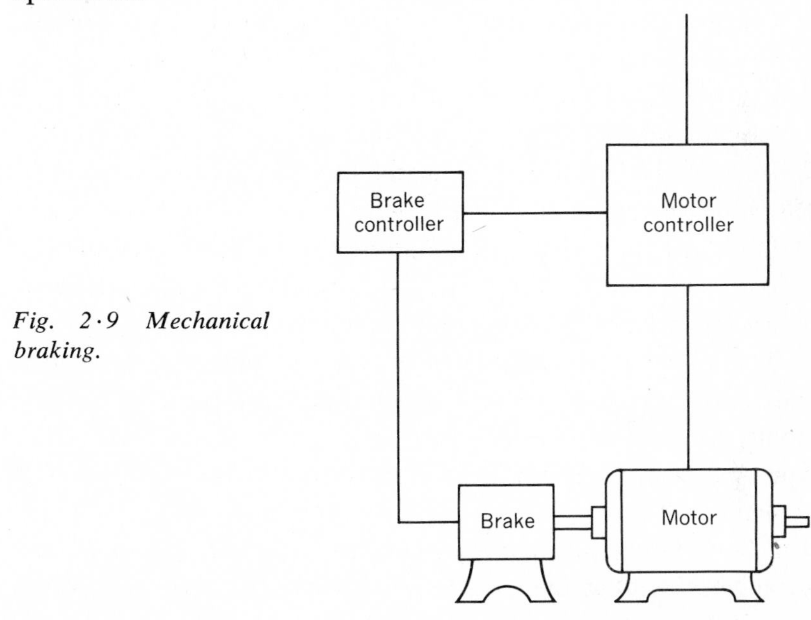

Fig. 2·9 Mechanical braking.

2·17 STARTING D-C MOTORS

Any device used to start a d-c motor over about ¼ hp must provide some means of limiting the starting current to approximately 150 percent of full-load value. An a-c motor offers a high impedance to the line which will limit the starting current. The d-c motor offers only the low resistance of the armature to limit inrush current until the motor begins to rotate.

Once rotation is started, the armature winding begins to cut the flux produced by the field, and a voltage is generated in the armature coil. The voltage generated in the armature coil is opposite in polarity to the applied voltage and is referred to as *counter electromotive force* (*cemf*). The value of the counter emf increases with speed, until at full speed it is 80 to 95 percent of the applied voltage.

Armature current is calculated by subtracting the counter emf from the applied voltage and dividing by the armature resistance. When the armature is at rest, the counter emf is at zero. Therefore, the armature current is equal to the applied voltage divided by the armature resistance, which is very low, generally 1 to 2 ohms or less.

If we assume an armature resistance of 0.85 ohm and an applied voltage of 110 volts, the initial inrush current would be 129.4 amp. At full speed the counter emf would be about 100 volts and would limit the current to 11.8 amp. The starting current should be limited to about 150 percent of full-load current by connecting a resistor of 650 ohms in series with the armature.

The series resistance must be removed in steps as the acceleration of the motor produces an ever-increasing counter emf and reduces the resistance required, until at rated speed all resistance is removed. A d-c motor develops its greatest power when the counter emf is at its maximum value.

When series or compound motors are used, the starting resistance is connected in series with the armature and series field. The shunt motor has no series field, and therefore the resistance is connected in series with the armature only (Fig. 2·10).

2·18 SPEED CONTROL OF D-C MOTORS

D-c motors are used chiefly because of their speed-control characteristics, which make them the best suited for many drive requirements.

When a d-c motor has its rated armature voltage and rated field voltage applied, it runs at its base speed. Speeds below base speed (underspeed) are achieved by maintaining the field voltage at rated value and reducing the armature voltage. Speeds above base speed (overspeed) are achieved by maintaining the armature voltage at rated value and reducing the field voltage.

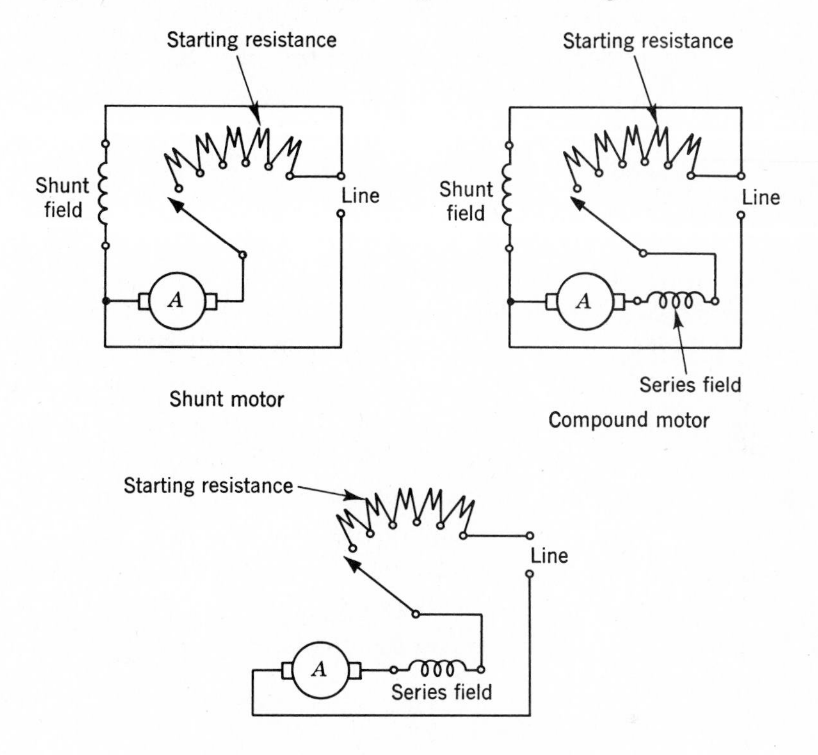

Fig. 2·10 Connections for d-c starting resistance.

The series motor's speed is controlled by the amount of resistance connected in series with the armature and series field. The resistors used for speed control must be rated for continuous duty rather than starting duty, since they are in the circuit whenever the motor is used at less than base speed.

The most common controller for speed control of series motors is the drum controller used with heavy-duty resistance grids.

Shunt and compound motors lend themselves well to applications where speed control is a major consideration. When the speed desired is over the base speed (overspeed control), resistance is added in series with the shunt field. When the speed desired is below the base speed (underspeed control), resistance is added in series with the armature.

The most popular manual controller for speed control of shunt and compound motors is the combination four-point starter and speed controller (Sec. 3·14).

Summary

A student beginning the study of motor control may feel that he will never learn all the functions that might be performed in the control of a motor or other device, and well he might. The advances made in this field are so rapid and far-reaching that new ones are developed almost daily. When analyzed completely, however, most of them are merely variations of the basic functions set forth in this chapter. It must be kept in mind that intelligent servicing, development, or installation of control equipment depends upon a thorough understanding of the requirements of the machine and the characteristics of the motor.

Review Questions

1. What is motor control?
2. What are the three basic types of motor control?
3. Name the two types of starting control.
4. How many types of protection are there for motors?
5. Do the running conditions of the motor affect the type of control to be used on a motor?

6. What is the difference between an automatic and a semi-automatic controller?
7. What factors need to be considered when selecting starting equipment?
8. What are the two basic methods of reduced-voltage starting?
9. How can the acceleration of motors be controlled automatically?
10. Reversing starters must be equipped with some form of .... to prevent both starters from closing at the same time.
11. When the motor is momentarily reversed to bring it to a stop, the function is called ....
12. Is the time it takes the overload relay to drop out the starter affected by the percentage of overload?
13. What is the difference between overload and short-circuit protection?
14. What is incomplete sequence protection, and on what two types of motors would it most likely be used?

3

CONTROL COMPONENTS

As soon as it has been decided what functions of control are needed for a machine, the components or devices to perform these functions must be selected. This selection should be made with care. For instance, if a float switch is needed and its duty cycle is only a few operations per day for a year or so, one of the cheaper competitive units might be satisfactory. If, however, the duty cycle is a few hundred operations per day on a permanent basis, then the best-quality unit available should be used. The small savings gained through the use of cheap components are usually soon offset by costly shutdowns due to failure of the components to function properly. In this chapter each of the basic types of control components will be discussed, how it works, both electrically and mechanically, and at least some of the functions that it can perform.

The student is strongly urged to obtain manufacturers' catalogs on control components as a further reference for use with this chapter. The more familiar the student becomes with the various manufacturers' equipment and the way it operates, the better prepared he will be to service it on the job.

3·1 SWITCHES AND BREAKERS

The one component common to all but the very smallest motors is a switch or breaker for disconnecting the motor. There are two types of switches in general use on motor circuits. The first of these is the isolation switch, which is rated only in voltage and amperage. This type of switch has no interruption-capacity rating and must not be opened under load. Quite often the switch used for this purpose is a nonfused type. Such switches are permitted by the national electrical code only for motors of 2 hp or less and over 100 hp.

The second type of switch is a motor-circuit switch, which is capable of interrupting the motor current under normal overloads. This type of switch is rated in horsepower and, when used within this rating, is capable of being used as a starting switch for motors (Sec. 2·1). When used for disconnecting and motor-circuit protection, this switch must be of the fused type.

Circuit breakers offer the same disconnecting features as switches and the circuit protection of fuses. The breaker operates on a thermal-released latch so that it may be reset and used again after an overload. Being built all in one unit and offering short-circuit protection, as well as serving for disconnecting, makes this unit more compact than a switch and fuse combination. Switches do offer more visual indication that the circuit is open.

Switches and breakers may perform the functions of start (Sec. 2·1), stop (Sec. 2·16), overload protection (Sec. 2·7), and short-circuit protection (Sec. 2·8), depending upon their rating and use in the circuit.

3·2 CONTACTORS

The contactor itself is not generally found alone in motor-control circuits. It is, however, the basic unit upon which the motor starter is built. Contactors are used to perform the functions of start and stop on many heavy loads such as electric furnaces, signs, and similar types of equipment that do not require running protection.

Perhaps the best way to describe a contactor would be to say that it is a magnetically closed switch. It consists of one set of stationary contacts and one set of movable contacts which are brought together by means of the pull of an electromagnet. The vast majority of contactors use an electromagnet and contact arrangement that falls into one of two general types. The first of these is the clapper type (Fig. 3·1). The contacts are

Fig. 3·1 Clapper-type contactor. (Square D Co.)

fastened to the pole pieces of the magnet and hinged so that they swing more or less horizontally to meet the stationary contacts.

The second is the solenoid type (Fig. 3·2). On this contactor the movable contacts are coupled to the movable core of a magnet. When the electromagnet is energized, the movable core is pulled to the stationary core, thus closing the contacts.

Regardless of whether the contactor is of the clapper or the solenoid type, the contacts themselves are broken by the pull of gravity or the force of a spring when the electromagnet is deenergized.

All that is necessary electrically to operate the contactor is to provide a voltage of the proper value to the coil of the electromagnet. When the voltage is switched on, the contacts close, and when the voltage is switched off, the contacts open.

Fig. 3 · 2 Solenoid-type contactor. (General Electric Company)

3 · 3 RELAYS

Automatic control circuits almost invariably contain one or more relays, primarily because the relay lends flexibility to our control circuits. The relay is by design an electromechanical amplifier.

Let us consider for a moment the meaning of the word *amplify*. It means to enlarge, increase, expand, or extend. When we energize the coil of a relay with 24 volts and the contacts are controlling a circuit of 460 volts, we are amplifying the voltage through the use of a relay. Relay coils require only a very low current in their operation and are used to control circuits of large currents. So again they amplify the current. The relay is inherently a single-input device in that it requires only a single voltage or current to activate its coil. Through the use of multiple contacts, however, the relay can be a multiple-output device which amplifies the number of operations controlled by the single input.

Suppose we have a relay whose coil operates on 115 volts at 1 amp, and the contacts of this relay control three separate circuits operating at 460 volts and 15 amp each. This relay then becomes a power amplifier in that it controls considerably more power in its output circuits than it consumes in its input circuit. It also becomes an amplifier in terms of the number of circuits, as its single input controls three separate outputs.

Relays are generally used to accept information from some form of sensing device and convert it into the proper power level, number of varied circuits, or other amplification factor which will achieve the desired result in the control circuit. These sensing devices used in conjunction with relays are commonly called *pilot devices* and are designed to sense or detect such things as current, voltage, overload, frequency, and many others, including temperature. The proper type of relay to be used in a given circuit will be determined by the type of sensing device which transmits the information to it. For instance, a voltage-sensing device must be connected to a voltage relay, and a current-sensing device must activate a current relay. Each of these types will be discussed individually.

Voltage Relay. This type of relay (Fig. 3·3) is probably the most common because it lends itself to so many applications and can be used to perform so many functions. The voltage relay is merely a small contactor (Sec. 3·2) which opens or closes its contacts, depending on whether they are normally closed or open whenever the proper voltage is applied to its coil. They are available with as many contacts either normally open or normally closed as needed. Voltage relays are used frequently to isolate two or more circuits controlled from one source (Fig. 3·4) or when the control voltage is different from line voltage.

It must be remembered that while a voltage relay is not a primary control device, it does require a pilot device (Chap. 4) to operate it.

Fig. 3 · 3 Voltage relay. (General Electric Company)

Current Relay. This type of relay (Figs. 3·7, 3·8, and 3·9) is used to open or close a circuit or circuits in response to current changes in another circuit, such as a current drawn by a motor (Sec. 2·1).

The current relay is designed so that when connected in series with the circuit to be sensed, it will close after the current through its coil reaches a high enough value to produce the necessary magnetic flux. There are a few terms used in connection with current relays that must be understood.

Pull-in current is the amount of current through the relay coil necessary to close or pull in the relay.

Drop-out current is the value of current below which the relay will no longer remain closed after having been pulled in.

Differential is the difference in value of the pull-in and drop-out currents for the relay in question.

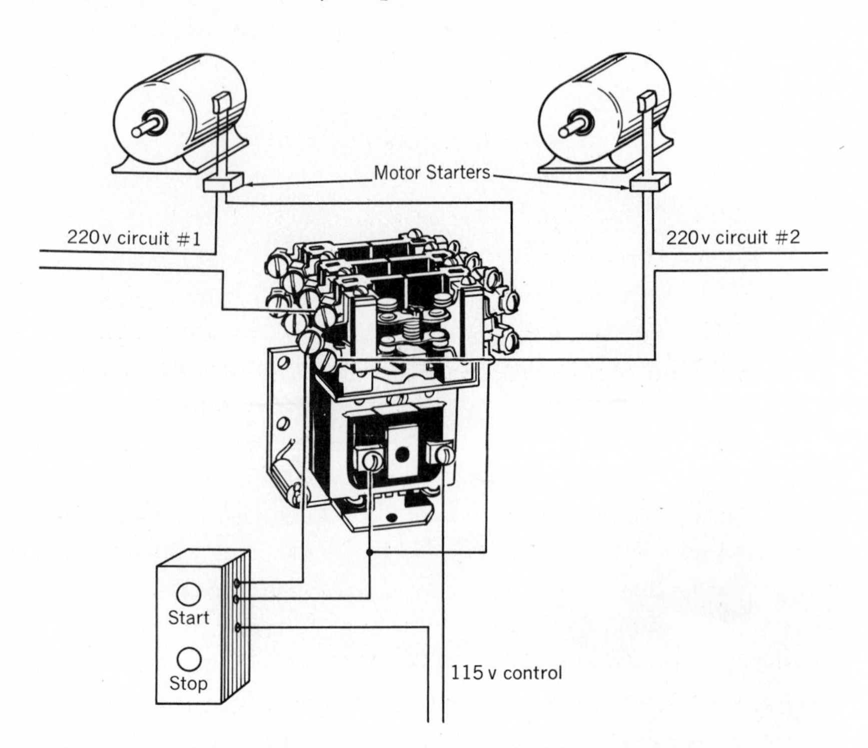

Fig. 3 · 4 Circuit diagram illustrating the basic use of voltage relays.

For example, if a relay is energized or pulled in at 5 amp and drops out at 3 amp, then the pull-in current is 5 amp, the drop-out current is 3 amp, and the differential is 2 amp.

Most relays of this type are provided with spring tension and contact spacing adjustments which allow a reasonable variation of pull-in, drop-out, and differential values. This type of relay should not be operated too close to its pull-in or drop-out values unless it is provided with some form of positive throw device for its contacts. This is important because the amount of contact pressure depends upon the difference in actual current and pull-in current for the particular relay. For example, when the above relay is operated with only 5.01 amp flowing through the coil, the contact pressure will be only that produced by the 0.01 amp of current.

Generally, true current relays are used only on circuits of very low current. For heavier current applications, a current transformer is used and its output applied to either a current relay or a voltage relay with the proper coil voltage.

Another type of current relay is the thermal type, in which a bimetallic strip or other device is heated by a coil connected in series with the circuit to be sensed. The bimetallic type depends upon the difference in expansion of two dissimilar metals when heated. It is constructed by riveting together two thin strips of dissimilar metals. When the current in the circuit produces sufficient heat, the bimetallic strip expands and releases the contacts. Motor overload relays and fluorescent starters are examples of this type of relay.

Frequency Relay. The frequency relay is used to apply field excitation to synchronous motors (Sec. 2·5) and for acceleration control on wound-rotor motors. Most of these units are specially designed for a particular application. One type consists of two balanced coils arranged on a common armature. These coils compare a reference frequency with that of the sensed circuit. The relay is closed one way when the frequencies are the same, or within a predetermined percentage, and is closed the other way when the frequencies differ by a given amount or more.

Time-delay Relay. This type of relay is often used for sequence control, low-voltage release, acceleration control, and many other functions.

Essentially, the time-delay relay is the voltage relay with the addition of an air bleed (Fig. 3·5) or a dashpot (Fig. 3·6) to slow down or delay the action of its contacts. This delay in action can be applied when the relay is energized or when it is deenergized.

If the delay is to be applied when the relay is energized, it is referred to as *on* DELAY (TDOE). If the delay is to be applied when the relay is deenergized, then it is referred to as *off* DELAY (TDODE). Both types are provided with an adjustment so that the time delay can be set within the limits of the particular relay. The contacts are always shown in the deenergized position and are either time opening (TO) or time closing (TC). These

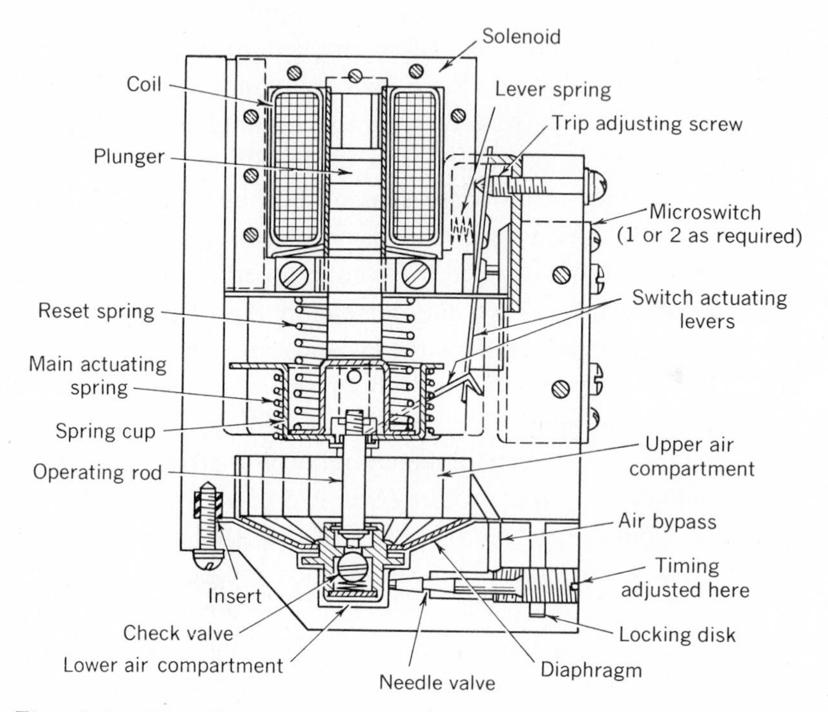

Fig. 3·5 Time-delay relay, air-bleed type. (Cutler-Hammer, Inc.)

units are built in various sizes depending upon the contact rating needed.

Overload Relay. The overload relay is found on all motor starters in one form or another. In fact the addition of some form of overload protection to an ordinary contactor converts it into a motor starter. This unit performs the functions of overload protection (Sec. 2·7) and phase-failure protection (Sec. 2·13) in motor circuits. The basic requirement for overload protection is that the motor be allowed to carry its full rated load and yet prevent any prolonged or serious overload. When a motor is overloaded mechanically, motor current increases, which in turn increases the temperature of the motor and its windings. The same increases in current and temperature are caused by the loss of one phase on polyphase motors or a partial fault in the motor windings. Therefore, to give full overload

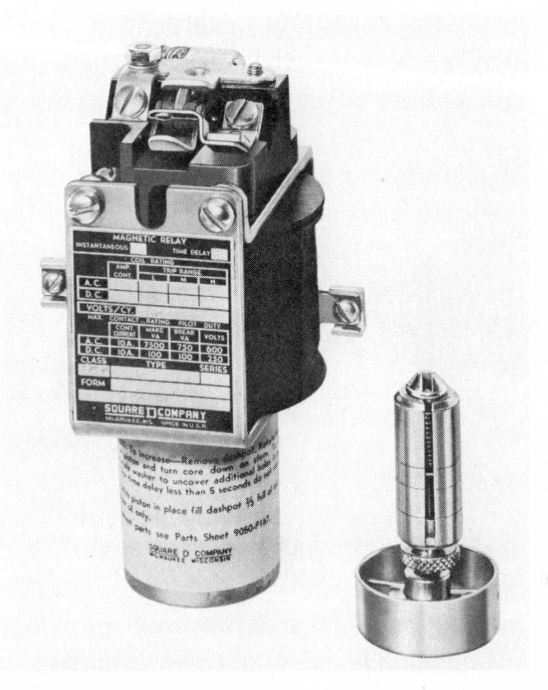

Fig. 3·6 Time-delay relay, dashpot type. (Square D Co.)

protection we need only to sense, or measure, the current drawn by the motor and break the circuit if this current exceeds the rated value for the motor.

There are three basic types of overload relays in general use on across-the-line starters. The first is a unit which employs a low-melting-point metal to hold a ratchet (Fig. 3·7*a*), which when released causes the opening of a set of contacts in the coil circuit of the starter. The second type uses a bimetallic strip (Fig. 3·7*b*) to release the trip mechanism and open the coil-circuit contacts.

Regardless of which type of device is used, it is activated by a heating element placed in series with the motor circuit. The amount of current needed to cause the relay to trip is determined by the size of the heating element used. When used for protection of small motors drawing low current, a coil of small wire or very thin metal is used as a heating element. On larger motors a heavier coil or strip of metal is used so that the same amount of heat is produced when the rated amount of current flows. Thermal units used in overload relays have an inherent

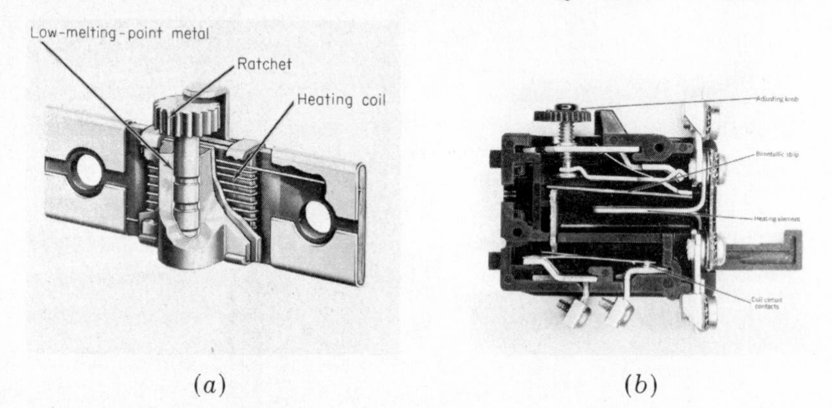

Fig. 3 · 7 Overload relays. (a) Low-melting point metal type. (Square D Co.) (b) Bimetallic-strip type. (General Electric Company)

time delay in their action that is inversely proportional to the amount of overload. This should be evident from a study of the curve shown in Fig. 3·8. When the overload is slight, the motor can go on running for some time without tripping the overload unit. If the overload is great, however, the overload relay will trip almost at once, thus removing the power from the motor and preventing damage.

Thermal relays trip on heat and heat alone, and they cannot

normally tell whether this heat is from the current to the motor or from the air that surrounds the starter. To offset this, it is sometimes necessary to install oversized heaters in high-temperature locations and undersized heaters in low-temperature locations. Some bimetallic units are designed to compensate for the ambient temperature change. This type of unit is called a *compensated overload relay*.

The third type of overload relay is magnetic (Fig. 3·9). This unit has a magnetic coil so connected that it senses the motor current either by the use of current transformers or by direct connection. When the current exceeds the rating of the motor, the overload coil lifts a plunger that forms its core and opens

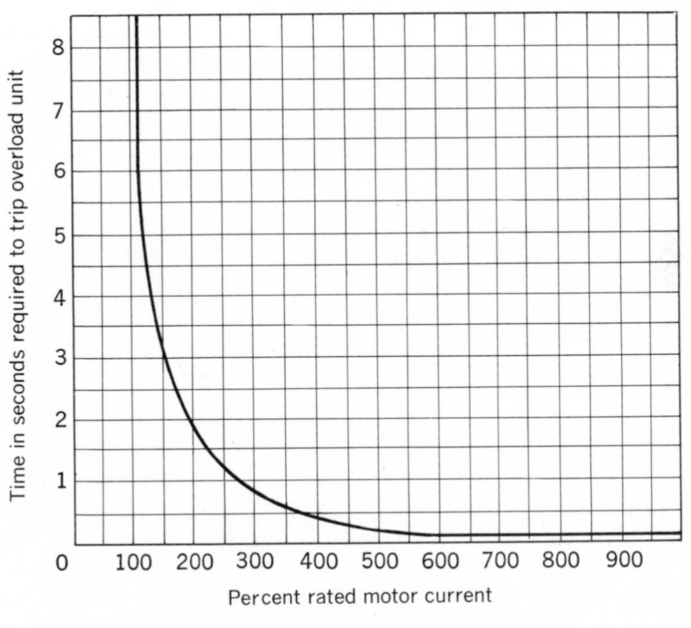

Fig. 3·8 Overload-relay current curve.

the contacts in the control circuit. Magnetic overload relays are generally found only on large motor starters.

Overload relays must be reset after each tripping, either automatically or manually. The automatic reset type should not be used except on equipment that is so designed, or used, that there can be no danger to life or equipment from the restarting of the motor. After the overload relay has been tripped, it requires a little time to cool, so that there is some delay before resetting can be accomplished.

3·4 MOTOR STARTERS

A motor starter in its simplest form consists of some means of connecting and disconnecting the motor leads from the line leads, plus overload protection for the motor. Many other refinements are added to this basic unit to achieve the desired degree of control and protection. There are many types and classifications of motor starters. Each of these types draws its name from the method or classification of operation of the motor that it

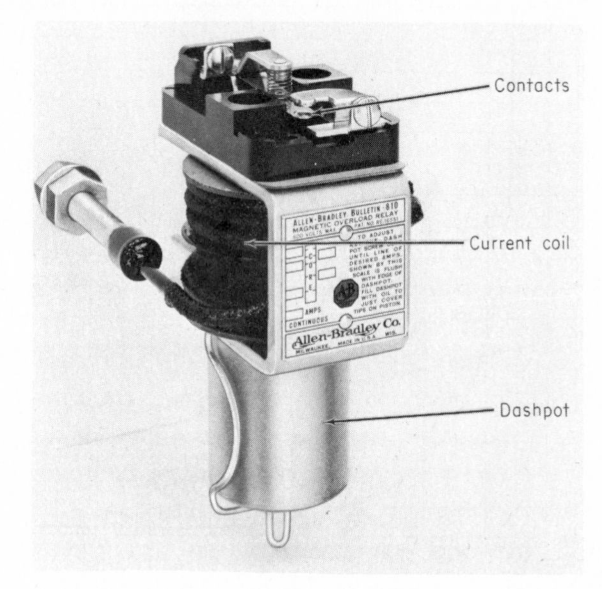

Fig. 3·9 Overload relay, magnetic type. (Allen-Bradley Company)

starts. Some of these classifications are manual or automatic, full-voltage, or reduced-voltage, single-phase or three-phase, and d-c or a-c. To describe a particular motor starter, it is necessary to use several of these terms or classifications. For instance, a particular motor might require a reduced-voltage, automatic, three-phase, a-c motor starter. Even this does not completely describe the unit, because it must be of a definite size for the motor and be rated for the proper voltage. Then we must know whether it is to be remotely controlled or to have a push button in the cover and many other features. In this chapter, we shall discuss many of these classifications, and the student should keep in mind that any particular starter might well be a combination of several of the types we discuss here.

Keep in mind that there is a difference between a motor starter and a motor controller. While it is difficult to draw a fine line between them, for the purpose of the discussion in this book the starter consists of the means of connecting the motor to the line plus providing the protection needed. By contrast, a controller contains not only the motor starter but at least a major part of the sensing devices and relays necessary for the complete control system.

Motor starters are built to NEMA (National Electrical Manufacturers Association) standards. These standards include such things as sizes, so that a purchaser may expect the equipment to be built to handle its rated load. For instance, the size 0 starter in 460-volt three-phase service is rated at 5 hp. At 660 volts the size one starter is rated at 10 hp, size two 25 hp, size three 50 hp, and size four 100 hp. For lower voltage service, each size starter has a smaller horsepower rating because of the increased current demand by the motor running at a lower voltage.

Also included in the NEMA standards are types of enclosures for starters to satisfy code requirements for atmospheric conditions existing in the place of installation. NEMA type 1 enclosures are for general purpose indoor use wherever atmospheric conditions are normal. They are intended primarily to prevent accidental contact with the control apparatus and energized circuits. NEMA type 3 enclosures are weather-resistant and protect against dust, rain, and sleet in outdoor applications. NEMA type 4 enclosures are watertight and suitable for outdoor applications on piers and in dairies and breweries. They may be washed down with a hose. NEMA type 7 enclosures are intended for use in hazardous gas locations such as are found in oil fields and satisfy the code requirements for class 1, group D, locations. NEMA type 8 enclosures are also intended for hazardous gas locations and have oil-immersed contacts. NEMA type 9 enclosures are built for use in hazardous dust locations, whose code classification is class 2, groups E, F, or G, such as flour mills. NEMA type 11 enclosures are corrosion-resistant enclosures that protect the enclosed equipment against corrosive effects of gases and fumes by immersing the equipment in oil. NEMA type 12 enclosures are dust-tight industrial enclosures and are designed for use where the enclosure is required to provide protection against dirt and oil.

If you were requested to order the starter for a 5-hp motor connected to a 230-volt three-phase line in ordinary service, you would need to specify several things. You would order a size 1 three-phase 230-volt across-the-line starter in a NEMA type 1 enclosure. Additional information would be needed as to whether the starter should be manual or automatic, depending upon the type of control to be utilized in the installation.

3·5 MANUAL MOTOR STARTERS

To be classed as manual, a motor starter must depend upon the operator closing the line contacts by pushing a button or moving a lever which is physically linked to the contacts in some manner. For an illustration, suppose we take the size 0 starter available in either manual or automatic type, having the push button in the cover. The manual type (Fig. 3·10) is so constructed that when the START button is pressed, a mechanical linkage forces the contacts to close. Once closed, the linkage is latched in this position. When the STOP button is pressed, or the overload units open, the linkage is tripped and the contacts open.

By contrast, when the START button is pressed on the magnetic starter, it merely energizes the starter coil, which in turn magnetically closes the line contacts. The STOP button or the overload relays break the circuit to the coil, thus allowing the line contacts to open.

The chief disadvantage of the manual starter is the utter lack of flexibility of control. It must be operated from the starter location, and it is definitely limited even as to protection-control possibilities. When the degree of control it offers is satisfactory for the installation, it does have the advantage of being less expensive. The vast majority of manual starters found in service will fall into one of three classes, namely, the thermal switch for use on very small single-phase motors, the size 0 and size 1 manual across-the-line starters in single- and three-phase motors, and the manual reduced-voltage compensator (Fig. 3·14) for large motors.

Fig. 3·10 Manual motor starter. (Square D Co.)

3·6 AUTOMATIC MOTOR STARTERS

The automatic starter, also known as a *magnetic starter*, consists of a contactor with the addition of protective control. This starter depends upon the magnetic pull of an electromagnet to close and hold its line and auxiliary contacts and offers unlimited flexibility of control. It is dependable and has a long life expectancy with reasonable maintenance. There are many mechanical arrangements used on this type of starter. They fall into two general classifications, however, depending on the movement of the magnetic coil core.

The first of these is the clapper type (Fig. 3·11*a*) which has the movable contacts attached to the hinge along with the magnetic core or a section of the core. The hinge is so arranged that the pull of the magnetic circuit swings the pole piece and

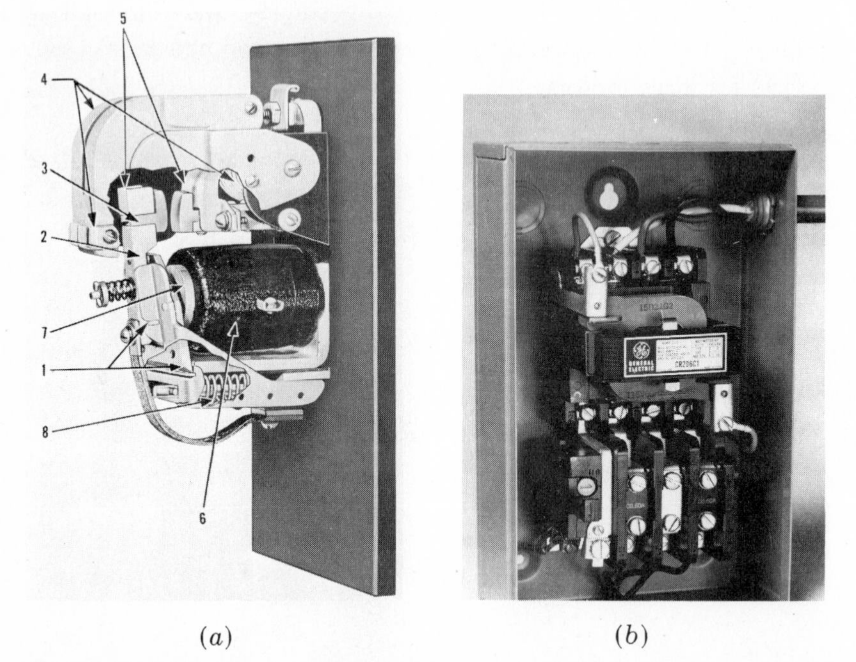

(*a*) (*b*)

Fig. 3 · 11 (a) Clapper-type construction as used for contactors and starters. (1) Hinge. (2) Contact arm. (3) Contact. (4) Arc shield. (5) Contact holders. (6) Coil. (7) Pole face. (8) Tension spring. (Square D Co.) (b) Solenoid-type construction as used for motor starters and contactors. (General Electric Company)

the contacts in a more or less horizontal direction, and the stationary line contacts are mounted on the vertical backboard of the starter.

The second is the *solenoid* type (Fig. 3·11*b*). With this type, the contacts are mounted to the movable portion of the core so that when the core closes, the movable contacts move to meet the stationary contacts.

Either of these basic types seems to be satisfactory, although each manufacturer has some reason for the particular type employed in his units.

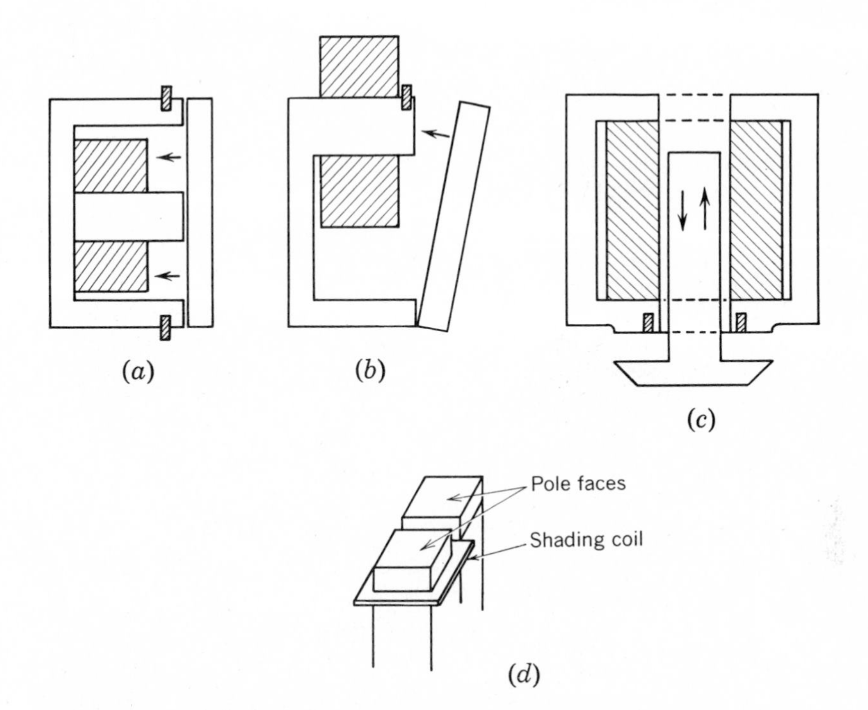

Fig. 3 · 12 The three basic magnetic shapes. (a) E type. (b) C type. (c) Solenoid type. (d) Magnetic pole piece showing shading coil.

The all-important magnetic circuit consists generally of an adaptation of one of the three basic magnetic shapes (Fig. 3·12). The E or C type is used on most clapper-type starters, and the modified E or solenoid type is used on vertical-action starters.

When a-c coil circuits are used, the pole faces of the magnet are equipped with a shading coil (Fig. 3·12*d*). This gives an out-of-phase flux to hold the magnet closed during the zero points of current, thus preventing chatter of the contacts. While this method of preventing chatter is effective, many starters for large motors employ a d-c coil circuit because of its constant magnetic pull and freedom from any tendency to chatter.

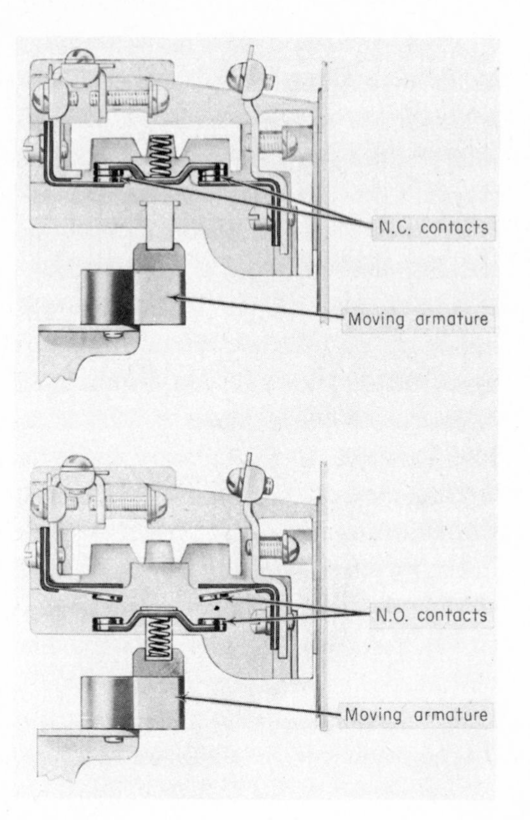

Fig. 3·13 Bridge-type contacts. (*Square D Co.*)

There are also two basic types of contacts in general use. Most small starters employ a bridge-type contact (Fig. 3·13). The bridge-type contact offers good contact alignment and a natural wiping action which helps to prolong contact life. Most large starters employ the required number of spring-mounted movable contacts that meet a corresponding number of rigid stationary contacts (Fig. 3·11*a*). The necessary wiping action is obtained by making the contacts in a curved shape that allows them to slide into alignment as they close and open. Good contact alignment is necessary in order to prevent excessive arcing and contact pitting.

The greatest single contribution to modern-day production machinery is the magnetic starter. The flexibility of control offered by the magnetic starter allows automation and automatic control accuracy never dreamed of, nor possible, with manual operation. This flexibility stems from the fact that all that is necessary to start a motor is to provide electric energy to the coil of the starter. The source of energy may be independent of the motor circuit and can be turned on and off from any point and by any means desired.

3·7 FULL–VOLTAGE STARTERS

Full-voltage, or across-the-line, starters (Fig. 3·11*b*) are the most widely used type. They are used on almost all three-phase, squirrel-cage, and single-phase motors. This type of starter is also used extensively as primary control on wound-rotor motors that employ manual secondary control. Rated for use on motors up to 600 hp and up to 600 volts, they can give full protection to the motor, the machine, and the operator. The limitation on the use of this starter for squirrel-cage motors is only the strain imposed on the wiring system and the machine by the starting current and torque of the motor. Across-the-line starters are available in a variety of enclosures to meet the needs of starter location conditions. These enclosures conform to the standards published by the NEMA to suit every condition of location. Availability also in either manual or magnetic types to suit the customer's needs adds to the flexibility of these units.

Any starter which connects the motor leads directly to the line voltage without any means of reducing the applied voltage or limiting the starting current would be classed as a full-voltage starter.

3·8 REDUCED–VOLTAGE STARTERS

As the name implies, the reduced-voltage starter contains some means of reducing the line voltage as it is applied to the motor during the starting period. This is done in order to limit the inrush of current during the starting cycle. The requirements for using reduced-voltage starting depend upon several factors (Sec. 2·1). These units are built in either manual or automatic

Fig. 3·14 Manual reduced-voltage starter, or compensator. (*General Electric Company*)

types, and, as with full-voltage starters, the manual type is cheaper but less flexible.

Manual reduced-voltage starters for squirrel-cage motors, more commonly referred to as *compensators* (Fig. 3·14), consist of a double-throw switch and an autotransformer. The START position of the switch applies power to the motor through an autotransformer. The operating handle is held in this position until the motor is running at the highest speed it will attain, and full voltage is then applied by throwing the handle to the RUN position. The switch mechanism is held in the RUN position by a latch which can be released by either the undervoltage release, the overload units, or the hand trip. Generally, these units are in a self-contained metal enclosure that is designed to be wall-mounted.

Automatic reduced-voltage starters (Fig. 3·15) take many forms and are generally designed for a particular type of motor

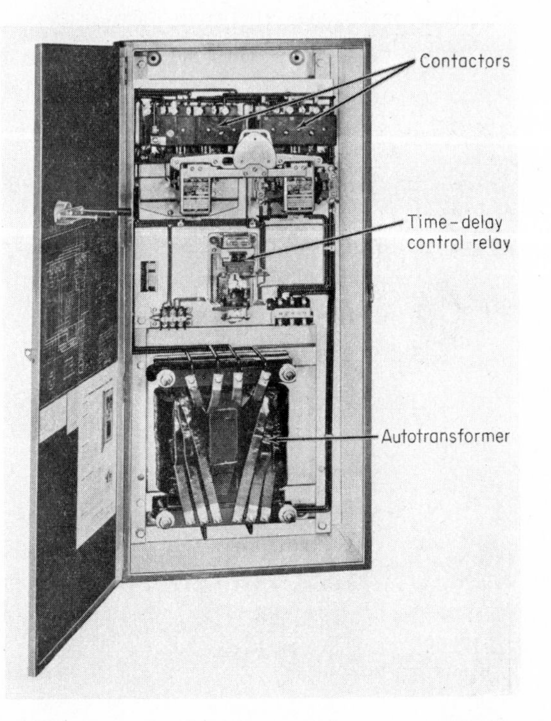

Fig. 3·15 Automatic reduced-voltage starter. (*Square D Co.*)

and for a particular application. The essential requirements are that some means be provided for connecting the motor to a source of reduced voltage and then automatically connecting it to the full line voltage after it has had time to accelerate.

The start contactor on a primary-resistance starter need be only a three-pole unit. This contactor connects the motor to the line in series with resistor units designed to limit the motor-starting current. The run contactor in this starter is also a three-pole unit which shorts out the resistors in order to apply full voltage to the motor.

The reactor type of reduced-voltage starter employs exactly the same contact arrangement as the primary-resistance starter. The only difference between a starter for primary resistance and a reactor type of reduced-voltage starter is in the use of reactors in place of the resistors.

The start contactor for an autotransformer type of reduced-voltage starter must be a five-pole contactor. The use of these

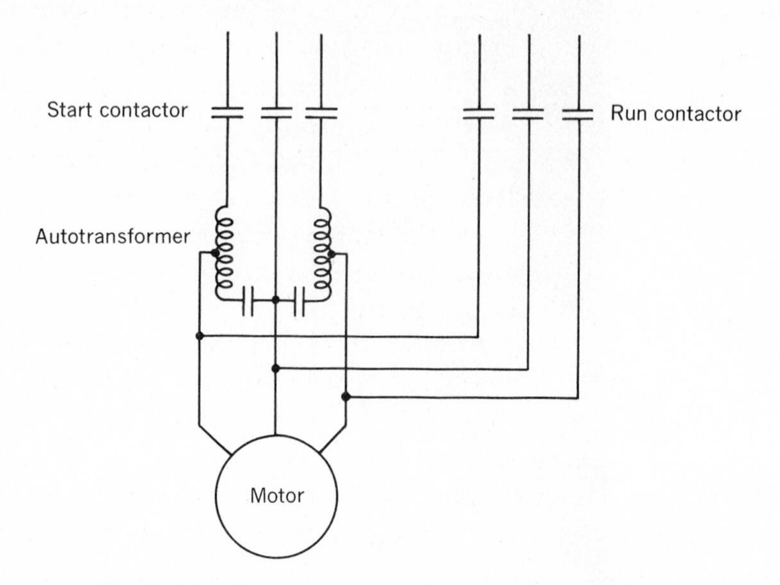

Fig. 3·16 Connections for starter autotransformer.

five contacts is shown in Fig. 3·16. These contacts connect the motor to the line through the autotransformer connected open-delta. The run contactor in this starter consists of three contacts which apply full line voltage to the motor.

Regardless of its type, an automatic reduced-voltage starter must contain some means of automatically changing from START to RUN at the proper time. Generally, this is accompanied by the use of a time-delay relay (Sec. 3·3). In the case of primary resistance or reactance starters, this relay is required only to energize the coil of the run contactor. When used on an autotransformer starter, this relay must break the circuit to the start contactor and then make the circuit to the run contactor. The use of a time-delay relay in this service gives definite-time control (Sec. 2·2). Another method which is sometimes used employs a form of current relay (Sec. 3·3) which will open the start contactor and close the run contactor when the motor current drops to a predetermined level. This gives current-limit control (Sec. 2·2).

When primary resistance or the reactor type of reduced-voltage starting is employed, there is no interruption of current to the motor. With the autotransformer type of reduced-voltage starting, however, the current may or may not be interrupted momentarily before the motor is placed across the line. When the current is not interrupted in the transition from reduced-voltage to full-voltage operation, it is referred to as a *closed-transition* start. When the motor is disconnected momentarily from the line and the current interrupted, it is referred to as an *open-transition* start. When the transition is of the open type, it is quite possible to have an inrush of current as high as twice the full-voltage starting current at the instant of voltage application. This surge of current is referred to as *transition current.*

Any of these controllers may contain almost any or all of the protection control functions, such as undervoltage release (Sec. 2·12), phase failure (Sec. 2·13), or incomplete sequence protection (Sec. 2·15).

One thing that you must keep in mind when choosing between manual and automatic controllers of this type is that when a manual unit is installed, it must be located so that the operator can not only see but also hear the motor, in order that he may properly judge when to apply full voltage. This limited selection of location can be overcome to some extent by installing a remote indicating tachometer. This enables the operator to determine the degree of acceleration of the motor from the control location.

Another method of obtaining the effect of reduced-voltage

starting is to use a wound-rotor motor with secondary control. This arrangement gives a higher starting torque with reduced starting current than does the squirrel-cage motor on primary reduced-voltage starting. The starter for such an arrangement would consist of an across-the-line starter connected to the primary of the motor. The secondary, or rotor, winding of the

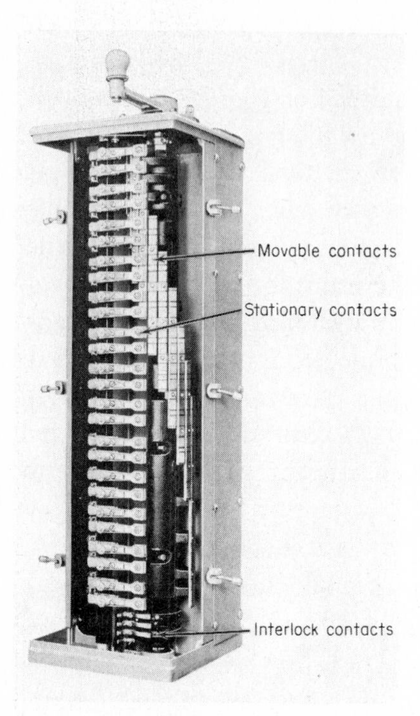

Fig. 3·17 Manual drum controller. (Cutler-Hammer, Inc.)

motor is connected to resistance grids by means of either a manual or automatic drum controller (Fig. 3·17). A second advantage to this arrangement is that it can offer speed control as well as limited starting current.

3·9 SPEED–CONTROL STARTERS

Besides the secondary speed control of wound-rotor motors, there are methods of controlling the speed of squirrel-cage motors, provided they are designed for multispeed operation.

One type of two-speed motor is wound with two separate stator windings and requires a starter with two sets of line contacts which can close only one set at a time. This interlock (Sec. 2·10) may be either mechanical or electrical or both. Two separate across-the-line starters may be used with electrical interlock when a special unit is not available. A reversing starter makes an excellent unit, provided the reverse phasing is eliminated. The two sets of contacts or starters are wired so that each set connects one speed winding to the line. As in other types of starters, they may be either manual or automatic.

Another type of two-speed motor is the consequent-pole motor, which has only one stator winding but gives two speeds by means of regrouping the stator coils to provide for a different number of poles. While the dual-stator-winding motor might have almost any ratio of high to low speed, the consequent-pole motor gives a 2 to 1 speed ratio. Three speeds may be obtained by using two stator windings. One of these windings gives one speed, and the other is regrouped for a second and third speed. To obtain four speeds, both windings must be regrouped. To obtain multispeed and variable torque, the high-speed windings are connected parallel-star and the low-speed windings are connected series-star (Fig. 2·3).

For constant horsepower and multispeed, the connection should be series-delta for high speed and parallel-star for low speed.

Because of the variety of possible connections, the starter for this motor must be designed for the particular type of motor to be used. One of the most popular manual arrangements requires the use of a drum-type controller to make the necessary changes in connections. This controller, however, must be preceded by an across-the-line motor starter which is interlocked through a set of contacts on the drum controller. The interlocks must disconnect the motor whenever a speed change is made by rotating the drum controller. The use of the across-the-line starter also provides the necessary protection for the motor which is not available on the drum controller.

Magnetic starters for multispeed motors must have a contactor for each speed (Fig. 3·18). The contacts on each contactor must be so arranged that they will make the proper connections to the stator windings for the motor to be used. These starters can be built to give any one of three types of control. The first and simplest of these is *selective* speed control. With selective speed control, the operator may start the motor at any speed

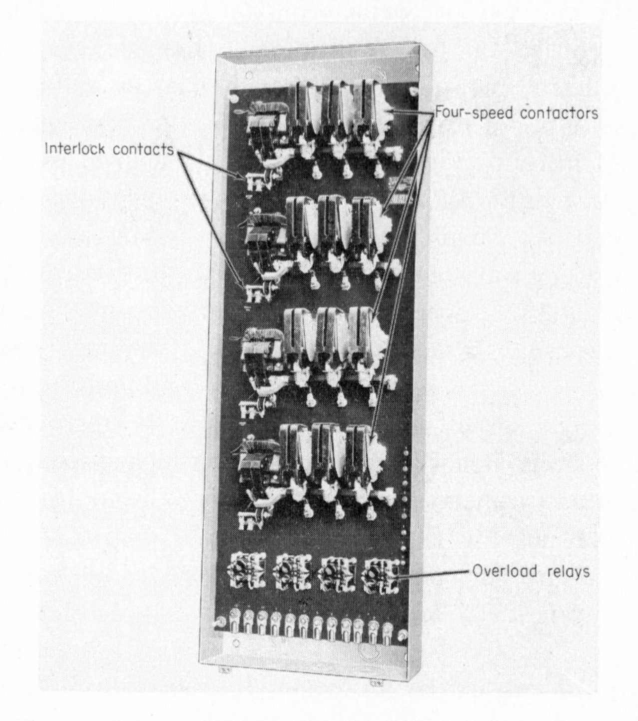

Fig. 3·18 Speed-control starter. (Cutler-Hammer, Inc.)

desirable and increase the speed merely by selecting any higher speed. To reduce speed, however, he must push the STOP button first and allow the machine to lose speed before the lower speed control is energized. This is done to prevent undue stress and strain on both the motor and the machine.

The second type is *sequence* speed control, which requires that the machine be started in its lowest speed and brought up to the desired speed through a set sequence. The acceleration to the desired speed requires that the operator push the button for each speed in proper sequence until the desired speed is reached. To reduce speed, the motor must be stopped and the sequence started over at the lowest speed.

The third type is *automatic* speed control, in which the operation is like sequence speed control, except that the operator need push the button only for the desired speed. The controller will automatically start in the lowest speed and accelerate through each successive speed to the one selected. To reduce speed, the STOP button must be pushed first. Then the button for the new speed should be pushed, which will recycle the controller to the new speed through each successive lower speed.

The choice of controller depends upon the type of load and the required operating conditions. Keep in mind that the basic difference lies in the fact that selective speed control allows starts in any speed, while the other two require a start at the lowest speed. It is not possible to describe properly the physical buildup of this type of unit in general terms, because of the many possible variations. In any case, however, a magnetic contactor will be required for each speed with the required number of contacts to make the connections for that speed plus the desired protection control. The possible automatic accelerations are discussed in Sec. 2·2, and any of these systems might be used.

3·10 COMBINATION STARTERS

The National Electrical Code requires a disconnect switch or breaker within sight of each motor. The combination starter includes this switch or breaker in the same enclosure with the starter itself. Combination starters are available in across-the-line or reduced-voltage and single-phase or three-phase types. In fact, almost any type of starter can be had in a combination form. The most common form of combination starter, however, includes a breaker or switch and an across-the-line starter (Fig. 3·19). A combination starter offers several mechanical advantages in that its compact size lends itself very well to a neat mechanical installation and reduces the wiring required at the job site. Electrically, the combination starter offers a protection for the operator or serviceman in that it generally includes a mechanical interlock which requires that the switch or breaker be in the OFF position before the door can be opened. This assures that the circuit is dead whenever the door to the starter is open.

The switch used in this type of unit may be fused or unfused. If the unfused switch is used, then the motor circuit must be

Fig. 3·19 Combination starter. (General Electric Company)

protected by another fused switch or breaker to give short-circuit protection. The use of a fused switch or circuit breaker in the combination starter adds short-circuit protection (Sec. 2·8) to the other control functions offered by the starter itself.

3·11 REVERSING STARTERS

The basic requirement of a reversing starter for three-phase motors is that it be capable of connecting the motor to the line in one phase rotation for forward and in the opposite phase rotation for reverse. A magnetic reversing starter (Fig. 3·20) incorporates two magnetic starters in one enclosure. The line sides of these starters are so connected (Fig. 3·21) that line 1 on starter 1 is connected to line 3 of starter 2 and line 3

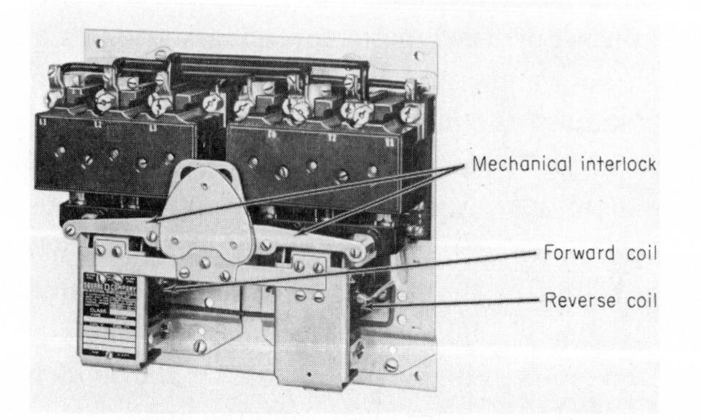

Fig. 3·20 Magnetic reversing starter. (Square D Co.)

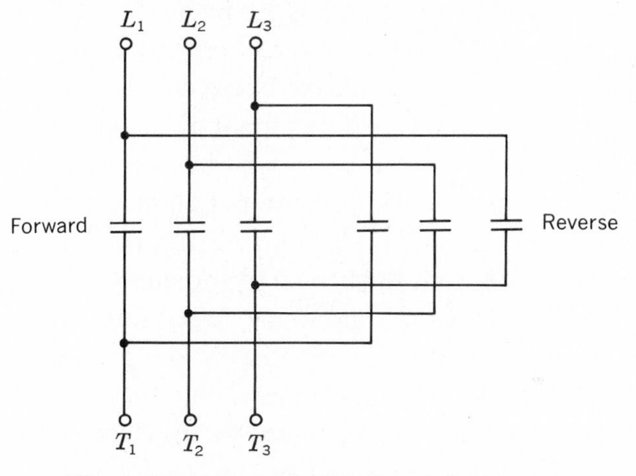

Fig. 3·21 Connection for reversing starter.

of starter 1 is connected to line 1 of starter 2. Line 2, or center phase, of both starters are connected.

Thus, when starter 1 is energized, *L*1 and *T*1 are connected and *L*3 and *T*3 are connected. When starter 2 is energized, however, *L*1 and *T*3 are connected and *L*3 and *T*1 are connected, thus accomplishing a reversal of phase rotation at the motor itself. These units are provided with a mechanical interlock consisting of a lever or arm which prevents either starter from closing when the other is energized. Many of these units also incorporate an electrical interlock to achieve the same purpose.

Remote control of a magnetic reversing starter requires only that the push button energize the coil of the starter, which gives a desired rotation of the motor. The STOP button must be wired so as to deenergize whichever coil is in use at the time. The normal wiring of a FORWARD, REVERSE, and STOP push-button station requires that the STOP button be pushed first when going from one direction to the other. This allows the motor to be disconnected from the line before being reversed and prevents plugging of the motor. Plugging of a motor is the sudden reversal of rotation without first removing it from the line.

If plugging is desirable, then the FORWARD and REVERSE buttons must be of the double-pole type, with one set of normally open and one set of normally closed contacts activated by one button. The normally closed contacts are so wired that the stop circuit is broken before the start circuit is made, regardless of which button is pushed. Caution should be exercised in employing plugging on any machine because not all machines can stand the severe strain imposed by the sudden reversal of the motor. Plugging can damage machinery. It may damage the motor and at times can endanger the personnel operating the machine. Plugging is used extensively in industry on presses, grinders, and many other pieces of machinery, but these should be so designed that they cannot be damaged by the stresses and strains encountered in this type of operation.

Manual reversal of three-phase motors of the squirrel-cage type is generally accomplished by use of a drum controller or switch between the line starter and the motor. This type of reversal requires that the motor be disconnected from the line before the drum switch is moved from forward to reverse or from reverse to forward. This prevents severe arcing of the drum-switch contacts. Proper wiring of the drum switch requires that it have a set of contacts interlocked with the line starter so that any time the handle of the drum switch is rotated, it will disconnect the motor from the line. This system of wiring also prevents plugging of the motor.

Reversal of single-phase fractional-horsepower motors may also be accomplished by the use of a drum switch or even a toggle switch. The reversal of this type of motor generally requires only that the starting winding be reversed in relation to the running winding. Suitable precautions must be taken to ensure that the motor comes to rest before attempting to reverse. Automatic or magnetic control for reversal of single-phase frac-

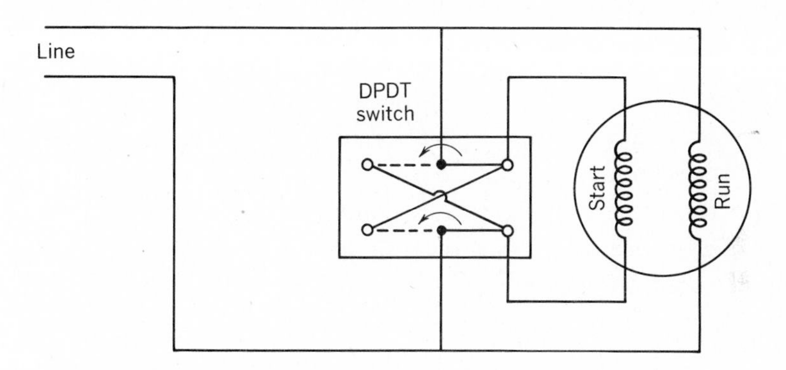

Fig. 3·22 Basic circuit for reversal of small motors.

tional-horsepower motors may be accomplished by the use of relays or starters. The possible connections for the use of drum switches, toggle switches, or relays for reversing this type of motor are shown in Fig. 3·22.

The reversing starter as such will offer the same control functions as any other starter of the manual or magnetic type and, in addition, provide the control function of reversing or reversal of the motor. Should a reversing starter not be available, two across-the-line starters properly connected may be used. When two starters are used in this service, electrical interlock must be used with them to prevent both starters from closing at one time. The electrical connection between the starters should be the same as that used in a reversing starter unit. Factory-built reversing starters generally are so wired that they require only one set of overload relays. When reversal is required on multispeed squirrel-cage motors or wound-rotor motors, it is generally accomplished by the use of a drum switch between the speed-control line starter and the motor itself.

3·12 WOUND–ROTOR MOTOR STARTERS

The starter for a wound-rotor motor consists of a full-voltage across-the-line starter used to energize the field, or primary, winding of the motor and some form of secondary control. The use of the primary starter provides the overload and undervoltage protection necessary for protecting the motor. The primary starter may be either manual or automatic and may be of the reversing type if primary reversal of the motor is desired. The primary starter should be interlocked with the secondary controller in such a manner that the motor cannot be started unless all of the secondary resistance is in the circuit.

There are several types of secondary control and controllers possible for use on wound-rotor motors. The most common component used for secondary control is the drum controller (Fig. 3·17). This unit is merely a set of rotating contacts operated by a handle attached to a shaft along with the movable contacts. The movable contacts engage a set of stationary contacts which short out resistance in the secondary circuit as needed for speed control. The drum controller may be of the reversing type or the nonreversing type. It may have several steps for speed control or may only have two or three steps used to give reduced-voltage starting. It is also possible to use a magnetic contactor similar to a speed-control contactor as a secondary controller.

Another type of secondary controller which gives the smoothest speed regulation, over the widest range of speed, for a wound-rotor motor is the liquid rheostat (Fig. 3·23). This type of secondary speed control is generally limited to very large equipment such as fans, blowers, and pumps where a constant torque or load is involved. The liquid rheostat consists basically of three tanks of water or other electrolyte. These cells or tanks must be made of insulating material. Inside each tank is a stationary contact or electrode and one movable electrode. The resistance is varied by increasing the distance between electrodes. The maximum resistance occurs when the electrodes are the furthest apart and the minimum resistance when the electrodes mesh. This change in resistance is affected both by the distance between the electrodes and the area of the electrode exposed to the electrolyte. One critical feature of this unit is that the water level must be maintained above the movable electrode

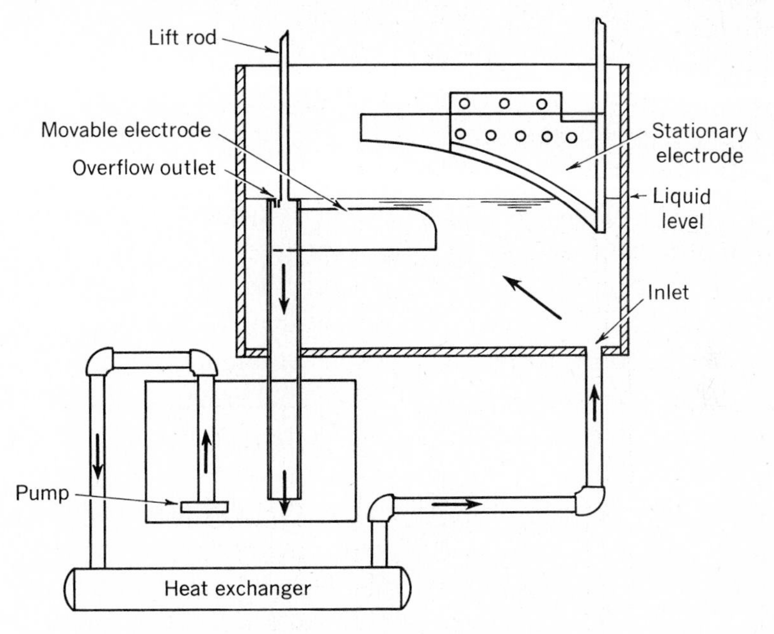

Fig. 3·23 Cross section of a liquid rheostat.

even though it moves up and down. This level is maintained by the use of a gate which is fastened so as to move up and down with the electrode. The water or electrolyte must be kept in constant circulation and provided with some means of cooling. Generally a heat exchanger is used for this purpose.

The various types of speed control such as definite-time control, sequence control, current-limit control, and frequency control, as discussed under speed-control starters, also apply to wound-rotor motor starters.

3·13 SYNCHRONOUS MOTOR STARTERS

The synchronous motor in its most commonly found form starts as a squirrel-cage motor. During the starting period, the stator is energized by alternating current and may be placed either directly across the line or through a reduced-voltage starter. It is necessary during the starting time to short out the d-c field winding through a field discharge resistor. This resistor protects the field from high induced voltages and also serves to increase the starting torque by serving as a secondary resistance. When the motor reaches synchronizing speed, which is usually between 93 and 98 percent of its synchronous speed, the starting resistance must be disconnected and the d-c voltage applied to the field winding of the motor. The application of the d-c field excitation will cause the motor to pull into step and synchronize.

The field application relay, or slip-frequency relay, as it is sometimes referred to, is probably the most critical component in a synchronous motor starter. Its function is to apply the d-c field excitation at exactly the proper time. The sensing of the proper time to apply field excitation is accomplished in the field application relay by sensing the induced alternating current that flows in the field winding during the starting period. This current is at a maximum when the motor first starts and diminishes in strength and frequency as the motor approaches synchronous speed. At approximately 95 percent of synchronous speed, the current induced in the field winding has reached a weak enough value to allow the field application relay to pull in and apply excitation to the motor. The use of a field application relay prevents application of excitation current when the motor is out of step more than approximately 75 to 80 degrees. This unit also applies the field discharge resistor after removing the excitation whenever an overload or other trouble causes the motor to pull out of synchronism.

The starter for a synchronous motor of standard design consists of a starter, either across-the-line or reduced-voltage type, similar to that required for a squirrel-cage motor, plus the necessary field control equipment (Fig. 3·24).

A manual starter consists of a manual compensator with overload and undervoltage protection, a field application relay and its contactor, a field rheostat to control excitation, and a field discharge resistor.

An automatic starter consists of an automatic reduced-voltage starter with definite-time control, a field application relay and contactor, a field discharge resistor, and overload and undervoltage protection.

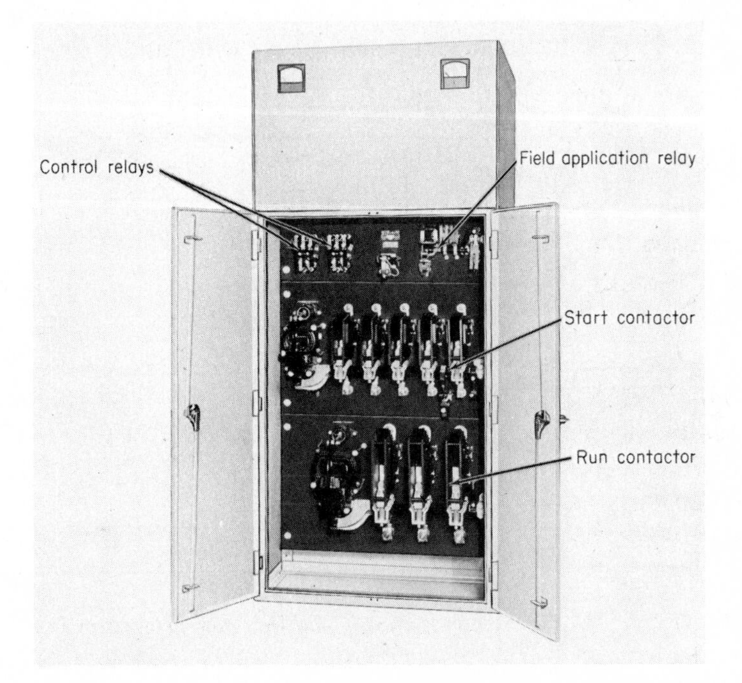

Fig. 3·24 Synchronous motor starter. (*Cutler-Hammer, Inc.*)

There are several factors involved in the starting of synchronous motors where extra-heavy loads are to be handled, such as in rubber mills and cement plants. Caution should be used in determining the starting requirements of this type of motor in its particular installation. It is beyond the scope of this book to go into the various possibilities involved in the starting and control of synchronous motors in special applications.

The manual starter for a series motor consists of a tapped resistor and a contact arm arranged so as to short out progressive steps of the resistance as the handle is moved from step to step. When all of the resistance has been cut out of the circuit, the motor is connected across the line. The handle must be held in the RUN position against the tension of a spring by the holding coil.

When the holding coil is connected in series with the motor (Fig. 3·25), it has a few turns of heavy wire and carries full

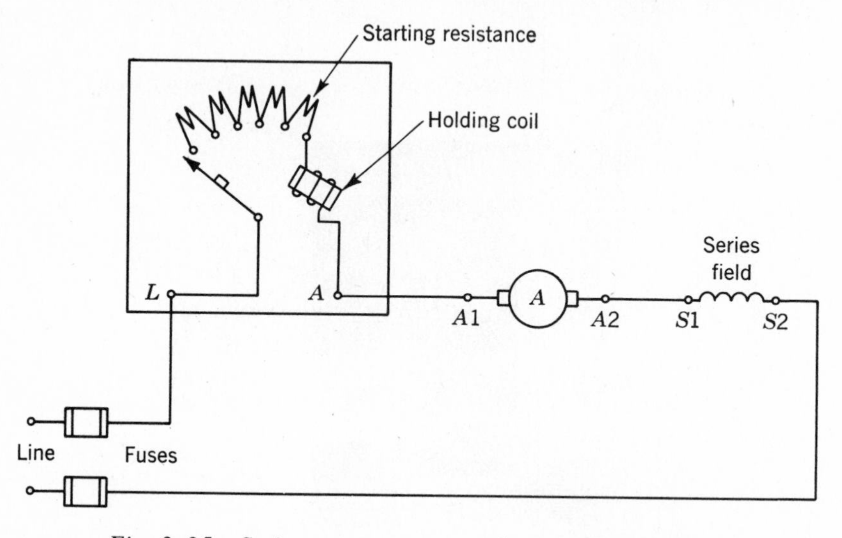

Fig. 3·25 Series motor starter with no-load protection.

motor current. When the load is removed or greatly reduced, the current drops to a low value which allows the spring to return the control arm to the OFF position. This arrangement provides no-load protection.

When the holding coil is connected across the line, it has many turns of fine wire and draws its own current from the line. Any serious drop or failure of the supply voltage will drop out the holding coil and allow the control handle to return to the OFF position. This arrangement provides no-voltage protection (Fig. 3·26).

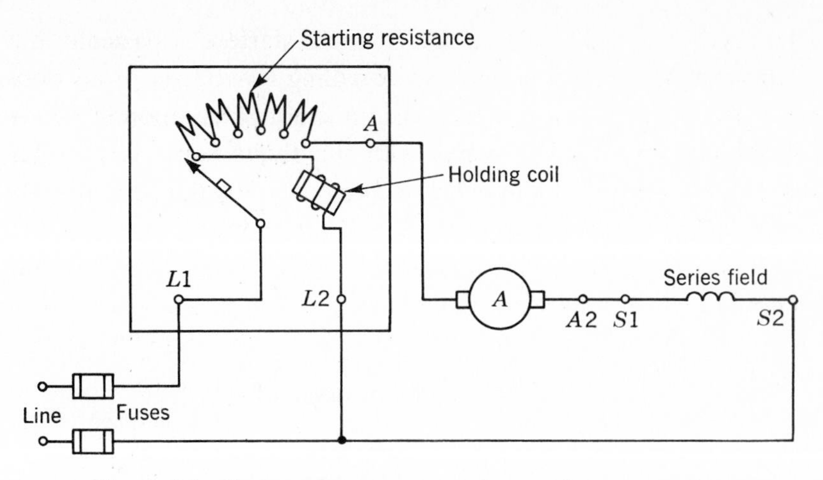

Fig. 3·26 Series motor starter with no-voltage protection.

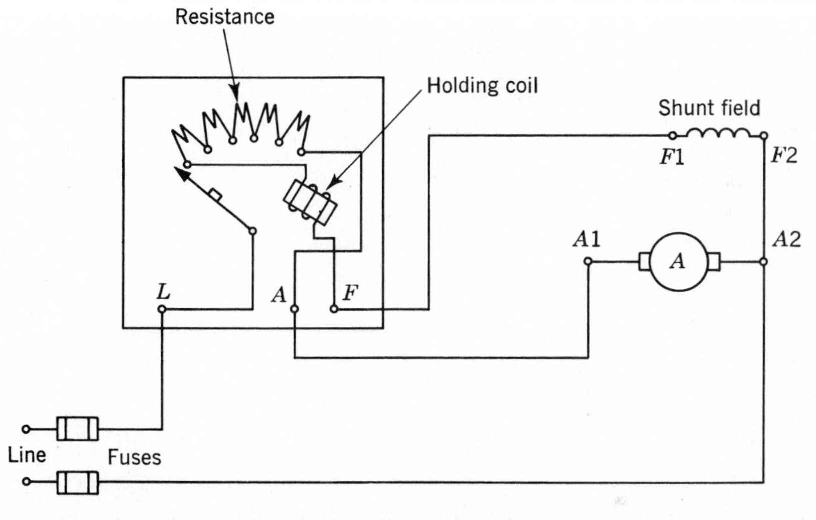

Fig. 3·27 Three-point starter.

D-c drum controllers used with resistance grids are generally employed for large series motors used on cranes, elevators, and other heavy loads which require speed and reversing control.

3·15 D-C SHUNT AND COMPOUND MOTOR STARTERS

When speed control is not desired, the basic starters for these motors are the three-point starter (Fig. 3·27) and the four-point starter (Fig. 3·28). The action of these starters is basically the same as that of the series starter providing no-voltage protection.

Speed control is accomplished in a shunt or compound motor by adding resistance in series with the shunt field. The starter used for this type of service to provide above-normal speeds is the four-point starter (Fig. 3·28). The motor is started in the same manner as the three-point starter. The field rheostat is used to adjust above-normal speeds.

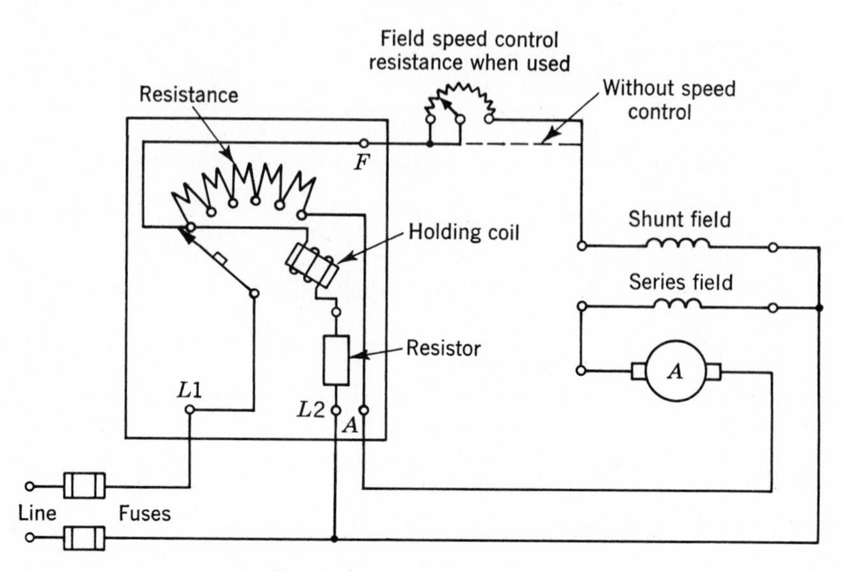

Fig. 3·28 Four-point starter.

Manual speed controllers are available which will provide both above- and below-base-speed control. These starters use mechanical linkage to position two tapped resistances, one in the armature circuit for low speeds and one in the field for high speeds. The armature resistance must be rated for continuous duty.

3·16 D-C REVERSING STARTERS

Reversal of a d-c motor is accomplished by reversing the direction of flow of current in either the armature or field winding, but not in both. When both windings are reversed, rotation remains the same as before. Generally, the practice is to reverse the armature leads by using a double-pole double-throw switch

(Fig. 3·29) in conjunction with a four-point starter. When a drum controller is used as a reversing speed controller, it has the ability to rotate in either direction from stop, and thus reverses the armature connections as needed for forward or reverse.

3·17 AUTOMATIC D-C CONTROLLERS

The requirement of an automatic controller for d-c motors is the same as that for manual controllers, except that contacts are used to replace the control arm of the manual unit.

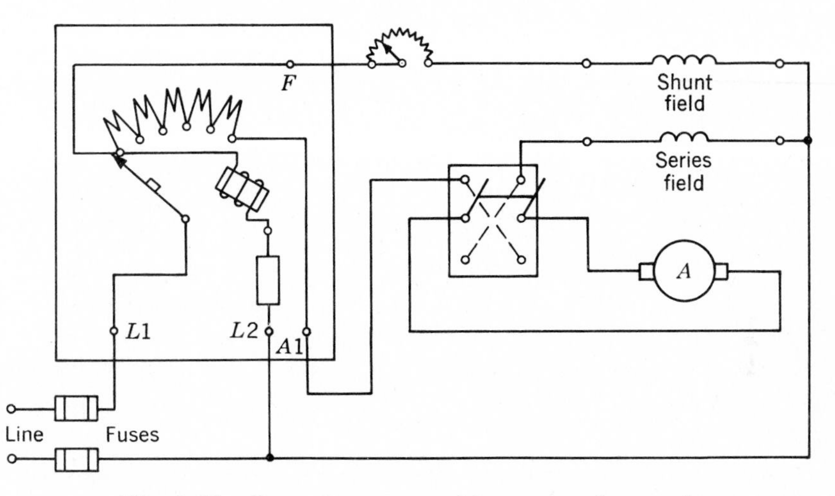

Fig. 3·29 Reversing starter with overspeed control.

For starting, we have learned that a tapped resistance must be progressively shorted out to bring the motor up to speed. When a relay contact is connected across each section of the starting resistance, we have a means of shorting out that resistance.

When the coils of the starting relays are connected to the line and the relays are time-delay relays, we have time-limit acceleration control. When the coils of the starting relays are connected across the armature, they will be subject to armature voltage changes. When these relays are adjusted to close on progressive steps of voltage, we have counter-emf acceleration control.

There are many other arrangements possible for automatic starting and speed control of d-c motors, but they are beyond the scope of this book.

Summary

This chapter on control components is intended to give the student an insight into the many variations in devices used to control the functions of motors. Always remember that a motor can perform only as well as the components in its control circuit. The fact that a particular component is a quality product does not necessarily mean that it will perform well in a particular circuit. The component must be selected to fit the motor to be controlled and the function to be performed.

Review Questions

1. What is the difference between switches intended for isolation purposes only and those intended for use as disconnecting means for motors?
2. What functions of control can be performed by switches and breakers?
3. What is the basic difference between a contactor and an across-the-line motor starter?
4. What is required to operate a voltage relay?
5. When speaking of current relays, what is meant by pull-in current, drop-out current, and differential?
6. Name two uses for a frequency relay.
7. The time-delay relay has two basic methods of delaying the closing or opening of its contacts. What are these two methods?
8. What are the two basic types of overload relays in general use on across-the-line motor starters?
9. Which of the two basic types of overload relays can have compensation built into it?

10. Basically, what is the difference between a manual motor starter and an automatic motor starter?
11. What is the chief advantage to using an automatic or magnetic motor starter?
12. What is meant by a clapper-type contact arrangement?
13. What is meant by a solenoid-type contact arrangement?
14. What are the three basic magnetic shapes used on modern motor starters?
15. What is the purpose of a shading coil used on magnetic pole pieces for a-c operation?
16. What is the basic difference between a motor starter and a motor controller?
17. What is a limitation on the use of across-the-line full-voltage starters for starting squirrel-cage motors?
18. What is the common name used for a manual reduced-voltage starter?
19. What are the three basic methods of achieving reduced-voltage starting?
20. What is meant by definite-time control?
21. What is meant by current-limit control?
22. Why must the autotransformer type of reduced-voltage starting remove the motor from the line momentarily before applying line voltage?
23. Which gives the higher starting torque, wound-rotor motors with secondary control or squirrel-cage motors with primary reduced-voltage starting?
24. Name several methods of obtaining speed control of motors.
25. What ratio of speed is offered by the consequent-pole motor?
26. What is meant by sequence speed control?
27. What is meant by selective speed control?
28. What is meant by automatic speed control?
29. What is a combination starter?
30. What is the basic requirement of a reversing starter?
31. Is interlocking a requirement on reversing starters?
32. What type of component is generally used for manual reversal of three-phase motors?

4

PILOT DEVICES

All components used in motor-control circuits may be classed as either primary control devices or pilot control devices. A primary control device is one which connects the load to the line, such as a motor starter, whether it is manual or automatic. Pilot control devices are those which control or modulate the primary control devices. Pilot devices are such things as push buttons, float switches, pressure switches, and thermostats.

4·1 DESCRIPTION OF PILOT DEVICES

An example of primary pilot control would be a magnetic across-the-line starter controlled by a simple toggle switch used to energize and deenergize the coil of the starter. When the switch is closed, the starter will be energized and will start the motor. When the switch is open, then the starter will be deenergized and will stop the motor. The starter, in that it connects the motor or load to the line, would be classed as a primary control device. The switch does not connect the load to the line but is used to energize and deenergize the coil of the starter. Therefore, it would be classed as a pilot control device.

For any given motor, generally there are two primary control devices used. These are the disconnecting means or switch and the motor starter. There may be many pilot devices used in parallel and series combinations to control the function of start and stop performed by the primary control device. The overload relays, for instance, which are included in the motor starter, are actually pilot devices used to control the primary device whenever the motor is overloaded.

The requirements of pilot devices vary greatly with their function and their proposed use. For instance, a float switch must open and close its contacts on the rise and fall of a liquid in some form of container. A pressure switch must open and close its contacts when the pressure in some vessel, pipe, or other container varies through the limits built into the pressure switch. Perhaps the best picture that can be drawn to show the difference between primary devices and pilot devices would be the comparison of a contactor and a voltage relay. The contactor is built to carry relatively large currents; therefore, it has heavy contacts capable of interrupting these currents. The relay, designed for pilot duty, has relatively small contacts because the current it is expected to interrupt is very small. In general, pilot devices might better be termed *sensing devices* because they are generally used to sense such things as pressure, temperature, liquid level, or the pressure applied to a push button. The function of these pilot devices is to convert the information that they sense into control of the primary control device with which they are connected.

4·2 FLOAT SWITCHES

Float switches take many forms in their physical or mechanical construction. Basically, however, they consist of one or more sets of contacts either normally open or normally closed, operated by a mechanical linkage. Many float-switch units, as well as other pilot devices, employ a mercury switch in place of metallic contacts. The simplest mechanical arrangement for a float switch (Fig. 4·1) would be a pivoted arm having the contacts fastened to one end and a float suspended from the other end. As the water level rises, it would lift the float, thus, moving the contact end of the lever downward and either making or breaking the contact, depending on whether the stationary contact were mounted above or below the arm. If a single-pole double-throw action of the contacts were desirable, then one stationary contact could be mounted above, and one below, the center of the arm. If the float were all the way up, it would

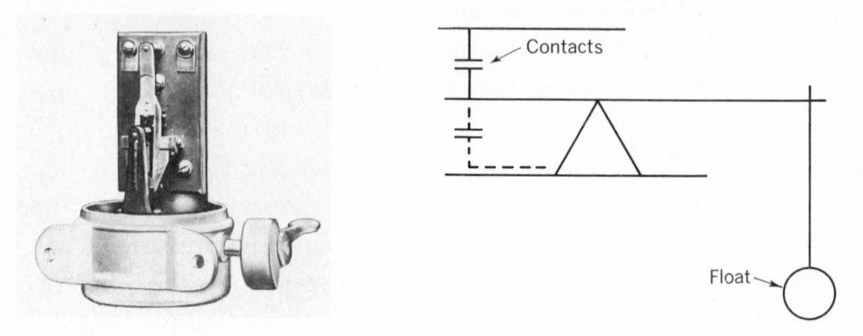

Fig. 4·1 Float switch, pivoted-arm type. (*Cutler-Hammer, Inc.*)

make the lower set of contacts, and if the float were all the way down, it would make the upper set of contacts.

Float switches require some means of adjusting the range of operation, that is, the amount of float travel between make and break of the contacts. In the simple float switch, this is usually accomplished by suspending the float on a rod which passes through a hole in the arm of the switch itself. Then if stops are placed above and below the arms on the float rod, the amount that the float travels before it operates the switch may be adjusted by moving the stops further apart or closer together.

Another system used in float-switch construction to give an even greater range of adjustment is to have the float suspended on a chain or cable which winds up on a reel. The action of the float is then transformed into a rotary motion which actuates a drum-type switch (Fig. 4·2). As may be noted from the accompanying photographs, these are only two of the many possible ways that a float may be made to actuate a set or sets of contacts. Any arrangement that will accomplish this may be properly classed as a float switch and used for pilot duty.

It should be noted here, however, that float switches are also made with heavy contacts and are suitable for primary control of small fractional-horsepower motors. When used for primary

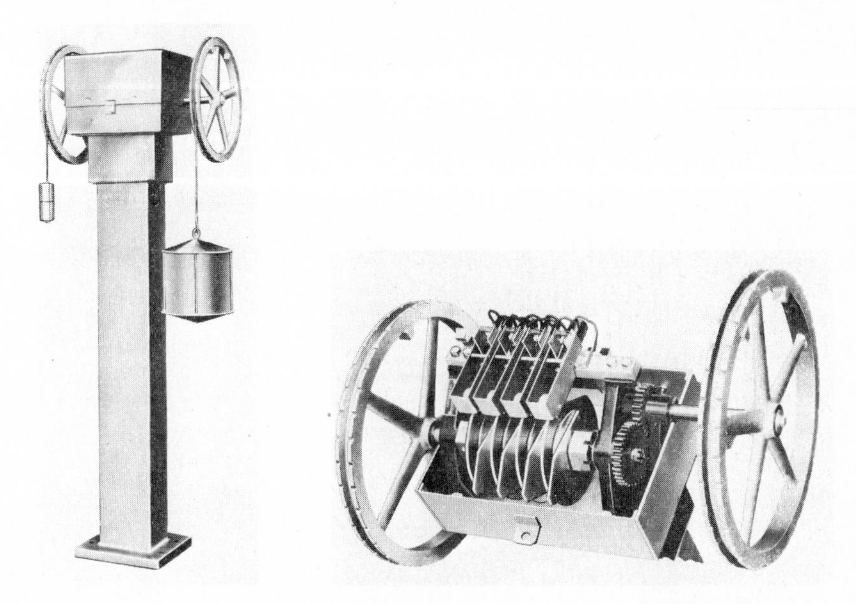

Fig. 4·2 Float switch, drum-switch type. (*Cutler-Hammer, Inc.*)

control, they are inserted in the line leads ahead of the motor and merely make and break the motor circuit in response to the action of the float.

It is highly desirable when studying pilot devices, if it is at all possible, to obtain several units made by different manufacturers and study the mechanical devices employed in their operation. The student will find that they vary greatly in actual mechanical design but fall into the same basic type of operation as described herein.

4·3 PRESSURE SWITCHES

Pressure switches, like float switches, are generally considered to be pilot devices. In the heavy-duty models, however, they are sometimes built for primary control of fractional-horsepower motors. Again, as with all pilot devices, they vary considerably in their mechanical design. Basically, they fall into three general classes according to the means of operation. The first of these is a bellows which is expanded and contracted in response to increase and decrease in pressure. The contacts are mounted

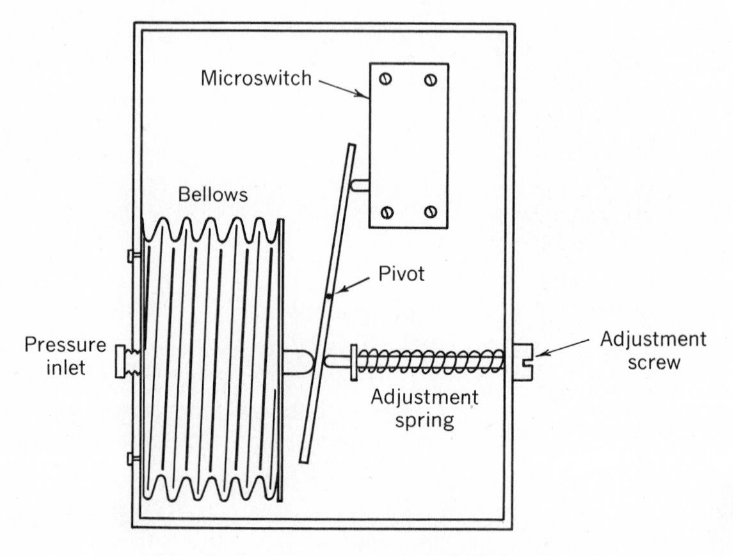

Fig. 4 · 3 Pressure switch, bellows type.

on the end of a lever, which is acted upon by the bellows (Fig. 4·3). The bellows expands, moving the lever, and either makes or breaks the contacts, depending on whether they are normally open or normally closed.

The second general type uses a diaphragm in place of the bellows (Fig. 4·4). Otherwise, the action of the switch is identical whether it contains a bellows or a diaphragm. The advantage of one type over the other depends greatly upon the installation and the pressures involved and would have to be decided in each installation.

It should be noted that pressure switches have a definite range of pressure where they are designed to operate. For instance, a pressure switch made to operate from a vacuum up to possibly several pounds of pressure would not be suitable for use on a line which normally would carry from 100 to 200 lb of pressure.

A third general type of pressure switch, the bourdon tube, employs a hollow tube in a semicircular shape so designed that an increase in pressure tends to straighten the tube. This action

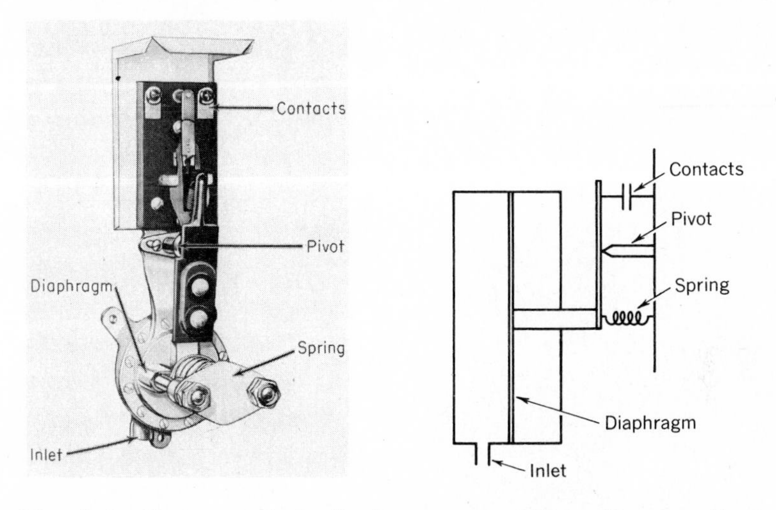

Fig. 4·4 Pressure switch, diaphragm type. (Cutler-Hammer, Inc.)

is transformed into a rotary motion by a linkage which trips a mercury switch mounted within the enclosure.

4·4 LIMIT SWITCHES

Limit switches are designed so that an arm, lever, or roller protruding from the switch is bumped or pushed by some piece of moving equipment. The movement of this arm is transformed through a linkage to a set of contacts. Movement of the arm causes the contacts to open or close, depending upon whether they are normally open or normally closed (Fig. 4·5).

There is a great variation in the internal design and action of these units, but, again, they fall into two general classifications as to mechanical design. The units intended for rugged use but not for precision control generally have metallic contacts operated directly from the lever action of the switch. Most manufacturers build a more accurate, or precision, unit which employs a microswitch to allow operation on very minute movements of the external lever of the switch. As with float switches, there are limit switches which are built so that a cable or chain is wound up on a reel which forms part of the limit switch. This

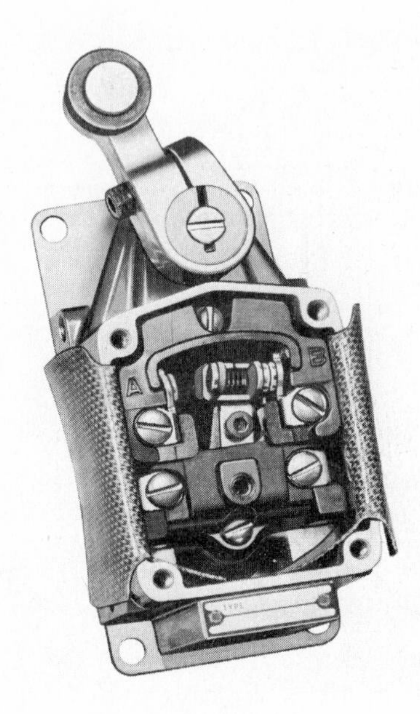

Fig. 4·5 Limit switch. (*Square D Co.*)

movement of the chain or cable is transformed to a rotary motion which actuates a drum-type switch. This type of limit switch is used where a great deal of travel must be allowed between actions of the switch.

Another type of limit switch (Fig. 4·6) which employs a drum-type switch is designed for direct shaft mounting where the rotation of the machine directly causes a rotation of the switch. The contacts in this type of limit switch must be designed to resemble a cam so that they can close and open with a continuous rotation in the same direction. Many limit switches of this type are coupled through a reduction gear so that many revolutions of the machine are required to produce one revolution of the limit switch, thus extending the range of control offered by the limit switch.

4·5 FLOW SWITCHES

The purpose of a flow switch is to sense the flow of liquid, air, or gas through a pipe or duct and to transform this flow

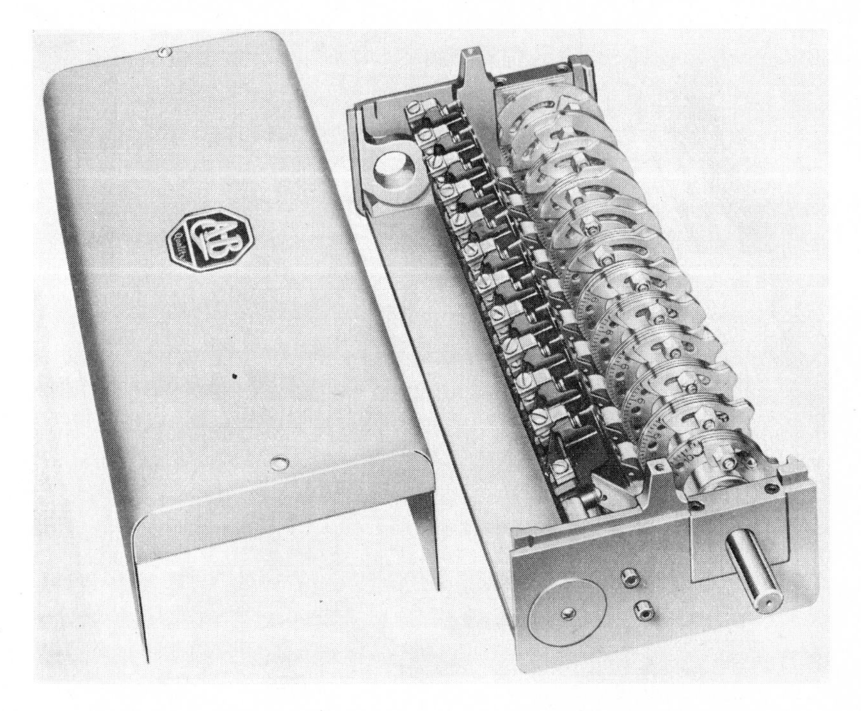

Fig. 4·6 Drum-type limit switch. (Allen-Bradley Company)

into the opening or closing of a set of contacts. One type of flow switch (Fig. 4·7) utilizes a pivoted arm having contacts on one end and a paddle or flag on the other end. The end with the paddle or flag is inserted into the pipe so that the flow of liquid or gas over this valve causes a lever to move and open or close the contacts.

Another type of flow switch uses a difference in pressure across an orifice flang which is installed in the pipe. A pipe

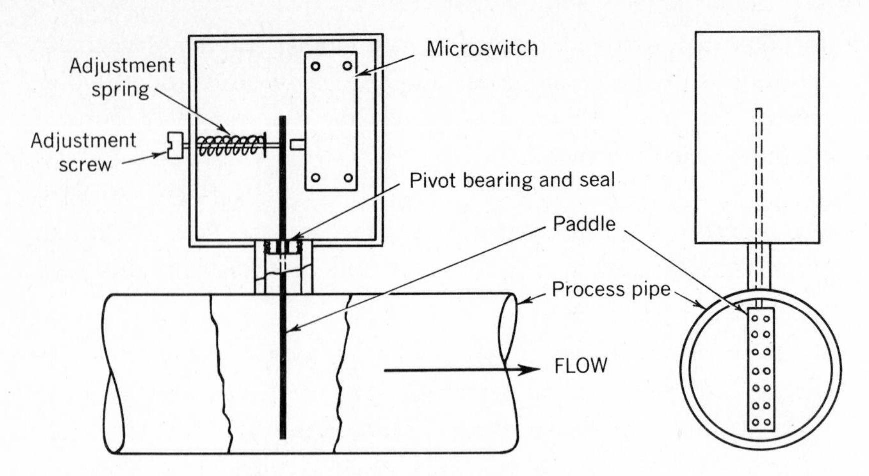

Fig. 4·7 Flow switch, paddle type.

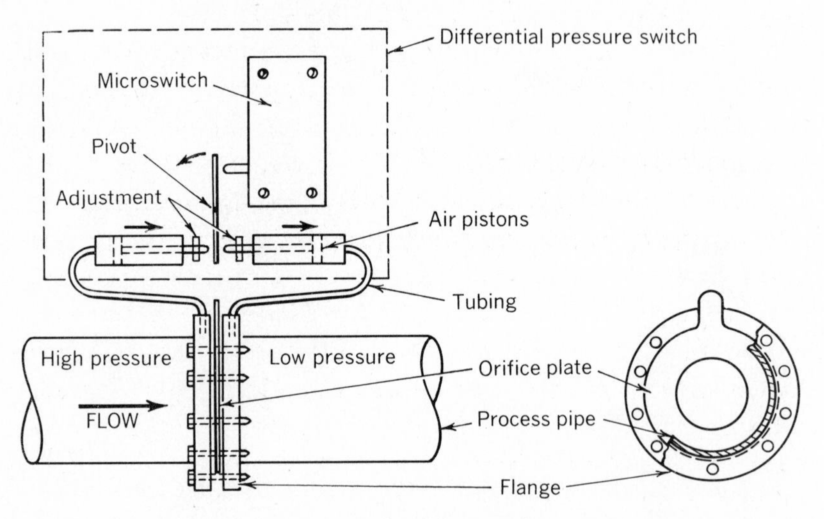

Fig. 4·8 Flow switch, differential pressure type.

is run from each side of the orifice to a pressure switch. The corresponding difference in pressure actuates the pressure switch in one direction or the other, opening or closing its contacts, depending upon their arrangement. Such a flow switch is illustrated in Fig. 4·8.

As with other types of pilot control, there are many other possible mechanical arrangements for flow switches. The student should consult manufacturers' catalogs and study the diagrams and illustrations in them to acquire a broader knowledge of the design and application of flow switches.

4·6 THERMOSTATS

Probably the thermostat is the pilot device which is built in the greatest variety of mechanical arrangements. Some are made to employ the action of the bellows to move the contacts. Some employ bimetallic strips to sense temperature and actuate the contacts. Many other arrangements are possible with this type of unit. A study of Fig. 4·9 will help the student to visualize a few of the arrangements found in everyday use on thermostats. Thermostats for use in motor-control circuits merely open or close a set of contacts in response to temperature changes, regardless of their mechanical construction and action.

The modulating thermostat moves a contact across a resistance in proportion to the change in temperature, thus varying the relative resistance of the circuit. When properly connected to a modulating motor (Fig. 4·10), it can control the position of the motor in direct response to the changes in temperature. The movement of an arm on the motor shaft is directly proportional to the amount of change in temperature. When connected to a damper, the modulating motor can control the amount of air flowing through a duct. When connected to a valve, the motor can control the flow of water or other liquids or gases through a pipe. While this type of thermostat is seldom, if ever, used for the direct control of a motor, it can initiate control through contacts mounted on the shaft of the modulating motor.

4·7 PUSH-BUTTON STATIONS

The push-button station (Fig. 4·11), while probably the simplest of all pilot devices, is the most commonly used in motor-control circuits. Push-button stations are of two general types: the maintained-contact type and the momentary-contact type. When the START button is pushed on the maintained-contact type, the contacts close and remain closed until the STOP button is pushed. This action is accomplished through a mechanical

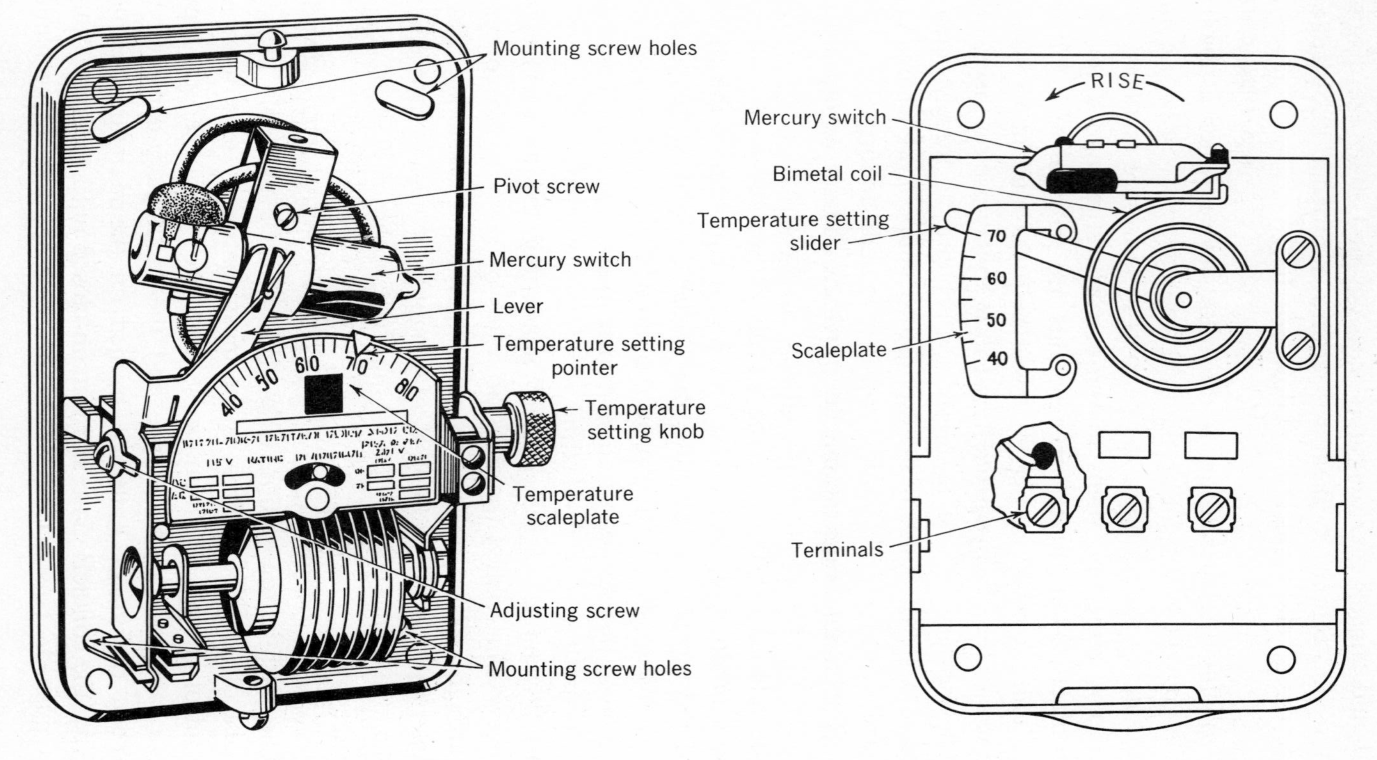
Mounting screw holes
Pivot screw
Mercury switch
Lever
Temperature setting pointer
Temperature setting knob
Temperature scaleplate
Adjusting screw
Mounting screw holes
40
50
60
70
80
RISE
Mercury switch
Bimetal coil
Temperature setting slider
70
60
50
40
Scaleplate
Terminals

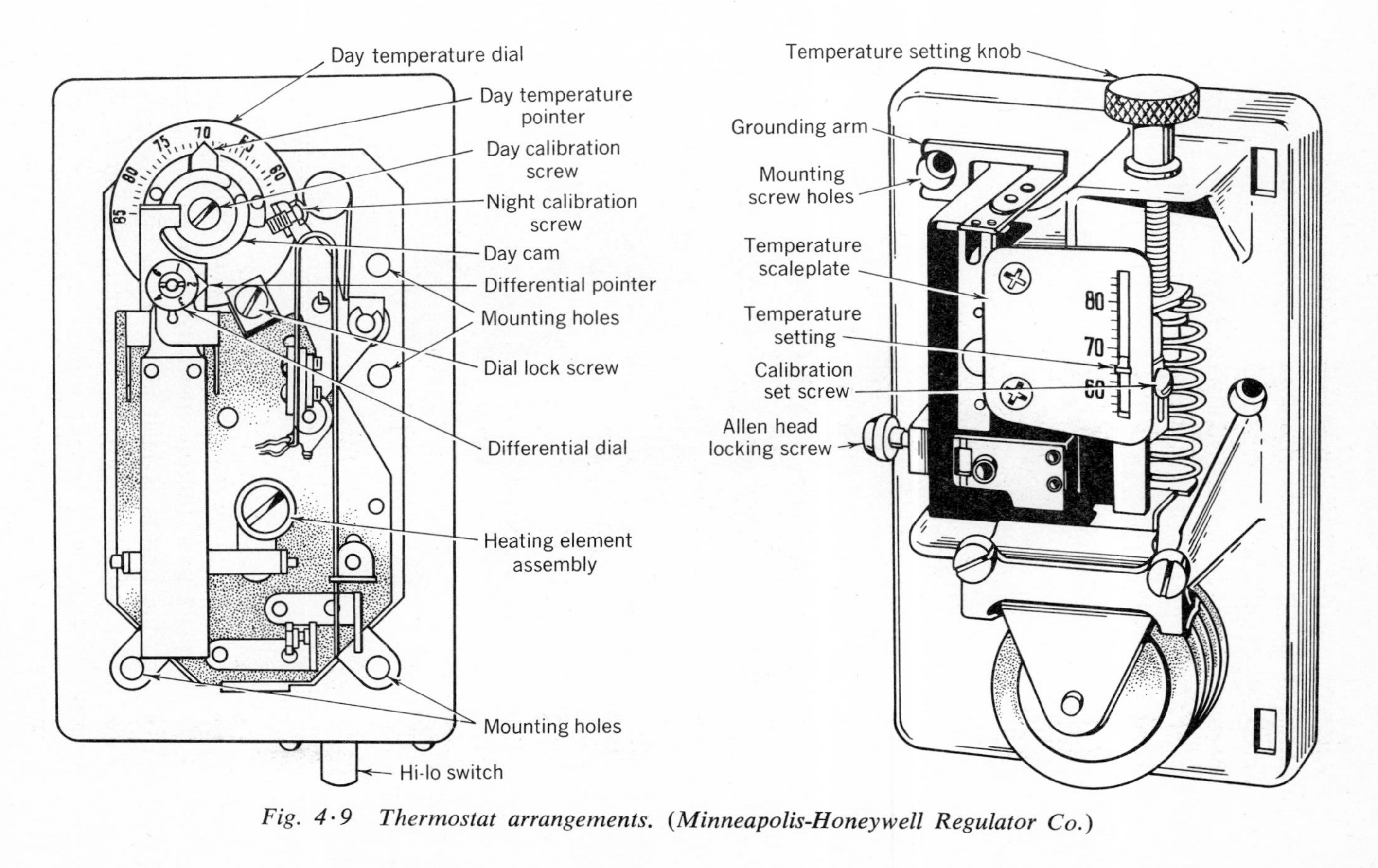

Fig. 4·9 Thermostat arrangements. (Minneapolis-Honeywell Regulator Co.)

linkage from the button to the set of contacts, which are so arranged that they will remain in either position until moved to the opposite position.

The momentary-contact normally open push button, such as used for START buttons, merely closes its contacts for whatever

Fig. 4·10 Modulating motor. (Minneapolis-Honeywell Regulator Co.)

period of time the button is held down. The normally closed momentary-contact push button opens its contacts for whatever period of time the button is held down. Push buttons also are available in the double-pole style. This push button has one set of contacts that are normally closed and one set that are normally open.

Push-button stations are made up of individual push buttons which may be normally open, normally closed, or double-pole units to give whatever combination of contacts that are needed. The most common push-button station is the start-stop station. Push-button stations are available, however, in most standard labelings to cover the normal control operations and are available with special labels to fit special needs. Also found on push-button stations are pilot lights to indicate when the motor is running, or possibly when it is not running, and selector switches.

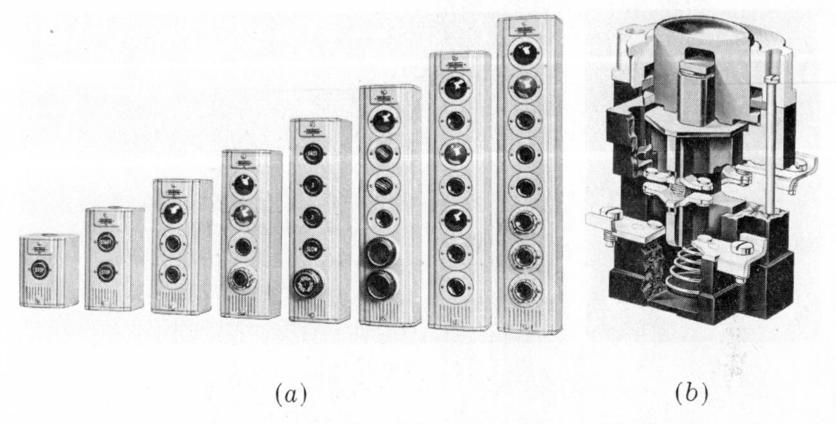

(*a*) (*b*)

Fig. 4·11 (a) Assortment of push-button stations. (b) Cutaway view of a single push button. (Cutler-Hammer, Inc.)

A selector switch may be used for hand-off-automatic control, or it may be simply an on-off switch.

4·8 PLUGGING SWITCHES

The plugging switch, sometimes referred to as a zero-speed switch (Fig. 4·12), is a special control device which is operated by the shaft of the motor or a shaft or pulley turned by some part of the machine. The rotation of the shaft causes a set of contacts to close, and when the power is removed these contacts cause a momentary reversal of the motor. When the motor is running in reverse, the forward set of contacts is closed. When the power is removed, this set of contacts momentarily energizes the motor in the forward direction. This sudden momentary reversal of direction of rotation of a motor is known as a *plugging stop*. Plugging is used on many precision machines such as presses, grinders, and other machine-tool operation. The purpose of plugging a motor is to bring it to an abrupt stop, so the plugging switch must not hold the starter closed for any appreciable length of time. Before a plugging switch is installed on a motor, it should be determined that the machine and motor

Fig. 4·12 Plugging switch. (Cutler-Hammer, Inc.)

are built for this rugged operation and that plugging will not endanger the operator.

4·9 TIME CLOCKS

The time clock as used for motor-control circuits consists of an electric clock to drive adjustable cams which open and close contacts at any present time. There are many versions of the time clock in general use. The simplest clock has a set or sets of contacts which can be adjusted to turn the external circuit on once and off once each day. There are several more elaborate versions between the simple time switch and the elaborate program clock. A program clock may be used to open and close several circuits independently at any desired time. This clock may also be set to skip undesired days so that a program for a period of time, generally consisting of a week, may be set up on the tape, and the clock will make and break the circuit by opening and closing its contacts at each predetermined time.

Summary

The discussion in this chapter is not by any means a complete list of all pilot devices that are manufactured. We have covered the most common and frequently encountered types. Most special pilot devices, such as aquastats, stack switches, relative-humidity controllers, airstats, high-pressure cutouts, and suction pressure controls are merely adaptations of the basic types that we have discussed. The control man must above all else be able to look at a strange control device and analyze its function both mechanically and electrically so that he will understand its operation in a control circuit. A thorough understanding of the basic types of control components will enable you to handle most, if not all, new components that you will find on the job.

The best method of becoming familiar with all types of control components is to make a study of the literature which manufacturers are happy to supply free of charge. This literature generally contains photographic illustrations showing the mechanical and electrical operation of the various devices and is usually accompanied by written description of the operation, ranges, and possible uses of these devices.

Quality should never be sacrificed in control components, particularly pilot devices, where the temptation to use cheap competitive units is the greatest. While a thermostat, float switch, or limit switch seems to be an insignificant part of an overall control system, the failure of one of these pilot devices can shut down the operation of a whole industrial plant. This is especially so if it is located on a critical machine.

Many times a mechanic is overwhelmed by the complexity in size of a control system for such things as central refrigeration plants, central heating plants, automatic manufacturing assembly lines, and other multiple-machine control systems. This fear of complexity is, in fact, groundless if you understand the basic functions of control and the operation of the basic types of control components. The overall system of complex control is made up of a series of individual control circuits involving these basic functions and basic components which are relatively easy to master. This subject will be pursued further in Chaps. 6 and 7, where it is hoped the student will be able to see that the most complex control circuits are merely groups of simple control circuits interlocked to give sequence or coordinated control of several machines.

Review Questions

1. What would be the difference between a float switch designed for primary control and one designed for pilot control?
2. What type of float switch should be used for great ranges of adjustment of float level?
3. Name the three general types of pressure switches in regard to their mechanical operation.
4. What is the advantage of the drum-type limit switch with a reduction gear?
5. What is the purpose of a float switch?
6. What is the difference between the ordinary thermostat and a modulating thermostat?
7. What is meant by momentary-contact push button?
8. What is the purpose of a plugging switch?
9. Basically, what is a time clock as used in motor controllers?
10. What is a pilot device?

5

CONTROL-CIRCUIT DIAGRAMS

If you were to find yourself in a foreign land and unable to read or speak the language, you would see the familiar things such as buildings, automobiles, newspapers, and people, but you would not be able to understand what was going on around you.

If you could speak and understand the oral language but could not read the printed words, you still would be dependent upon someone else for a full understanding. The same thing applies to control work. If you have mastered the first four chapters of this book, you can now speak and understand the oral language of controls. Until you master the control diagrams, however, you will be dependent upon someone else for most of your information.

This chapter deals with the written language of control and control circuits. Do not be satisfied until you can read and understand control prints readily and with reasonable speed. When you learned to read English, you first learned the 26 letters of the alphabet that are arranged into combinations to form all the words we use. The same thing is true in the language of control. There are only a few basic symbols that are used to express the meaning and purpose of the control circuit. The chief difficulty is that while there is a standard for symbols,[1] there is no real standard usage and sometimes it is necessary to do a little guessing as to what a symbol means.

The symbols used in this chapter are those in most common use.

5·1 SYMBOLS

With reference to Fig. 5·1, symbol 1 represents a normally open contact that is automatically operated. It might represent a line contact on a starter, the contact of a limit switch, the contact of a relay, or any other control device that does not require manual operation. Symbol 2 represents a normally closed automatic contact, and all that applies to symbol 1 applies to symbol 2 also, except its normal position. The method of telling what operates this type of contact will be discussed under Sec. 5·2.

Symbol 3 represents a manually operated, normally open contact of the push-button type. Symbol 4 represents the same type of contact except that it is normally closed. Symbol 3 for the normally open push button should be drawn so that there is space between the dots and the cross bar, but it is not always drawn with such care. If the cross bar is above the dots, the symbol is for a normally open contact, even though the bar may be touching the dots. Symbol 4 should be drawn so that the cross bar just touches the bottom of the dots, but again this is not always done. If the cross bar is below the dots, then it is normally closed even though the bar does not touch the dots. A good way to remember this is to picture the symbol as being a drawing of a push button, which it is. If you push on the button part, represented by the vertical line, the cross

[1] NEMA ICS-1970, part ICS 1-101.

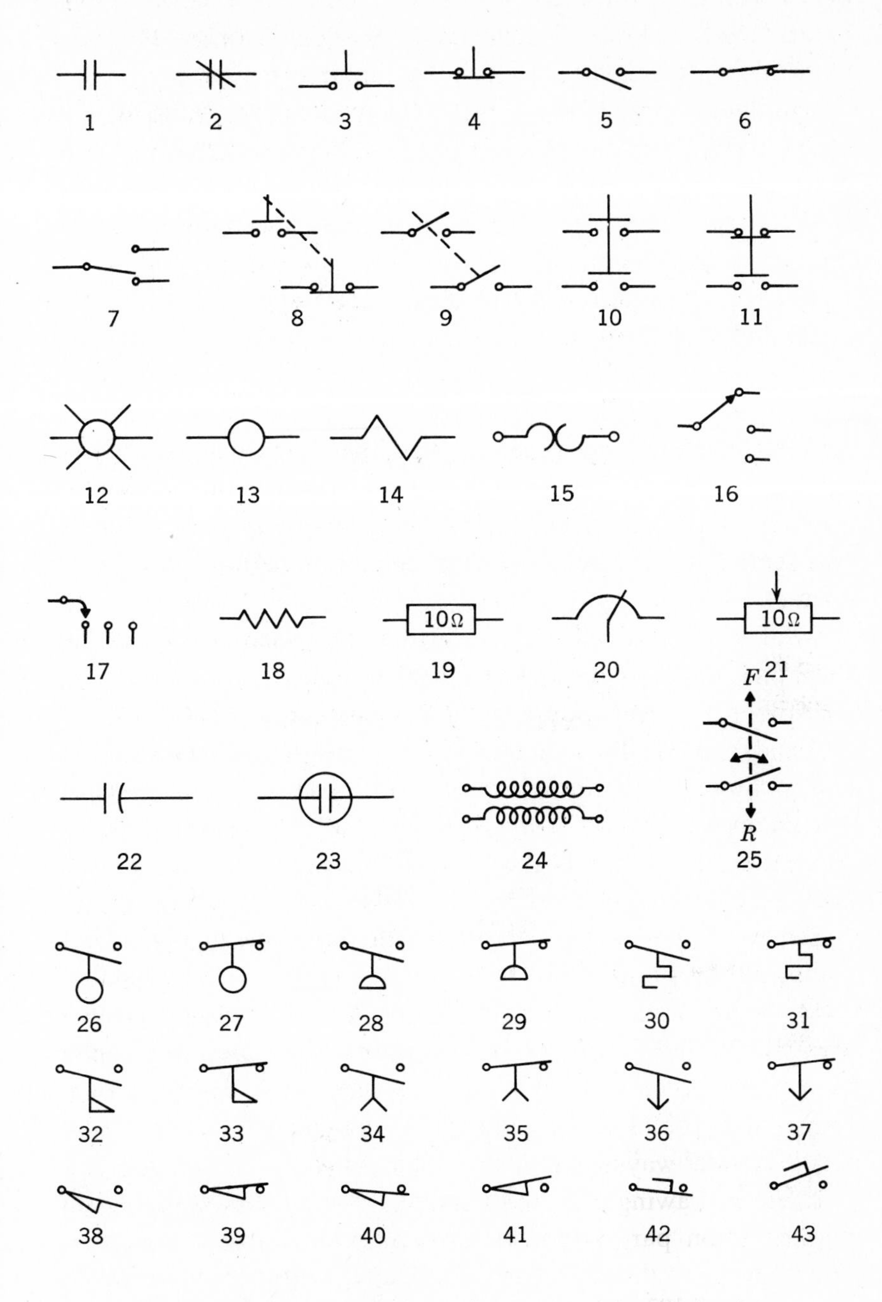

Fig. 5 · 1 Basic symbols used in motor-control circuits.

bar will move downward. When it is above the dots or contact points, the pressure will close them. When it is below the dots, it will move them apart and open the circuit.

Symbols 5 and 6 represent manual contacts of the toggle-switch type, 5 being normally open and 6 being normally closed.

Symbol 7 is a toggle switch of the single-pole double-throw (SPDT) type, where one contact is normally open and the other normally closed.

When more than one set of contacts are operated by moving one handle or push button, they are generally connected by dotted lines, as in symbols 8 and 9. The dotted lines represent any form of mechanical linkage that will make the two contacts operate together. One other method that is used frequently to show push buttons that have two sets of contacts is shown in symbols 10 and 11. Symbol 10 has two normally open contacts, and symbol 11 has one normally open and one normally closed contact.

Symbol 12 is a pilot light which is identified chiefly by the short lines radiating out from the center circle.

Symbol 13 represents a coil. It might be a relay coil or a solenoid coil or the closing coil on a starter. Later we shall discuss how to tell which it is. Symbol 14 also is used to represent an operating coil.

Symbol 15 represents the heating element of an overload relay.

Symbol 16 is a rotary selector switch. The same type of switch is shown by symbol 17.

Symbols 18 and 19 show two ways that resistors are drawn. Symbols 20 and 21 are variable resistors. Capacitors are shown in symbols 22 and 23. Symbol 24 is used to represent a transformer. Symbol 25 shows a plugging switch, which stops plugging action after drive has practically come to rest.

There is a need in some diagrams for clarification of specific devices and their operation; the additional symbols 26 to 43 are NEMA approved. Symbols 26 and 27 represent normally open and normally closed liquid-level switches. Symbols 28 and 29 represent normally open and normally closed vacuum or pressure switches. Symbols 30 and 31 represent temperature activated switches. Symbols 32 and 33 represent flow switches.

Symbols 34 and 35 represent timer contacts which have delay on energizing (TDOE). Symbols 36 and 37 represent timer contacts which have delay on deenergizing (TDODE).

Symbols 38 and 39 represent direct-actuated limit switches. Symbol 40 represents a normally open limit switch which is held closed. Symbol 41 represents a normally closed limit switch which is held open.

Symbols 42 and 43 represent foot switches.

With English or any other language, the meaning of a word depends to some degree upon how it is used, and so it is with the language of control symbols. As we progress through the study of control circuits, we shall develop these few basic symbols into words and sentences that will tell the story of what functions are to be performed by the control components represented in the diagram by symbols.

5.2 DIAGRAMS

The control diagram is the written language of control circuits, and it takes several different forms to fit the particular needs for which it is to be used. As with all languages, the same form will not suit all needs. Some things are better expressed by poetry, while others are best written in prose. There are three general types of control diagrams in use.

The first of these is the wiring diagram (Fig. 5·2*a*), which is best suited for making the initial connections when a control system is first wired or for tracing the actual wiring when troubleshooting.

The second type is a schematic or line diagram (Fig. 5·2*b*), which is by far the easiest to use in trying to understand the circuit electrically. Most circuit diagrams are first developed by drawing a schematic diagram.

The third type is the wireless-connection diagram (Fig. 5·3), which has very few claims to usefulness except that it is compact and saves confusing lines when many wires must be shown. Its chief advantage is for installing already-formed wiring harnesses for factory assembly lines.

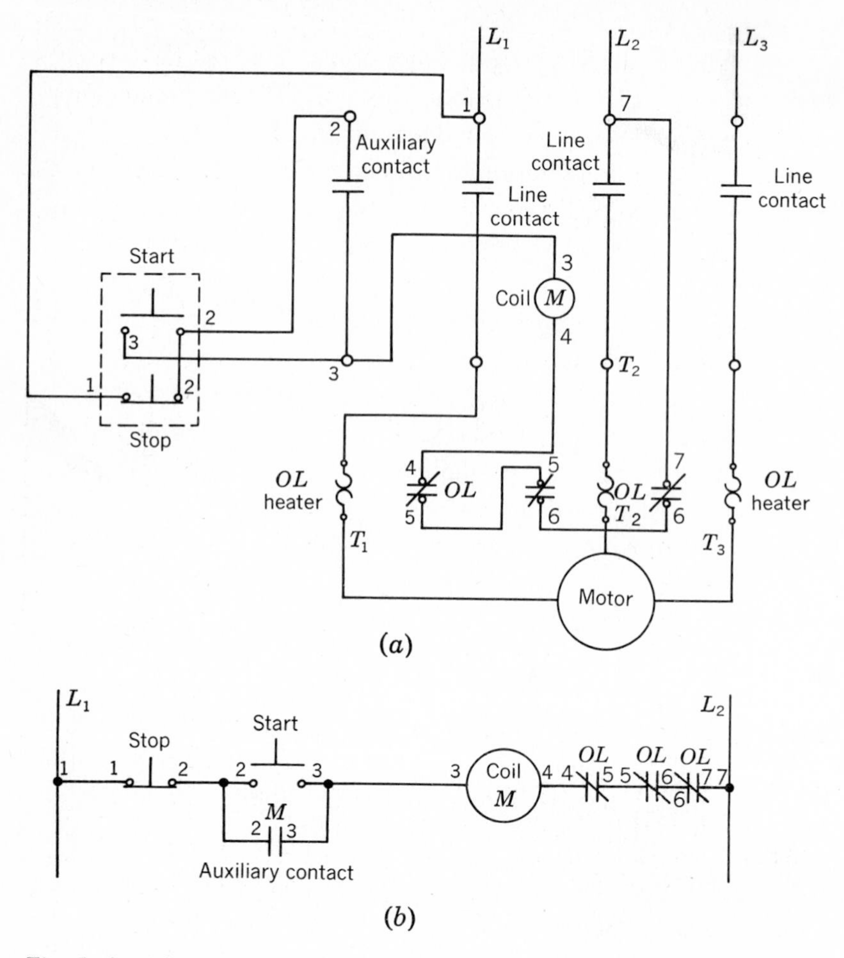

Fig. 5 · 2 (a) A typical wiring diagram. (b) A typical schematic diagram.

5·3 WIRING DIAGRAMS

Wiring diagrams (Fig. 5·2*a*) are developed by drawing the symbol for each component in its proper physical relationship to the other components and then drawing the wires between the proper terminals. In other words, it is a drawing of the equipment and wires more or less as they will be run on the job. Therefore, we can say that the wiring diagram is a representation

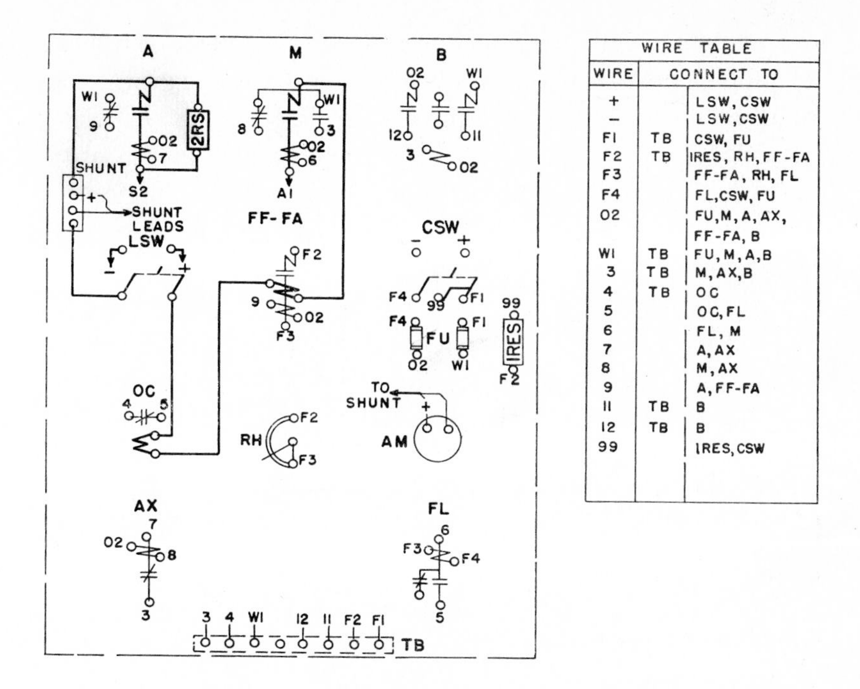

WIRE TABLE

WIRE	CONNECT TO	
+		LSW, CSW
−		LSW, CSW
F1	TB	CSW, FU
F2	TB	1RES, RH, FF-FA
F3		FF-FA, RH, FL
F4		FL, CSW, FU
02		FU, M, A, AX, FF-FA, B
W1	TB	FU, M, A, B
3	TB	M, AX, B
4	TB	OC
5		OC, FL
6		FL, M
7		A, AX
8		M, AX
9		A, FF-FA
11	TB	B
12	TB	B
99		1RES, CSW

Fig. 5 · 3 A typical wireless-connection diagram.

of the control circuit in its proper physical relationship and sequence. Its chief advantage is that it helps to identify components and wires as they are found on the equipment. Symbols as used in the wiring diagram (Fig. 5·2*a*) are usually a pictorial representation of the components with the contacts and coils in their proper physical relationship.

5·4 SCHEMATIC DIAGRAMS

The schematic diagram (Fig. 5·2*b*) is a representation of the circuit in its proper electrical sequence. Assume that you have wired a part of a control circuit beginning at line 1 and continuing through each contact, switch, and coil until you reached line 2. If all the contacts, switches, and coils are free of their mountings and the wire is out in the open, you can take each end of the wire and stretch it tight. What you would see would be a straight wire, broken in places by the contacts, switches, and coils. This is what you see in a schematic diagram. Each line from line 1 to line 2 represents a wire and its associated components as it would appear if stretched out in the above manner. A careful study of the diagrams in this chapter will show you that the more complex circuits have several of these wires or lines stretched out and that each of them is a small circuit within itself.

The chief advantage of the schematic diagram lies in the fact that it shows the circuit in its proper electrical sequence. Each component is shown where it falls in the electric circuit without regard to its physical location. There is no diagram that can compare with the schematic diagram for obtaining an understanding of a control circuit or for locating trouble in a control circuit.

To read a schematic diagram, start at the left-hand side of the top line and proceed to the right. If a contact is open, the current will not go through; if it is closed, the current will go through. In order to energize the coil or other device in the circuit, you need every contact and switch closed to form a complete path. In other words, if there is an open contact, the coil will be dead; if not, it will be energized. Remember that contacts and switches are shown in their normal, or deenergized, position.

The symbols used in schematic diagrams must have some means of telling you what operates them and on what component they will be found. Since we have put them in their electrical instead of their physical position in the circuit, the several contacts of a relay might be scattered from one end of the diagram to the other. In order to identify the relay coil and its several contacts, we put a letter or letters in the circle that represents the coil (Fig. 5·4). Each of the contacts that are operated by this coil will have the coil letter or letters written next to the symbol for the contact. Sometimes, when there are several contacts operated by one coil, a number is added to the letter to indicate the contact number, generally counted from left to right across the relay.

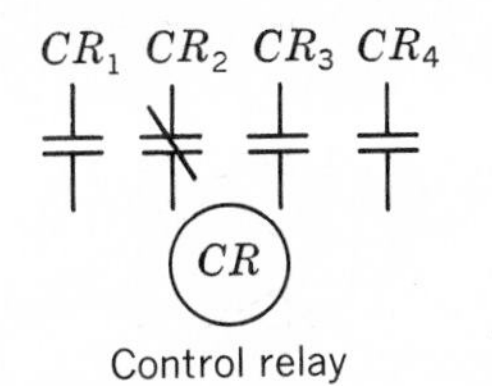

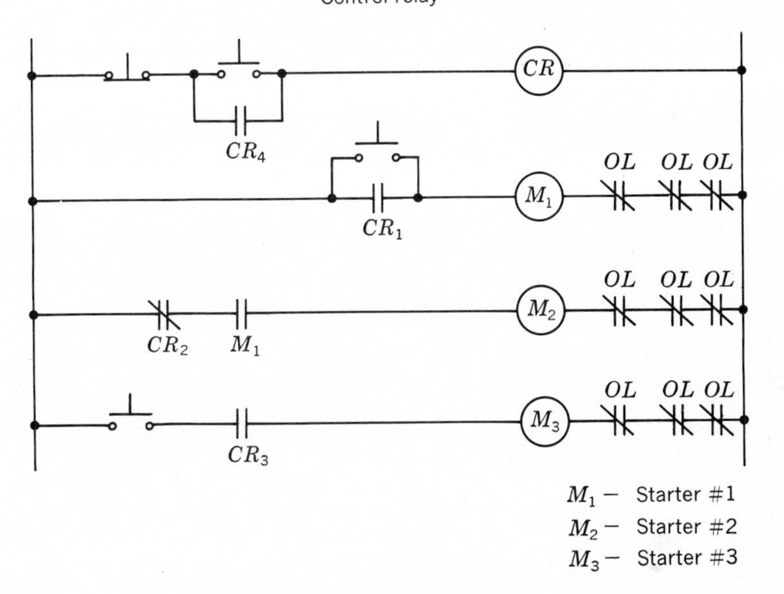

Fig. 5 · 4 Identification of contacts and coils.

While there is a standard[1] for the meaning of these letters, most diagrams will have a key or list to show what the letters mean, and generally they are taken from the name of the device. For instance, the letters *CR* generally are used to indicate the coil of the control relay. The letters *FS* are used frequently to show a float switch. The letters *LS* are used to show a limit

[1] NEMA ICS-1970, part ICS 1-101.

switch. Quite frequently, when several motor starter coils are shown on one control diagram, such as a circuit for sequence operation of several motors, the starter coil may be shown with the letters *M*1, *M*2, *M*3, etc., for the total number of motors.

5·5 DEVELOPING A SCHEMATIC DIAGRAM

In order to see the relationship between the schematic and the wiring diagrams, suppose we develop a schematic diagram from a wiring diagram (Fig. 5·2). This method of development is highly recommended for use in the field when a schematic is needed but not available. The first step is to number the wires of the control circuit. Start where the control wire leaves *L*1, and number each end of each wire. Change numbers each time you start another wire until you reach *L*2, or the end of the control circuit. In Fig. 5·2*a* we numbered the wire from *L*1 to STOP button 1 and placed the numeral 1 at each end. The wire going from the other side of the STOP button to the START button and to terminal 2 of the starter auxiliary contact is numbered at each of its three ends with the numeral 2. The wire that connects the other side of the START button to terminal 3 of the starter auxiliary contact and the starter coil is numbered 3. A wire from the starter coil to the first overload contact is numbered 4. The wire between the two overload contacts is numbered 5. The wire from the second overload contact to *L*2 is numbered 6.

To draw the schematic of this circuit, the first step is to draw two vertical lines, one on each side of the paper (Fig. 5·2*b*). Now draw a short horizontal line to the right from *L*1 and number each end with numeral 1. This represents wire 1 of the wiring diagram, which ends at the STOP button. Draw the symbol for the STOP button at the end of this line. Now draw wire 2 from the STOP to the START button and down to the auxiliary contact. Note that this is shown as an automatic contact operated by coil *M*; therefore it is labeled *M* to show what operates it. Continue the circuit horizontally across the paper, following the numbers on the wiring diagram, until you reach *L*2. Be sure to show each contact in its normal, or deenergized, position. When a wiring diagram is needed and only a schematic is available, the reverse of the above method should be used.

To read the schematic diagram in Fig. 5·2*b*, start at *L*1, which is a hot line, and follow the circuit across the page. First we come to the STOP button. It is normally closed so that the current can flow through, and we can proceed to the START button and auxiliary contact *M*. Both of these are normally open, so the current cannot go any farther. Contact *M* closes when coil *M* is energized, so we cannot complete the circuit that way. The START button can be pushed, which will close its contacts and allow current to flow to the coil *M* and on through the two normally closed overload contacts, marked *OL*, to *L*2. This completes the circuit to coil *M*, and it closes the starter and contact *M*. When we release the START button, thus opening its contacts, the coil does not drop out, because contact *M* is now held closed by coil *M*. The motor is now running and will remain so until the control circuit from *L*1 to *L*2 is broken.

To stop the motor manually, all that is needed is to push the STOP button, which interrupts the circuit at this point, causing an interruption of current to coil *M* and dropping out the contacts of the starter. Contact *M* being operated by coil *M* is now open, so that when we release the STOP button, the coil is not energized again. Note that when the motor draws too much current, one or both overload contacts will be opened, thus interrupting the circuit between coil *M* and *L*2. The result of opening the circuit at this point is the same as that of pushing the STOP button.

While this is a simple circuit and fairly easy to follow on either diagram, the same system for the development of the schematic diagram and analyzing the control circuit will work regardless of the complexity of the circuit.

Suppose now that we add one more START-STOP station to the circuit in Fig. 5·2*a*. The new circuit is shown in Fig. 5·5*a*. If you look at the numbering of the wires in the wiring diagram, you will see that it goes from one to seven. This increase in total numbers is caused by the insertion of the extra STOP button

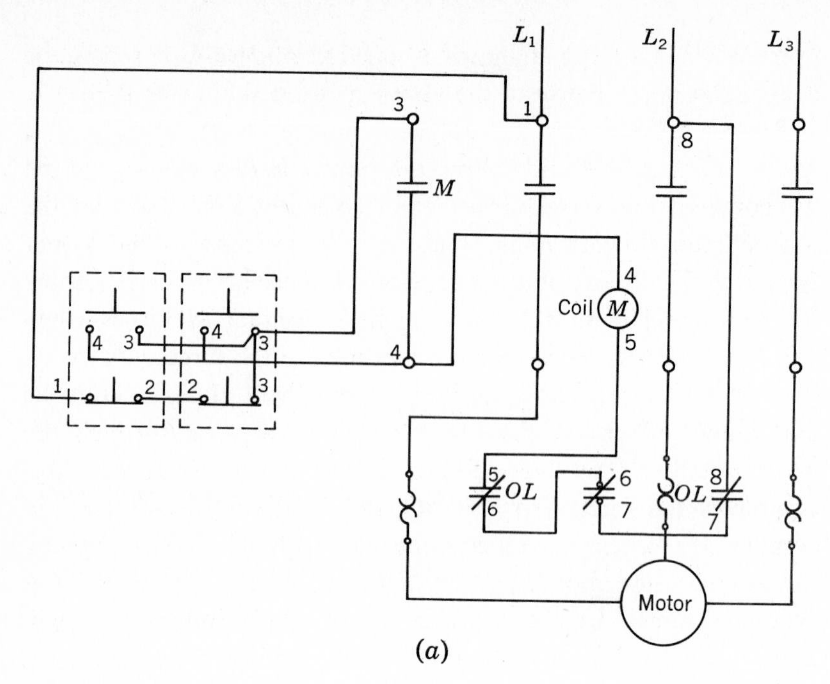

(a)

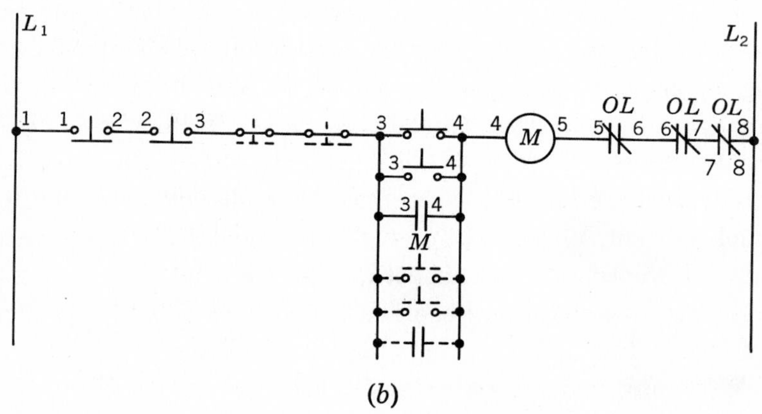

(b)

Fig. 5 · 5 The basic T *formation.*

in the circuit. If we follow the same technique used on Fig. 5·2*b*, we shall develop the schematic diagram of Fig. 5·5*b*, which starts at *L*1 and proceeds horizontally through the first STOP button, the second STOP button, and the first START button, which is paralleled by the second START button and auxiliary contact. From there it proceeds to the coil of the starter marked *M* and thence to the first overload contact and the second overload contact to *L*2.

As in the preceding circuit, the STOP buttons will be closed in their normal position so that current can flow from *L*1 as far as the parallel group of START buttons and the auxiliary contact. Current can flow from *L*2 through the normally closed overload contacts to the coil *M*, so that now all that is needed to start the motor is to close one of the START buttons. Since the START buttons are in parallel, either of them will complete the circuit from *L*1 to coil *M*, so that it makes no difference which one is pushed in order to energize the coil.

It should be noted that there are two additional STOP buttons, two additional START buttons, and an additional contact shown dotted on this diagram. They indicate additional controls as they would be added to this circuit. Careful note should be taken of this diagram, since it is the basic T formation which is assumed by any circuit with multiple control components used to control a single coil. You will note that all the STOP buttons are connected in series from one side of the line or the other. The start components, in this case consisting of two or more START buttons and one or more contacts, are in parallel. The value of this T formation lies in understanding that if any control component, regardless of its type, is to be used to stop the motor, it will be placed in series with the STOP button; if it is to be used to start the motor, it will be placed in parallel with the START button. In short, if you can draw, read, and understand the circuit of Fig. 5·2*b*, then you can develop more complex circuits by the addition of components to perform the function of stop in series with the original STOP button and to perform the function of start in parallel with the original START button.

5·6 ADDING CIRCUIT ELEMENTS

Suppose that you are instructed to add a limit switch, float switch, or push button to an existing circuit. Then, if this component is to be used to stop the motor, all that is necessary is that you locate the wire connecting $L1$ to the STOP button or other components and break it at some point through the new control component.

Suppose instead that you are required to install a control component such as a limit switch, float switch, or push button to perform the function of start for the motor. Then all that is required is that you parallel the new component with the existing start component. These additional components are represented in Fig. 5·5*b* by those components shown dotted.

It should be noted also that components used to perform the function of stop are normally closed components. That is to say, their contacts are in the closed position whenever the component is deactivated. Those components which are to perform the function of start are normally open components. In other words, their contacts are open in their deactivated state. There is no limit to the number of components that can be added in series with the STOP button of the simple circuit shown in Fig. 5·2*b* to perform the function of stop, nor is there any limit to the number of components that can be added in parallel with the START button to perform the function of the start for the coil M.

Consider the circuits shown in Fig. 5·6. The top circuit is the same as that shown in Fig. 5·5*b*. The current can flow through both of the normally closed STOP buttons as far as the START buttons. All that is required to energize the coil is to push the START button, thus closing its contacts and energizing coil $M1$. Coil $M1$, in turn, closes contact $M1$ in parallel with the START buttons, thus maintaining the circuit to coil $M1$.

Now look at the bottom circuit, and you will see that the current can flow from $L1$ through the normally closed contact only as far as the normally open contact $M1$. This contact must be closed in order to energize coil $M2$ through the START button. This contact has an identification $M1$, which indicates that it would be closed whenever coil $M1$ is energized. This means, then, that the motor which is energized by coil $M1$ must be running before we can start the motor which is energized by coil $M2$. If we start motor $M1$ by pushing the START button, coil $M1$ is energized, thus closing both contacts labeled $M1$. The contact in parallel with the START buttons is used to maintain the circuit to coil $M1$. The contact in the lower circuit, labeled $M1$, will be closed and will allow current to flow as far as the START button. When this START button is pushed, current can reach coil $M2$. Energizing this coil closes contact $M2$, maintaining the coil circuit and permitting the second motor to run.

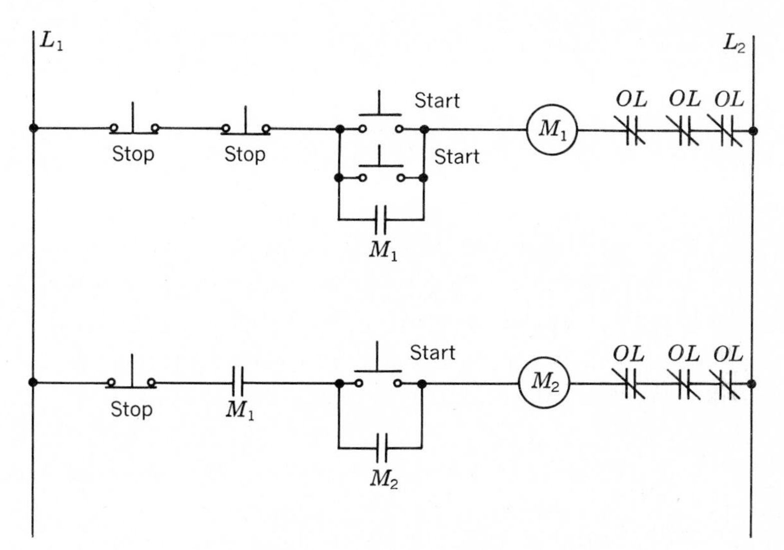

Fig. 5·6 Interlocking.

Consider what happens if we push the STOP button of coil $M1$. This will break the circuit and deenergize coil $M1$, thus dropping out all its contacts. This will open the maintaining contact in parallel with its START buttons and the contact in series with the STOP button of coil $M2$. Opening of the contact in series with this STOP button will deenergize coil $M2$, which will drop out its contact $M2$, and both motors will be stopped, even though the button we pushed was in the circuit for motor $M1$. Circuits of this type are frequently used for multiple conveyor-belt operation, where the first conveyor must not run unless the following conveyor is running, thus preventing material from piling up where the two conveyors converge.

Actually we have considered only three fairly simple basic control circuits. These circuits, however, represent the majority of conditions that are found in the most complex control circuits. The same type of analysis of the operation of the electric circuit will enable you to understand many circuits which now might puzzle you considerably. In Chap. 7 we shall consider many more complicated circuits and develop a system for analyzing their operation.

Summary

In this section we have discussed in some detail the schematic diagram. The emphasis has been put on the schematic diagram because it is the type of diagram which transmits to the reader the most concise and understandable electrical information about the control circuit. The same procedure for reading and understanding the control functions of a circuit from the schematic diagram also applies to the wiring diagram. When it is necessary to use a wiring diagram to analyze or understand the control circuit, it is necessary that you trace each wire, beginning at the source of power and noting each component or contact that is in the circuit and what its function might be. It is highly recommended that on the more complicated circuits, if a schematic diagram is not available, you develop such a diagram using the methods put forth in this chapter. Understanding the circuit will be much easier when this procedure is followed.

To read a wireless-connection diagram, the same principles apply except that you must find the proper component by comparing the number of the wire indicated as it leaves a terminal of each piece of equipment. Caution is needed to be sure that you have found all of the places where a particular wire is connected by finding all the points labeled with the same numeral. Again it is suggested that you develop a schematic diagram from the wireless-connection diagram before attempting analysis of the control circuit if it contains more than a very few components.

Should it be desired to have a wiring diagram when only a schematic diagram is available, one can be developed by applying the reverse of the procedure outlined for the development of a schematic diagram from a wiring diagram. Draw each component to be used in the circuit in its proper physical relationship. Now number each wire on the schematic diagram as we have been doing. Then number each terminal of each component on the wiring diagram as it is numbered on the schematic. All that is left is to connect corresponding numbers by wires or lines on the drawing, and you will have a wiring diagram which represents the electric circuit shown on the schematic diagram.

In order to understand the symbols as found on drawings or diagrams made by various people, you should study manufacturers' booklets and control-circuit diagrams that can be obtained from the manufacturers of control equipment. This study will enable you to become familiar with the many types of symbols used to represent a single component just as your understanding of the spoken or written word in English depends upon the size of your vocabulary. The knowledge of words and phrases in English makes it easier for us to understand the spoken and written word as it is presented by various people, and so it is with control circuits. The greater your knowledge of the symbols in use and the components that will be used to perform the functions of control, the better will be your understanding of the various diagrams and circuits as drawn by the many people engaged in this work.

Review Questions

1. Draw a symbol for a START push button.
2. What method is used to show two contacts which are mechanically connected so that they operate simultaneously?
3. When a push button is intended to be normally closed, is the cross bar drawn above or below the contact dots?

4. What is the chief advantage of a wiring diagram?
5. What is the chief advantage of a schematic diagram?
6. What is the chief advantage of a wireless-connection diagram?
7. Components which are to be used to perform the function of stop are connected in__________with each other.
8. Components which are to perform the function of start are connected in__________with each other.
9. How are contacts identified to show what operates them?
10. How is a schematic diagram developed from a wiring diagram?
11. How is a wiring diagram developed from a schematic diagram?

6

DEVELOPMENT OF CONTROL CIRCUITS

Control circuits are very seldom drawn or designed as a complete unit. Rather, they are developed one step at a time to provide for each control function that they will be expected to perform. It is very much like writing a letter when the writer has the general subject that he wishes to convey in mind. He proceeds, sentence by sentence, to put this idea on paper. The same procedure should be followed in developing a control circuit. You must have all of the control functions in mind when you start, so as to provide for each function in its proper relationship to all other functions as the circuit is developed.

6·1 TYPES OF CONTROL CIRCUITS

There are two basic types of control circuits: three-wire circuits and two-wire circuits. These designations stem from the fact that for three-wire circuit control only three wires are required from the ordinary across-the-line motor starter to the control components, and in the two-wire circuit only two wires are required (Fig. 6·1).

The three-wire control circuit requires that the primary pilot-control components be of the momentary-contact type, such as momentary-contact push buttons. Maintained-contact devices, such as limit switches and float switches, may be used in various parts of the circuit to supplement the primary start and stop control devices. This type of control is characterized by the use

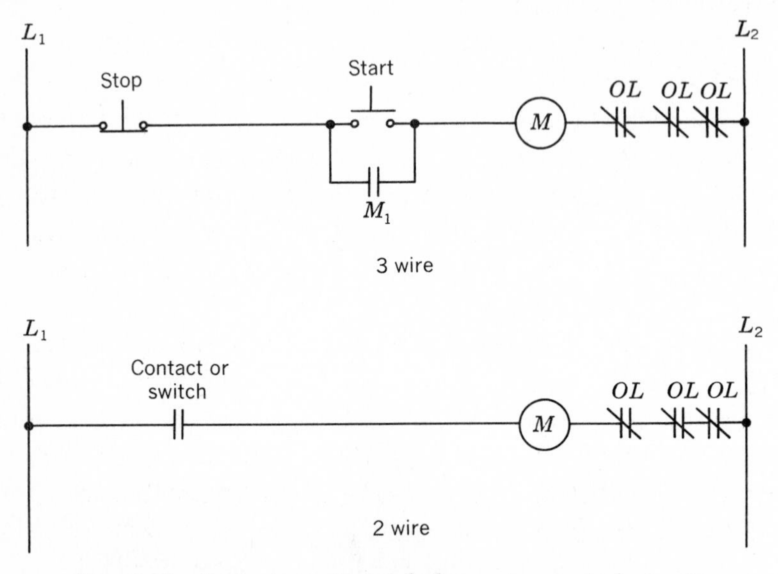

Fig. 6·1 Basic two-wire and three-wire control circuits.

of the auxiliary contact on the starter to maintain the coil circuit during the time that the motor is running.

The two-wire control circuit uses a maintained-contact primary pilot-control component, which may be a simple single-pole switch, a maintained-contact push-button station, or any type of control component which will close a set of contacts and maintain them in that position for as long as the motor is to be running. The opening of this contact or contacts stops the motor by dropping out the coil of the starter.

All control circuits, regardless of how complex they may be, are merely variations and extensions of these two basic types. It is the purpose of this chapter to show how each of these basic circuits can be developed into the necessary control of a motor or motors by the addition of contacts or push buttons operated by one or more of the various control components to perform the control functions desired. We shall use the schematic diagram for development of all control circuits because it is the type of diagram which lends itself most readily to the development of control circuits.

The simplest method for the development of a control circuit is to start with the coil and the overloads. Add the primary start and stop control device, which generally, in a three-wire circuit, consists of a STOP and START push button used in conjunction with the auxiliary contact of the starter. To this circuit all contacts or push buttons that are to be used to perform the additional control functions are added one at a time until the final circuit has been developed.

Keep in mind, when considering a three-wire control circuit, that all devices intended to perform the function of stop must be normally closed devices and will be located in series with the original STOP button. All devices which are to perform the function of start must be normally open control devices and will be connected in parallel with the original START button.

Sometimes a circuit requires that two or more normally open contacts or push buttons must be closed before the function of start can be performed. These contacts or push buttons would be connected in series, and then the series connection paralleled with the original start components. If it is desired that several contacts or push buttons will have to be opened before the function of stop is performed, then these normally closed contacts or push buttons will be wired in parallel and then connected in series with the line to perform the function of stop.

When there is a definite sequence to the action of the various control components, you should add them one at a time to your control circuit in the same order as their operating sequence.

Be sure to check the circuit for proper electrical operation after each contact or push button has been added to be sure that you have not interfered with the proper function of any control component which has already been placed in the circuit.

If you have mastered the preceding part of this book, you should have the necessary knowledge of control functions, control components, and circuit diagrams to begin to learn to develop control circuits. Until you can develop a circuit to perform desired functions, it is doubtful that you will be able to interpret or analyze someone else's control circuit as to its proper operation and the functions which it is designed to perform.

6·2 DEVELOPMENT OF CIRCUIT 1

In order that the step-by-step method of circuit development can be made more clear to you, we shall consider our first circuit as a series of jobs done at different times to improve the operation of the original circuit.

The existing control circuit (Fig. 6·2*b*) is to control a pump which pumps water from a storage tank into a pressure tank. The physical arrangement of the pump and the two tanks, along with the final control components, is illustrated in Fig. 6·2*a*. As the original circuit stands, it is a manual operation requiring that the START button be pushed whenever the water is too low in the pressure tank. The pump is allowed to run until the tank is observed to be full. The operator then pushes the STOP button, securing the pump and stopping the flow of water into the pressure tank.

The owner now desires that a float switch be installed in the pressure tank near the top so that the operator need only push the START button, thus energizing the pump and starting water to flow into the tank. When the level of the water has reached float switch 1, its contacts will be opened, thus stopping the pump and the flow of water. The function to be performed by float switch 1 is that of stop. Therefore, it must be a normally closed contact and must be connected in series with the original STOP buttons, as shown in Fig. 6·2*c*.

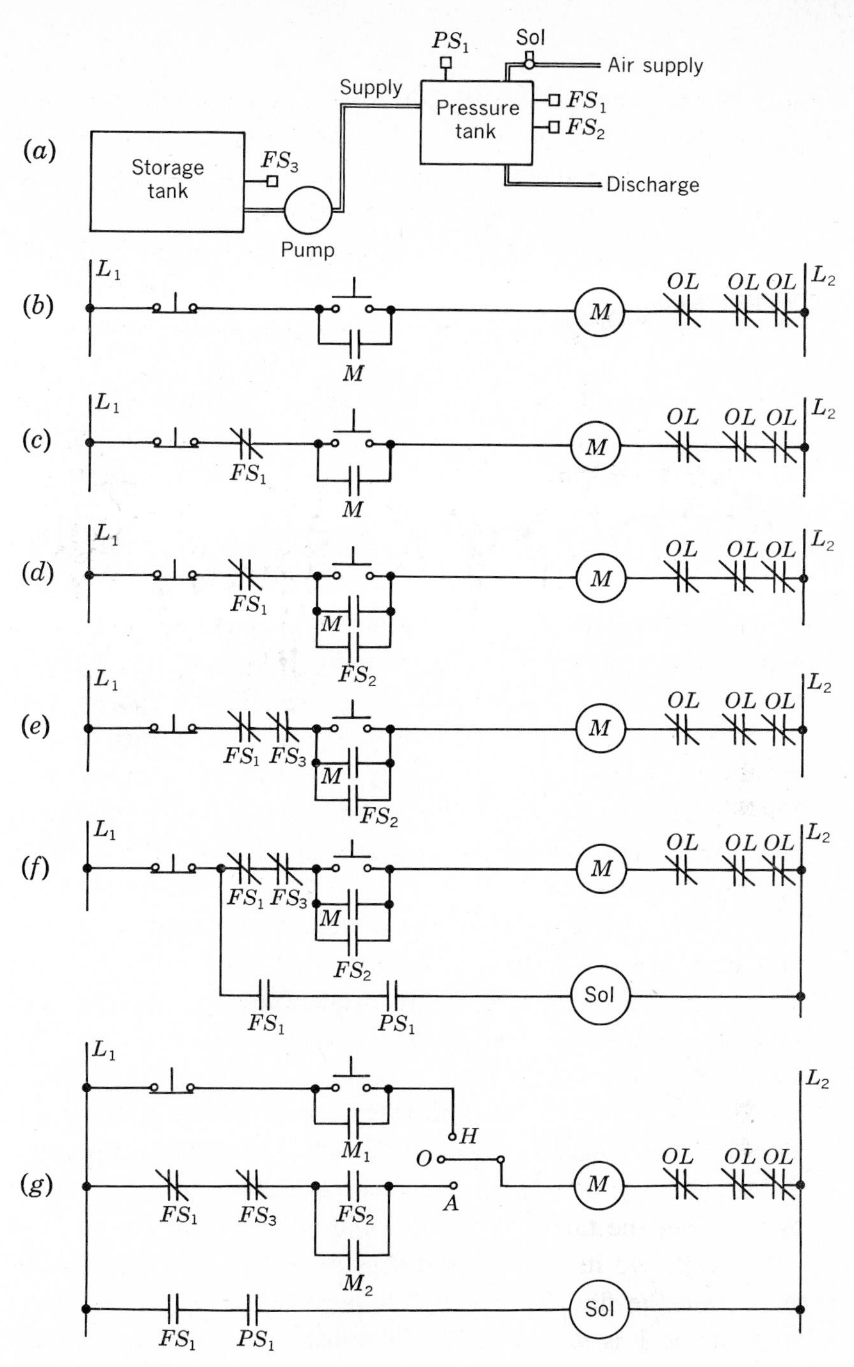

Fig. 6 · 2 Development of circuit 1. Automatic control for a water pump.

After operating with this control for some time, the owner decides that it will be more convenient if the pump is started automatically as well as stopped automatically. He requests that another float switch be installed to maintain the lower level of the tank. This version of the control circuit requires that the pump be started whenever the water reaches a predetermined low level. The control function desired is that of start, so the float switch must have a set of normally open contacts that will be closed whenever the water drops to the lowest desired level. These contacts must be connected in parallel with the original START button so as to perform the function of start for the motor. This connection is shown in Fig. 6·2*d*.

After some time of operating with the new control circuit, it is discovered that occasionally the storage tank drops so low in water level that the pump cannot pick up water. The owner requires a control to prevent the pump from starting whenever the storage tank is low in water. Even though this control does not necessarily stop the pump while it is running, it must prevent its starting whenever the water is low. It must also stop the pump if it is running and the water reaches this low level in the storage tank. Thus, the new control performs the function of stop for the pump.

This function of control can be obtained by the installation of a float switch to sense the extreme low level of water in the storage tank. Float switch 3 was installed and adjusted to open a set of contacts whenever the water in the storage tank reached the desired low level. Because the control function to be performed is that of stop, float switch 3 must have normally closed contacts which will be opened whenever the water level drops to the set level of the float switch. It is wired in series with the other stop components, as shown in Fig. 6·2*e*.

Later it is decided that the pressure placed on the line by the pressure tank when it is full is insufficient for the needs of the plant. The owner requests the installation of the necessary components and controls to maintain a pressure on the tank by the addition of the proper amount of air to the top of the tank. In order for the proper balance of water level and air pressure to be maintained at all times, air must be let into the tank only when the water level is at its highest position and the pressure is under the desired discharge pressure of the tank.

In order to achieve this, suppose that we install a solenoid valve in the air-supply line which will allow air to flow into the tank only when the coil of the solenoid valve is energized. Now we can install a pressure switch in the top of the tank which will sense the pressure in the tank at all times. This pressure switch will perform the function of start for the solenoid valve. When the pressure is lower than the set point of the pressure switch, its contacts must close and complete the circuit through it to the solenoid. If, however, the water is below its top level when the pressure drops, we do not want the solenoid valve to open. Therefore, we require the function of stop in regard to water level, to prevent air being put into the tank when it is not desired.

If float switch *FS*1 is of the double-pole variety, having one normally open and one normally closed set of contacts, we can wire it into the circuit as shown in Fig. 6·2*f*. The circuit for the solenoid valve is a two-wire control circuit requiring that both *FS*1 and pressure switch *PS*1 be closed in order that air will be placed in the tank by the energizing of the solenoid valve. When the water level reaches its highest point, *FS*1 will be activated. The normally closed contact in the pump circuit will open and the normally open contact in the solenoid circuit will close. If the air pressure is low, the contacts of *PS*1 will be closed and air will flow into the tank until either the water level drops and opens *FS*1 or the pressure increases to normal and opens *PS*1, thus satisfying the requirements of the control circuit as specified by the owner of the plant.

While the circuit of Fig. 6·2*f* gives a degree of hand operation because the push buttons were left in the circuit, it will be preferable to have either a definite hand operation or a definite automatic operation, as desired by the operator. The necessary changes required to give hand, off, and automatic operation to the circuit are shown in Fig. 6·2*g*.

If you had been charged with the responsibility of developing the final circuit of Fig. 6·2*g*, you would have had certain specifications or requirements as to the proper function or operation of the completed circuit. The first of these probably would have been that it have hand, off, and automatic control selection, the second that the pump be controlled so as to maintain the water level in the pressure tank between a high and low point, third, that the pump be prevented from running whenever the water levels in the storage tank were below a given point, and fourth, that the pressure on the pressure tank be maintained by adding air to the tank whenever necessary. To develop this circuit properly from this set of specifications, the procedure would be the same as that we have followed, assuming that the circuit was built up a little at a time by going back and adding control components to the original manual circuit.

6·3 DEVELOPMENT OF CIRCUIT 2

Our second circuit will be for the control of three conveyors so arranged that conveyor 1 dumps material onto conveyor 2, which in turn dumps its material onto conveyor 3, which is used to load trucks or other vehicles at the shipping dock or a warehouse.

The specifications for the operation of the circuit are:

1. One push button is to start all conveyor motors in sequence from 3 to 1 or, in other words, from the last to the first.
2. An overload on any conveyor will stop all conveyors.
3. One STOP button will stop all conveyors in sequence from 1 through 3 or, in other words, from first to last.

An additional requirement is that there be a 2-minute delay between the stopping of each conveyor in the sequence in order that the material on the following conveyor might clear each conveyor before it is stopped.

If we are to develop this circuit step by step, then our first step is to meet the requirements of specification 1 that a single push button start all conveyors in sequence starting with conveyor 3. The circuit for this will be found in Fig. 6·3*a*. Here you will find a control relay which is started and stopped by a three-wire push-button control. It is maintained during the run operation by a set of contacts on the control relay, identified on the drawing by the letters *CR*1. Since conveyor 3 is required to be the first conveyor to start, the contacts identified on the drawing by *CR*2, which are closed by the control relay, are connected between the starter coil and the line, giving two-wire control for conveyor 3. This conveyor will start when the control relay is energized and stop when the control relay is deenergized.

In order that conveyor 2 be prevented from starting until conveyor 3 is running, we can use the auxiliary contacts on starter *M*3 for conveyor 3 to energize the coil of conveyor 2. These contacts are identified as *M*3 to indicate that they are closed by energizing coil *M*3. The use of this contact satisfies the condition that conveyor 2 start in sequence, following conveyor 3.

By the same token, if we use the auxiliary contact of the starter for conveyor 2 to energize the starter of conveyor 1, it must follow in sequence behind conveyor 2. The contacts identified on the drawing as *M*2 are connected in series with the coil *M*1 for conveyor 1, thus satisfying the condition that conveyor 1 start after conveyor 2 in its proper sequence. We have now satisfied the conditions of specification 1 and are concerned with specification 2, which requires that an overload on any one conveyor will stop all conveyors.

The conditions of specification 2 can be obtained by series connection of all overload contacts between the line and each of the starter coils and the coil of the control relay, as indicated in Fig. 6·3*a*. If any one or more of these six overload contacts open, the control circuit to all coils is broken, thus deenergizing the coils and stopping all the conveyor motors at the same time. We have now fulfilled the requirements of specification 2.

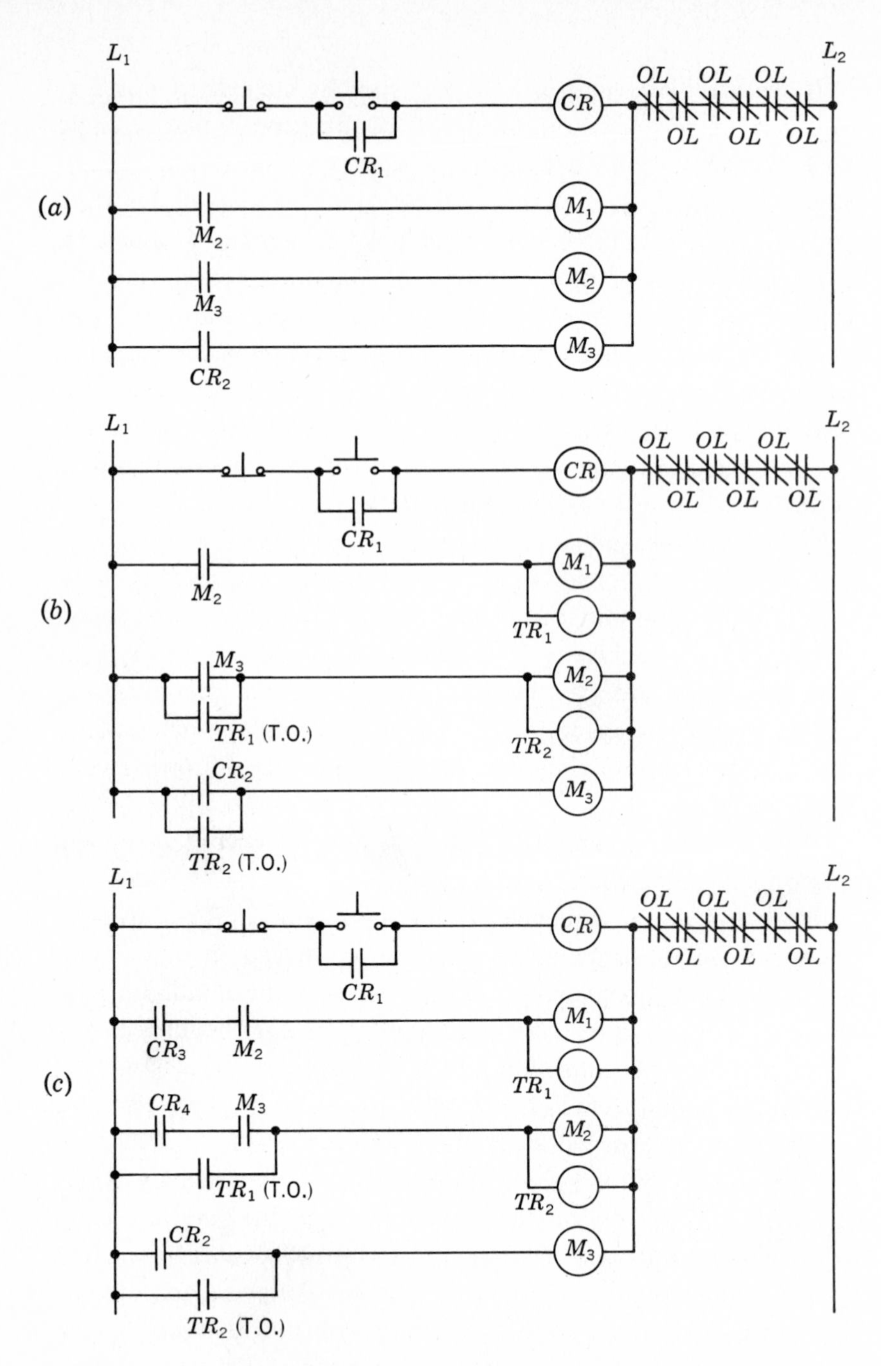

Fig. 6·3 Development of circuit 2. Sequence control for three conveyors.

While the circuit of Fig. 6·3*a* satisfies the first and second specification, it does not meet the conditions of specification 3 that conveyors stop in the reverse order. The requirement that the conveyors stop in reverse order and that they have a 2-minute time delay between the stopping of each conveyor in sequence indicates the use of timing relays with time opening (TO) contacts. The first inclination is to connect them as indicated in Fig. 6·3*b*. A careful study of this circuit, however, should reveal to you that when the STOP button is pushed, the control relay will drop out, opening contacts *CR*1 and *CR*2, which will only result in the control relay being deenergized, because contact *M*2 is still closed, maintaining the circuit to coil *M*1, contact *M*3 is still closed, maintaining the circuit to coil *M*2, and contact *TR*2 is closed, maintaining the circuit to coil *M*3. Thus, all conveyor motors continue to run. A modification of this circuit must be made in order that the conveyors may be stopped by the pushing of the STOP button.

In order to satisfy condition 3 of the specifications, the circuit, Fig. 6·3*b*, must be modified as indicated in Fig. 6·3*c*. In this circuit we have added two contacts, normally open, operated by the control relay and identified on the drawing as *CR*3 and *CR*4. Now when the STOP button is pushed, the control relay drops out, opening all its contacts and isolating each conveyor-motor starter from the line except for the time-delay relay contacts, which are held closed by time-delay relay 1 and time-delay relay 2. The opening of contact *CR*3 breaks the circuit to coil *M*1, thus stopping conveyor 1. The contact identified as *TR*1 is time opening; therefore the circuit to coil *M*2 is maintained for a period of 2 minutes, which is the setting of time-delay relay 1. At the end of this 2-minute period, the contact *TR*1 will open and drop out coil *M*2, thus stopping the second conveyor in accordance with the specifications. This starts the timing action of time-delay relay 2, and after a period of 2 minutes its contact (*TR*2) will open and drop out coil *M*3, thus stopping conveyor 3.

We have now satisfied all the specifications for this circuit. The conveyors will start in sequence, beginning with 3 and progressing to 1 by the pushing of the single START button. Any overload on any conveyor will drop out all starter coils, thus stopping all conveyors. When the STOP button is pressed, the conveyors will stop in the reverse order to that in which they started, with the delay of 2 minutes between the stopping of each conveyor.

This circuit performs the functions of start, stop, sequence control, overload protection, and time-delay action. The use of the control relay with its three-wire control circuit provides low-voltage protection not possible with the two-wire control circuit to each of the conveyor starters.

6·4 DEVELOPMENT OF CIRCUIT 3

This circuit will be for forward and reverse control of a motor. The specifications state that it must have three-wire control to give low-voltage protection. It must have electrical interlock, and the STOP button must be pushed in order to change direction of rotation of the motor. The first step in the development of this circuit is to provide start and stop in the forward direction. The circuit for this is shown in Fig. 6·4*a*. You will notice that this is the ordinary three-wire push-button control circuit and satisfies the requirement that the motor start and stop in the forward direction with three-wire control.

The second provision of the circuit is that it start and stop in the reverse direction, which is accomplished by the addition of a START button and auxiliary contact as shown in Fig. 6·4*b*. The START button is wired behind the STOP button so that only one STOP button will be required to stop the motor, regardless of the direction in which it is running.

The requirement that electrical interlock be used is satisfied by the addition of the contacts shown in Fig. 6·4*c* and identified by the letters *R*2 and *F*2, which are auxiliary contacts on the forward and reverse starters. The normally closed contact *R*2 will be opened whenever coil *R* is energized, thus preventing coil *F* from being energized at the same time. Contact *F*2 will be open whenever coil *F* is energized, thus preventing the reverse starter from being energized at the same time. This circuit satis-

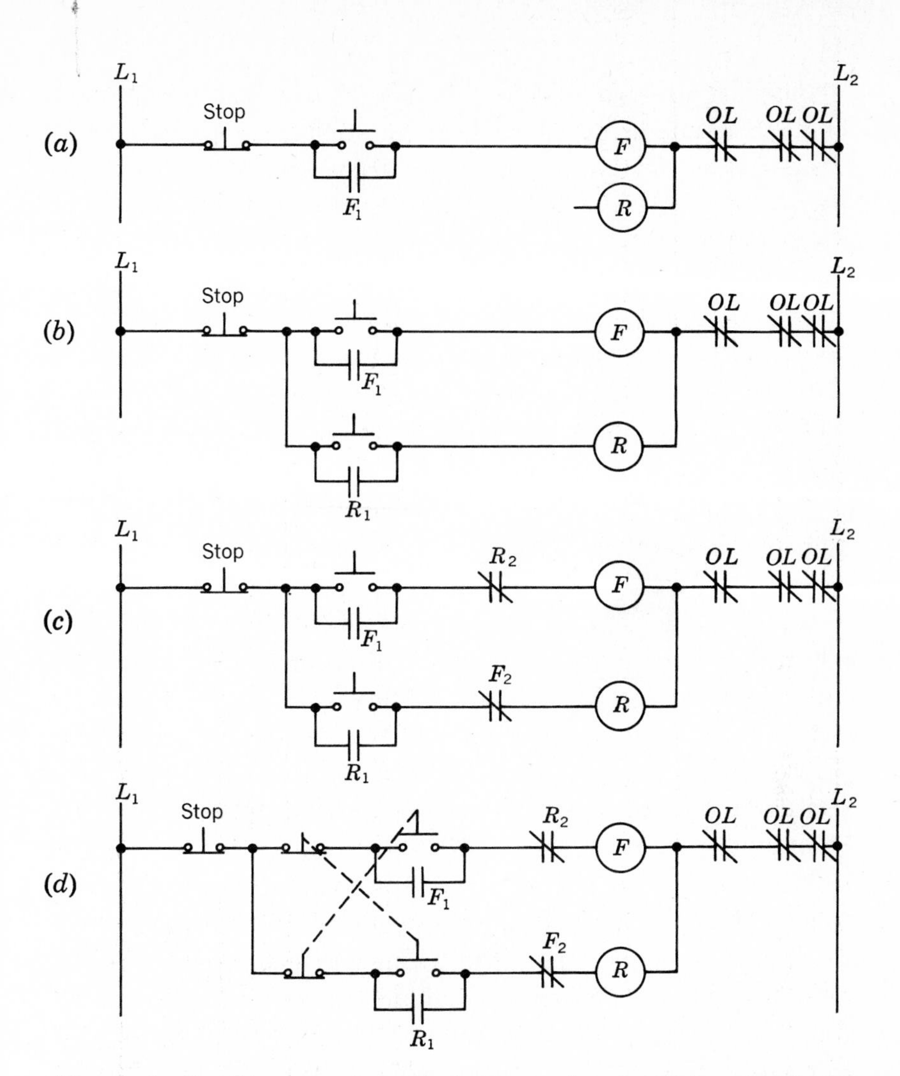

Fig. 6 · 4 Development of circuit 3. Forward and reverse control of a motor.

fies fully the specifications that the motor be able to start and stop in either the forward or reverse direction and that the STOP push button must be pressed in order to change from forward to reverse or from reverse to forward. Electrical interlock has been provided so that both starters cannot be energized at the same time.

Suppose now that plugging reversal is required on this machine. The circuit would have to be modified as shown in Fig. 6·4*d*. The START push buttons for forward and reverse would need to be of the double-pole type, having one set of normally open and one set of normally closed contacts. When we push the forward START button, it closes the circuit for the forward starter and at the same time breaks the circuit for the reverse starter. When the reverse START push button is pushed, it not only will complete the circuit to the reverse starter but also will break the circuit to the forward starter, thus giving plugging action.

This circuit performs the control functions of forward-reverse control, start, stop, interlock, overload protection, plugging, and low-voltage protection.

6·5 DEVELOPMENT OF CIRCUIT 4

The specifications for this circuit are as follows. It must give limit-switch control for forward and reverse running of the motor by the use of momentary-contact limit switches. It must also provide low-voltage protection. The initial start and stop for the control system will be by momentary-contact START and STOP push buttons.

The requirement that START and STOP push buttons be used to initiate a control of the circuit by limit switches would indicate the use of a control relay. The wiring for this is shown in Fig. 6·5*a*. Contact *CR*1 is used to maintain the circuit to the control relay during the running operation of the circuit. Contact *CR*2 is used to make and break the line circuit to the forward and reverse control circuit, thus satisfying the provision that the START and STOP buttons initiate and terminate the automatic control of the motor by limit switches. The use of the control relay and its START and STOP buttons also provides low-voltage protection.

The specifications call for the use of momentary-contact limit switches, which would require a three-wire control circuit for forward and reverse. These limit switches necessarily have two sets of contacts, one normally open and the other normally

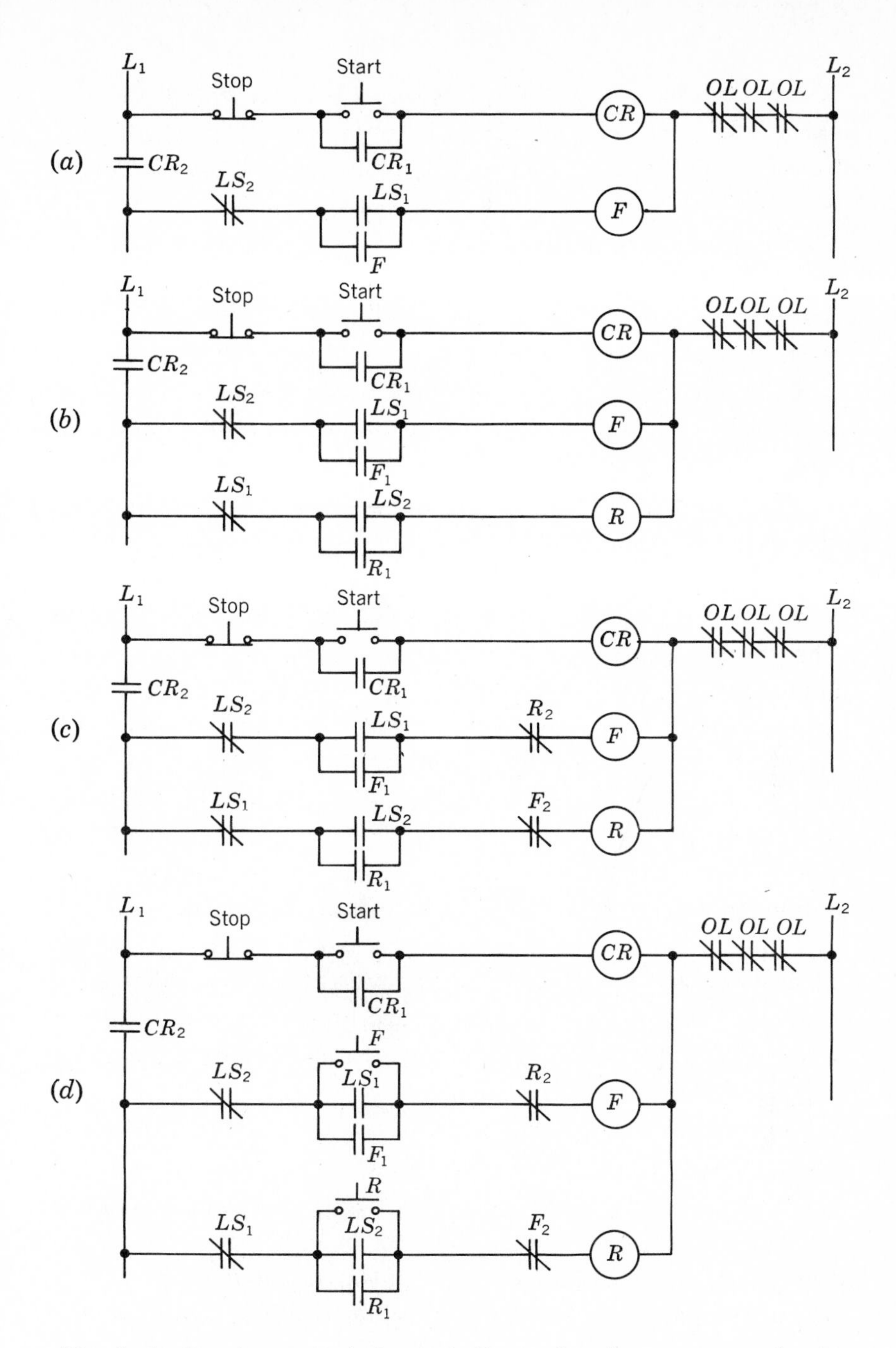

Fig. 6 · 5 Development of circuit 4. Forward and reverse control using limit switches.

closed. When wired as shown in Fig. 6·5*a*, the normally closed contact of limit switch 2 would act as the stop for the forward controller, and the normally open contact of limit switch 1 would act as the start contact for the forward controller. The auxiliary contact of the forward starter must be connected in parallel with the normally open contact of limit switch 1 in order to maintain the circuit during the running of the motor in the forward direction.

Figure 6·5*b* shows the additional wiring required for the reverse starter. The normally closed contact of limit switch 1 is wired as a stop contact for the reverse starter, and the normally open contact of limit switch 2 is wired as a start contact for the reverse starter. The auxiliary contact on the reverse starter is wired in parallel with the normally open contacts of limit switch 2 to maintain the circuit while the motor is running in reverse.

This circuit satisfies all the requirements of the specifications with the exception of electrical interlock, which is shown in Fig. 6·5*c*. This electrical interlock is accomplished by the addition of a normally closed contact in series with each starter and operated by the starter for the opposite direction of rotation of the motor.

Plugging reversal is provided in this circuit by the action of the limit switches themselves. When limit switch 1 is moved from its normal position, the normally open contact closes energizing coil *F* and the normally closed contact opens and drops out coil *R*. The reverse action is performed by limit switch 2, thus providing plugging in either direction.

The circuit of Fig. 6·5*c* would work perfectly and satisfy all the conditions of operation if it were always to stop in a position that would leave either normally open limit switch contact closed. This is not very likely to be the case, however, and therefore we must provide some means of starting the motor in either forward or reverse in order that the limit switches can take over automatic control. The circuit additions necessary to accomplish this are shown in Fig. 6·5*d*. Here we have added a push button in parallel with the other start components in the forward and reverse circuits. The function of these push buttons is to start the action of the motor in the desired direction so that it can run until the first limit switch is actuated and then will continue to operate automatically until the STOP button is pushed.

Circuits similar to this are used frequently for the control of milling machines and other machine tools which require a repeated forward and reverse action in their operation.

6·6 DEVELOPMENT OF CIRCUIT 5

The requirements of this circuit are to add jogging control in both forward and reverse to the circuit of Fig. 6·4*d*. In order

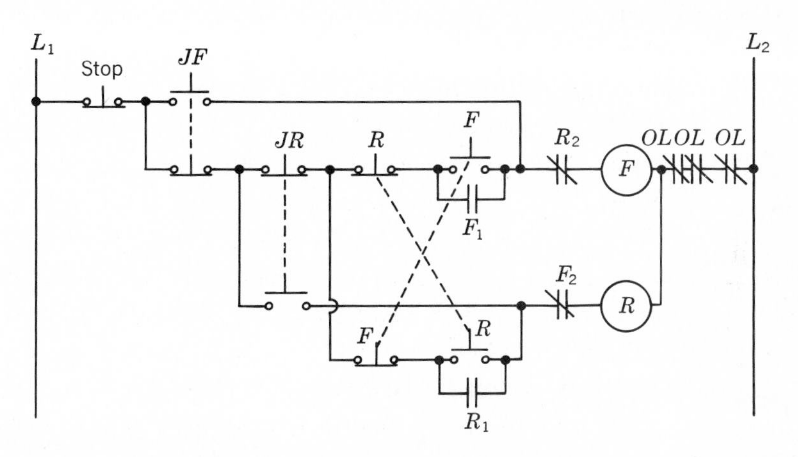

Fig. 6·6 Development of circuit 5. Forward and reverse control with jogging.

to jog a motor, the push button must connect the line to the starter coil while it is held down, causing the motor to run. It must also prevent the auxiliary contact of the starter from maintaining the circuit when the JOG button is released.

The circuit of Fig. 6·6 shows two JOG buttons identified as *JF* and *JR*. If you follow this circuit, you will see that the JOG-FORWARD button has a normally open contact which is connected from the STOP button to the coil side of the auxiliary contact for the forward starter. The normally closed contact of this push button is connected between the STOP button and all the other control devices. When the JOG-FORWARD button is pressed, the circuit is made from line 1 through its normally open contacts to the coil of the forward starter through the normally closed electrical interlock. At the same time, the normally closed contact of this JOG button breaks the circuit between the STOP button and all the other push buttons and contacts in the circuit, thus preventing the motor starter from sealing in when the JOG button is released.

The installation and wiring of the JOG-REVERSE button are identical with those of the JOG-FORWARD button, except that it is connected to the reverse starter; its action electrically and mechanically is the same as that for the JOG-FORWARD button.

This circuit incorporates many of the functions of control. It has start and stop in both forward and reverse, manual plugging duty, jog service in forward and reverse, electrical interlock, low-voltage protection, and overload protection.

6·7 DEVELOPMENT OF CIRCUIT 6

The assignment here is to add automatic plugging reversal to the circuit of Fig. 6·4*c*. The easiest method of obtaining automatic plugging reversal is by the use of a plugging switch. The connection for this switch is shown in Fig. 6·7. The action of this switch is such that when the motor is running in the forward direction, the movable arm of the switch is held in the direction shown by the arrow marked with the letter *F*. When the STOP button is pressed, the circuit is broken to the forward starter, allowing it to drop its contacts and thus closing the interlock contact marked *F*2. At this instant, the circuit is made from line 1 through the plugging switch through the normally closed interlock contact *F*2 to the coil *R* on the reverse starter. This will plug the motor in the reverse direction. The closing of the maintaining contact *R*1 will energize coil *R* which assures the running of the motor in the reverse direction. Should the STOP button be held down, the flow of current through *R*1 to coil *R* would not be possible and the result is plugging stop for the motor. The action of the plugging switch when the motor is running in reverse is such that its arm is in the position marked R. When the STOP button is pressed, the circuit functions in exactly the same way that it does when the motor is running in forward, except that now the motor is plugged by the energizing of the forward starter.

6·8 DEVELOPMENT OF CIRCUIT 7

This circuit is to control a three-speed motor, and the requirement is that it provide selective speed control (Sec. 3·9). To satisfy the requirement that the circuit give selective speed control would indicate the use of three simple start circuits, one

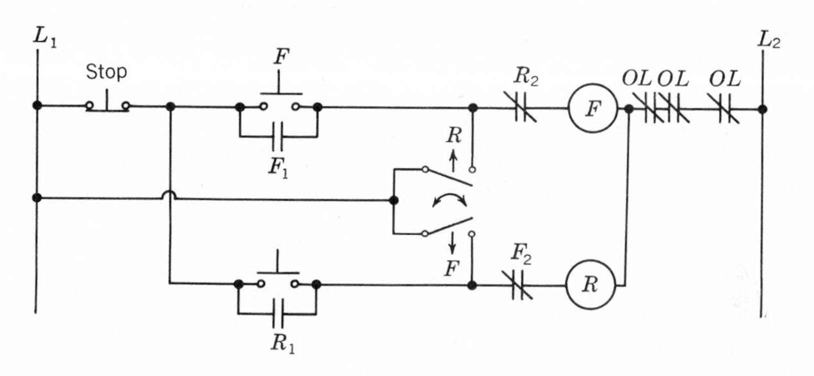

Fig. 6·7 Development of circuit 6. Forward and reverse control with automatic plugging.

circuit for each speed, so that the operator can start the motor in any desired speed. To increase speed, he need only to press the button for the desired higher speed. Such a circuit is shown in Fig. 6·8*a*.

This circuit, however, ignores any form of interlock which will prevent two speeds from being energized at the same time unless such interlock is provided mechanically in the starter. The necessary electrical interlock has been added in Fig. 6·8*b*. A careful study of this circuit will reveal that it is possible to increase speed by merely pushing the button for the next speed. For instance, if the motor is running in its first speed and it

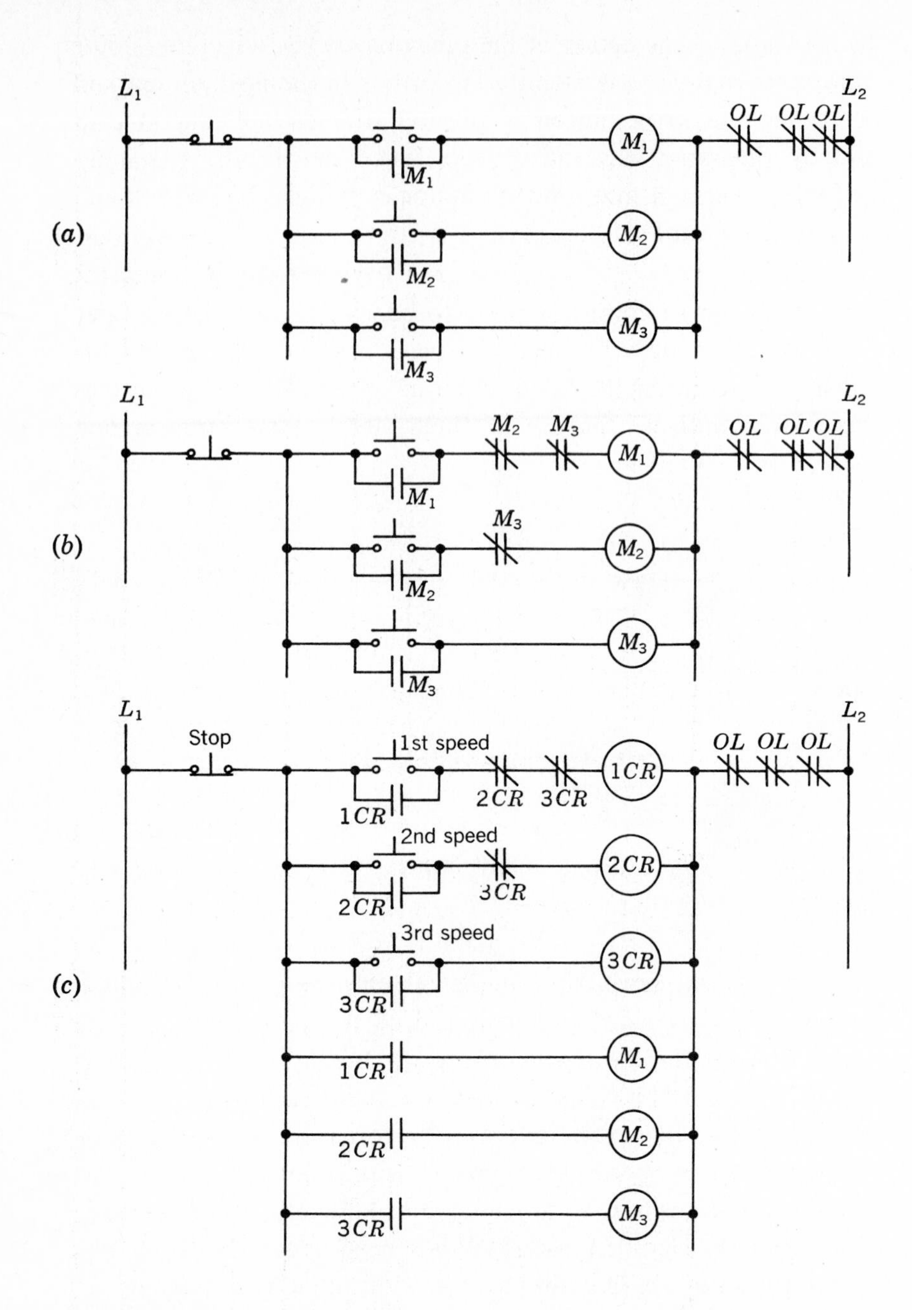

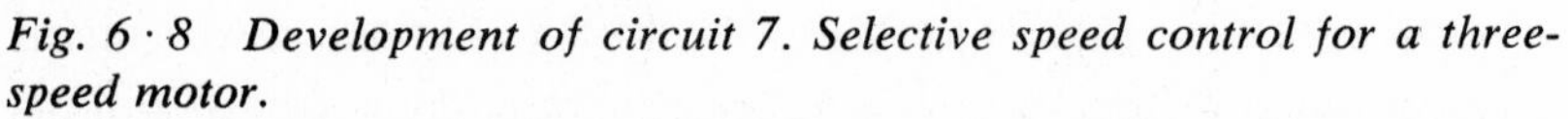

Fig. 6 · 8 Development of circuit 7. Selective speed control for a three-speed motor.

is desired to increase to the second speed, the normally closed interlock contact identified as *M*3 will be closed and the coil *M*2 will be energized. This will break the normally closed contact identified as *M*2, thus dropping out coil *M*1 and deenergizing the contactor for speed 1.

If this circuit is to function properly, then contact *M*2 must be built so that it will break before the line contacts of contactor *M*2 are made. If this is not done, then the starter will energize two speeds at one time, causing damage to the motor and the wiring. The action of the circuit for speed 3 is similar, in that by energizing coil *M*3 the normally closed contact *M*3 is broken before the line contacts for speed 3 are made, thus dropping out either speed 1 or speed 2, whichever is energized at the time.

In order to reduce speed, the STOP button must be pressed first. Analyzing the circuit, we see that if we try to go from speed 3 to speed 2, the pressing of the START button for this speed will result in current flowing only as far as contact *M*3. The same is true if we try to reduce from speed 3 to speed 1. If we try to reduce speed from speed 2 to speed 1, current can flow only as far as contact *M*2. The pressing of the STOP button drops out any coil that is energized, thus returning all contacts to their normally closed position and allowing the circuit to be energized in any desired speed.

Chiefly because it is difficult to obtain contact arrangements on starters which will provide the break-before-make action necessary in this circuit for interlocking, starters of this type generally employ control relays for each speed, which in turn energize the proper coils. The circuit for use of control relays is shown in Fig. 6·8*c*. It should be noted that the circuit for the three control relays is identical with the circuit of Fig. 6·8*b*, with only the addition of a contact on the relay for each contactor coil. Thus, it gives a three-wire control to the control relays and essentially a two-wire control to the contactor coils.

While this circuit has been developed for speed control of a single motor, it is equally applicable to sequence control of three motors. If coils *M*1, *M*2, and *M*3 were coils of individual starters for individual motors, they would start in selective sequence. This means that the operator could start any motor he desired and could progress upward in the sequence of motors at will. To go backward in the sequence, however, he must first stop whatever motor is running and then select the motor desired. This would be selective sequence control of three motors and could be expanded to any number of motors desired.

6·9 DEVELOPMENT OF CIRCUIT 8

This circuit will be a modification of circuit 7 to give sequence speed control (Sec. 3·9). The requirement of sequence speed control is that the motor be accelerated by pressing the START button for each successive speed in order until the desired speed is reached. Figure 6·9 is a circuit to accomplish sequence speed control of a three-speed motor using control relays. The contact arrangement on these relays in this type of service is critical, and it must be pointed out that normally closed contact *CR*2*b* must break before normally open contact *CR*2*c* makes. Also, contact *CR*3*b* must break before contact *CR*3*a* makes.

This circuit will not be developed step by step, because a similar circuit for speed control was developed as circuit 7. Rather we shall analyze the operation of this revised circuit. Suppose that the operator wishes to run the machine in its third speed. Then he must first press the button for speed 1, which will energize coil *CR*1. Energizing the coil causes contacts *CR*1*a* and *CR*1*b* to close. The closing of contact *CR*1*b* energizes the contactor for speed 1, and the motor starts and accelerates to this speed. The closing of contact *CR*1*a* sets up the start circuit for speed 2, and when this button is pressed, the circuit is complete to coil *CR*2, thus energizing this coil and closing contacts *CR*2*a* and *CR*2*c*. Also, it opens contact *CR*2*b*. The opening of contact *CR*2*b* drops out the contactor for speed 1.

Immediately thereafter, contact *CR*2*c* closes, energizing the contactor for speed 2 and allowing the motor to accelerate and run at the second speed. The closing of contact *CR*2*a* sets up the start circuit for the third speed. When the start button for the third speed is pressed, the circuit is complete to coil *CR*3, which in turn first opens contact *CR*3*b*, which drops out the contactor for speed 2. Immediately thereafter, contact *CR*3*a* closes, thus energizing the contactor for speed 3, which will allow the motor to accelerate and run in speed 3.

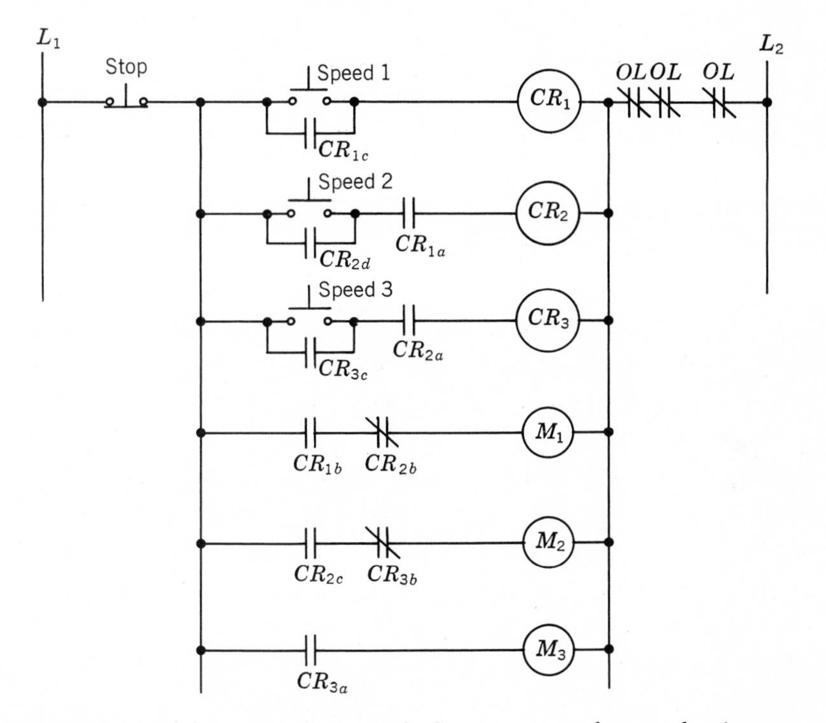

Fig. 6 · 9 Development of circuit 8. Sequence speed control using control relays.

It should be noted that the control relays remain energized until the stop button is pressed and that the only way to reduce speed is to press the STOP button and then progressively accelerate the motor, starting with speed 1 and increasing speed as desired. While this circuit is designed for only three speeds, it could be extended to include as many speeds as desired for the motor in question.

This circuit is not presented as the only or most desirable method of providing sequence speed control. There are many factors involved in a design of a control circuit for a given motor and controller. The control man will find many variations of circuits to accomplish the same purpose and should try to develop an overall understanding of the operation of the components and circuits which might be used to accomplish an end. Attempting to memorize a circuit to perform any particular function is a detriment to the student of control.

6·10 DEVELOPMENT OF CIRCUIT 9

This circuit will be a magnetic control for a wound-rotor motor. The customer desires four steps of automatic definite-time acceleration when the RUN button is pressed. He also desires the option of running the motor at any one of the four reduced speeds by pressing a push button for that speed. He also wishes to be able to change speeds at will, either up or down, after the motor has accelerated to run speed.

In order to visualize what contacts and control relays will be needed, an elementary drawing of the secondary circuit for the motor should be considered (Fig. 6·10*d*). This circuit provides the essentials of a four-step acceleration or four independent speeds, provided the contacts are properly controlled. To provide definite-time control, these contacts would be time closing (*TC*).

The first step in development of this circuit would be to provide the four-step definite-time acceleration, using a RUN button to initiate the control process. Figure 6·10*a* shows our circuit as it is developed. The provision of definite-time acceleration requires the use of time-delay relays for each speed. A control relay with three-wire control seems required for the run condition.

When the RUN button is pressed, the circuit is complete to the coil of *CR*, thus closing *CR*1 to maintain the circuit. Contact *CR*2 closes and energizes the primary contactor *PC*, thus energizing the primary of the motor. Some form of interlock should be provided to prevent the motor from starting unless all the resistance is in the secondary circuit, speed 1; however, this can best be provided after the balance of the speed control is developed.

A third contact on the control relay (*CR*) (Fig. 6·10*b*), *CR*3, can be used to energize a time-delay-on-energizing (*TDOE*) relay *S*2 which has two *TC* contacts connected to short out

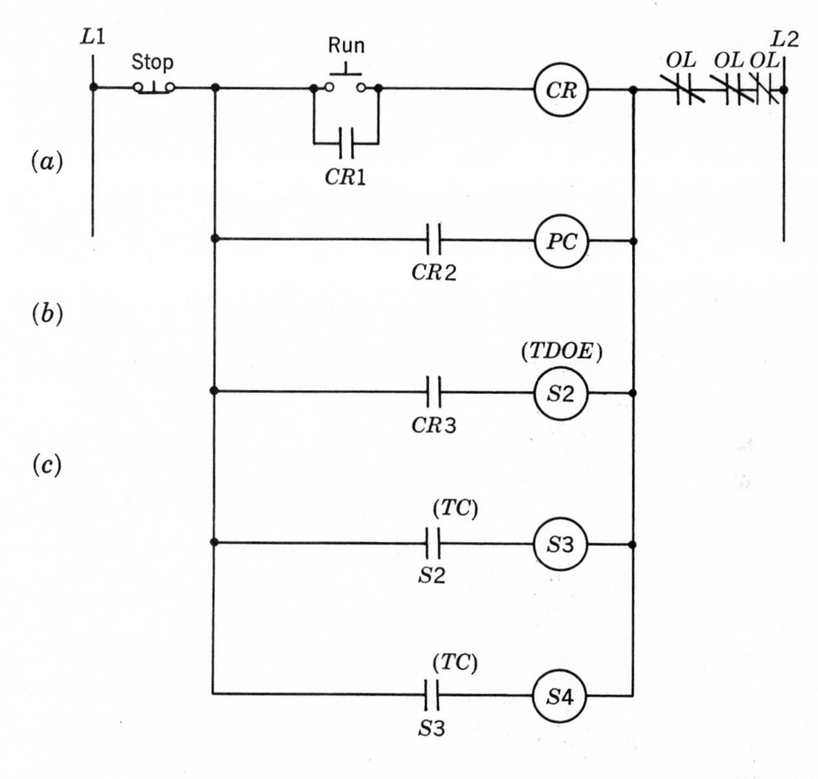

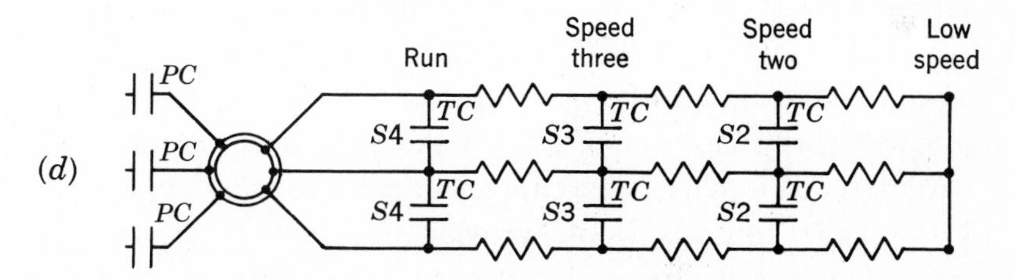

Fig. 6 · 10 Development of circuit 9. Definite-time acceleration control for wound-rotor motor.

the first section of the resistance grid (Fig. 6·10*d*), thus providing acceleration to the second step when *S*2 times out.

To provide the third step of acceleration (Fig. 6·10*c*), it is obvious that a second *TC* contact on *S*2 will be required to energize a second *TDOE* relay *S*3 which has two *TC* contacts connected to short out the second section of the resistance grid (Fig. 6·10*d*). The fourth step of acceleration is provided by a similar circuit.

The circuit of Fig. 6·10*c* provides a satisfactory degree of interlock in that the contacts of *S*2, *S*3, and *S*4 open whenever the STOP button is pressed and restore all resistance to the secondary circuit of the motor. This circuit satisfies the first specification for the circuit development.

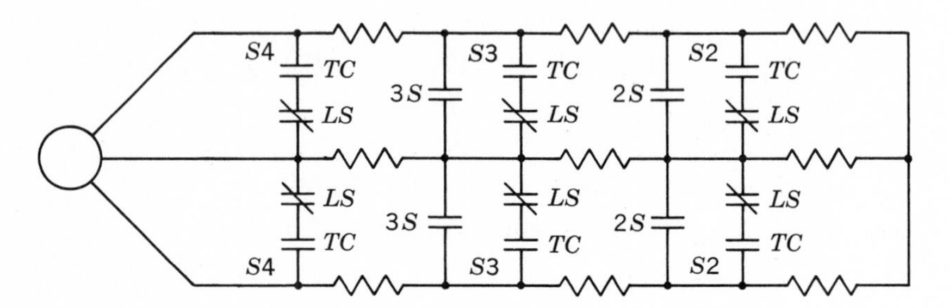

Fig. 6 · 11 Wound-rotor-motor secondary circuit.

The second specification for independent speed control after acceleration will require some modification of the circuit of Fig. 6·10*d*. To be able to select any speed at will, the operator must be able to open any closed contacts and close any open contacts in the secondary circuit as required for that speed.

Adding a normally closed contact in series with each *TC* contact in the circuit of Fig. 6·10*d* would allow the equivalent of opening any closed contacts in the secondary circuit.

Adding a normally open contact in parallel with each group of two series contacts would provide the effect of closing any open contacts in the secondary circuit. These contacts are shown in Fig. 6·11.

Figure 6·12 provides push-button control of the added contacts of Fig. 6·11 and satisfies the second specification for the circuit. A time closing contact on *S*4 should be used in a stop function for the push buttons to assure that the motor has accelerated to run speed before it can be operated at a lower speed.

The low-speed push button energizes a relay *LS* in a three-wire control circuit. Six normally closed contacts on *LS* are connected in series with the *TC* contacts in the circuit of Fig. 6·11. When the low-speed button is pressed, relay *LS* is energized and opens all six normally closed contacts, which restores all resistance in the secondary circuit and results in low-speed operation.

To increase speed from low or first speed to second speed, relay 2*S* must be added in a three-wire circuit with a normally open push button. One normally open contact 2*S* is used to seal in *LS* when moving up in speed and to energize *LS* when moving down from run to second speed. Two normally open contacts on 2*S* are used to short out part of the resistance grids (Fig. 6·11).

The third speed is added by duplicating the second speed components using contacts indicated by 3*S* on the circuits of Figs. 6·11 and 6·12. There is no need for interlock when going from second speed to third speed because the contacts 2*S* in Fig. 6·11 do not affect the operation even though they remain closed.

The fourth or run speed can be restored by merely dropping out any and all instantaneous contacts and relays associated with the lower speeds, since all *TC* contacts in Fig. 6·11 are now closed. This can be done by adding a normally closed push button as indicated in Fig. 6·12.

The circuit as developed up to this point would work as long as the operator always wanted to go from a low speed to a higher speed. To reduce the speed, it is necessary to open the contacts 2*S* and 3*S* of Fig. 6·11. Interlocking by the use of contacts could involve more relays in the circuit, but normally closed contacts as part of the speed-control push buttons, as indicated in Fig. 6·12, provide positive drop out of all higher speeds whenever a lower speed is desired.

This circuit may seem slightly impractical in its requirements, but is used to indicate that any requirements of the system can be met by a systematic development.

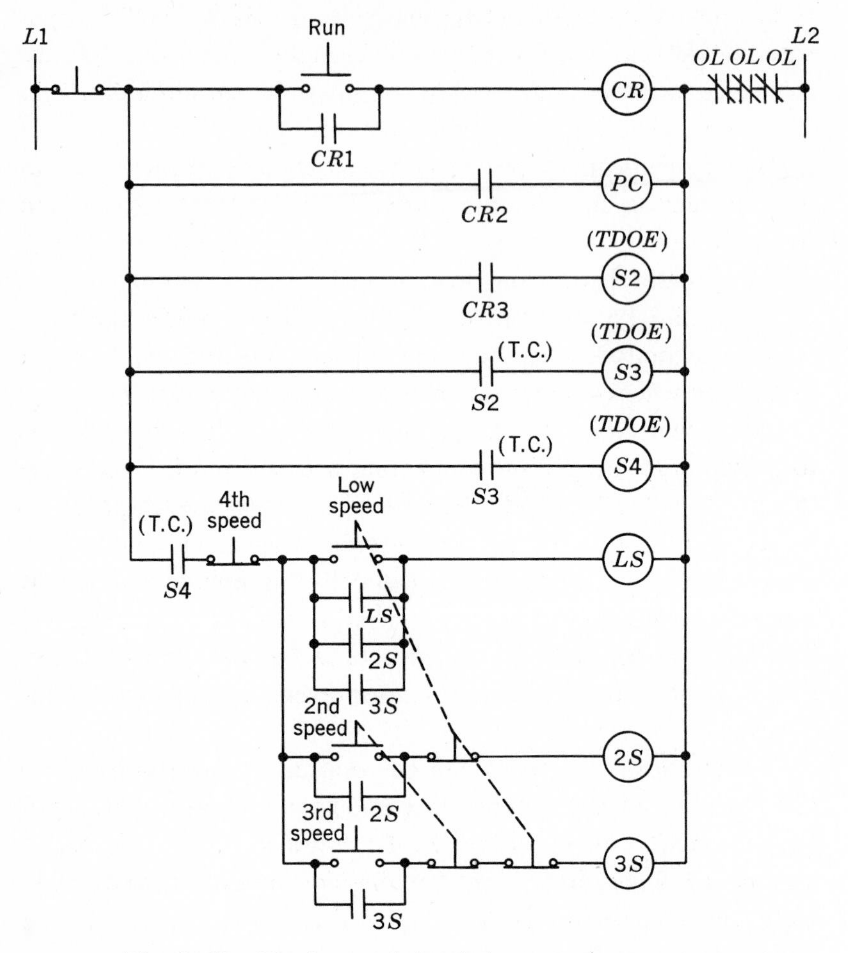

Fig. 6 · 12 Final control circuit for wound-rotor motor.

6·11 DEVELOPMENT OF CIRCUIT 10

The requirements for this circuit are to develop a control system which will provide definite-time speed control for a d-c shunt motor. The owner wishes to be able to start the motor and have it accelerate to any one of five preselected speeds. Speeds 1 and 2 are below base speed (underspeed control). Speed 3 is base speed, and speeds 4 and 5 are above base speed (overspeed control).

The requirement of two speeds below base speed indicates three resistance grids in series with the armature (Fig. 6·13*a*). One of the resistances would be short-circuited for speed 1, the second for speed 2, and the third for base speed by contacts *S*1*a*, *S*2*a*, and *S*3*a*.

The requirement of two speeds above base speed indicates two resistance grids in series with the shunt field and shorted out by normally closed contacts *S*4*a* and *S*5*a* until base speed has been reached and overspeed is required.

The addition of an overload heating element and line contacts completes the motor circuit. This leaves only the development of the control circuit to activate the contacts properly.

Step 1 in the development of the control circuit would be to provide for speed 1 by connecting a push button to energize the first *TDOE* relay *S*1. This circuit must remain energized; therefore, it requires a three-wire control.

When a d-c motor is started, considerable resistance is required in the armature circuit until it begins to accelerate. This requires a time delay before *R*1 is shorted out. If contact *S*1*a* is a time closing contact, it will provide time for the motor to reach the desired speed.

The line contacts must be closed before any current can flow through the motor circuit. If the line contact is an instantaneous contact on *S*1, this requirement will be met. An overload contact and STOP button must be provided to protect and stop the motor. Contact *S*1*c*, used to maintain the circuit to coil *S*1, must be an instantaneous closing contact on *S*1.

At this point it is obvious that *S*1 must be energized regardless of the speed selected; therefore, we can add a maintaining contact for each speed in parallel with *S*1*c*. All these contacts must be of the instantaneous closing type.

The circuit of Fig. 6·13*b* will start the motor and allow it

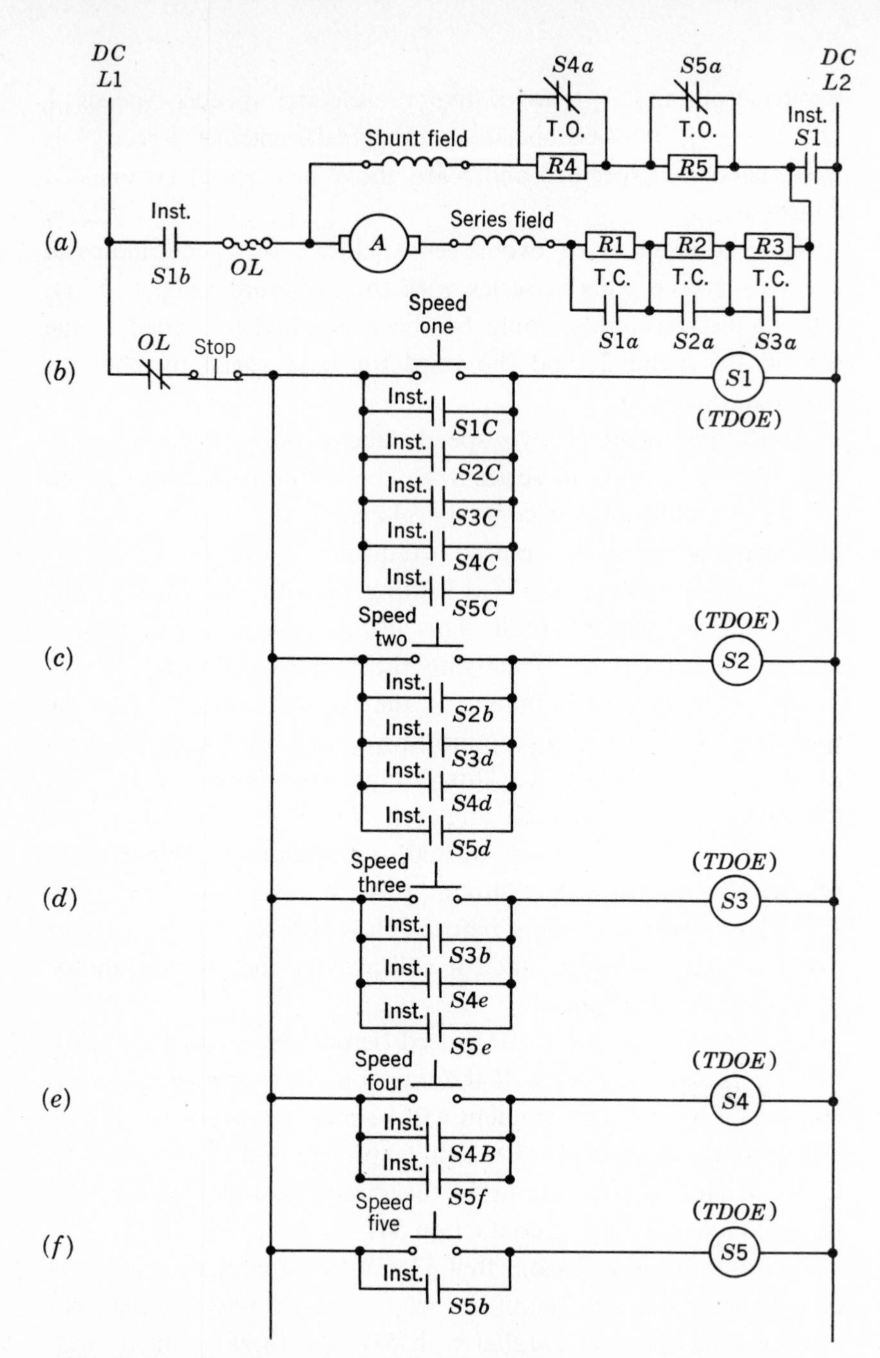

Fig. 6 · 13 Development of circuit 10. Definite-time preset speed control for a d-c shunt motor.

to accelerate to the desired speed, at which time $S1a$ times out and removes $R1$ from the circuit. The motor will run at this speed unless other buttons are pressed.

The requirement of a push button for each speed indicates a three-wire circuit and a *TDOE* relay for each speed. Figure 6·13*c* is a three-wire circuit to energize and maintain $S2$. The instantaneous contact $S2c$ will energize $S1$ and thereby close the line contact $S1b$. The relay $S2$ must provide a time delay twice as long as that of $S1$ so that $S1$ may time out and allow acceleration time for speed 2. Instantaneous contacts must be provided parallel to $S2b$ in order to bring in speed 2 whenever a higher speed is selected.

The other speeds are controlled by circuits similar to the one for speed 2. Each higher speed must have a longer setting on the time-delay relay and provide contacts to energize all lower speeds.

There are other circuits which could provide this control; however, this was chosen to illustrate the degree to which a circuit can be dependent upon proper interlocking.

6·12 DEVELOPMENT OF CIRCUIT 11

This circuit will be one for a chilled-water air-conditioning system. The compressor is a 500-hp squirrel-cage motor driving a centrifugal compressor. Head pressure is controlled by condenser water furnished by a condenser-water pump. The chilled water is pumped to the heat exchanger by a chilled-water pump. Protective control must be provided to shut down the compressor whenever flow ceases in either the chilled water or condenser water.

The starter for the compressor is a reduced-voltage autotransformer type using air circuit breakers requiring that a coil be energized to close the breaker and a second coil be energized to trip the breaker. Definite-time control of reduced-voltage starting is provided by a *TDOE* relay in the starter.

The first step in the system is to establish chilled-water flow and to sense by means of a flow switch that there actually is flow in the pipe. The flow switch (CWF) is usually located in the return water line to assure that flow is complete around the system. Figure 6·14*a* provides the START and STOP buttons for the system and energizes the chilled-water pump.

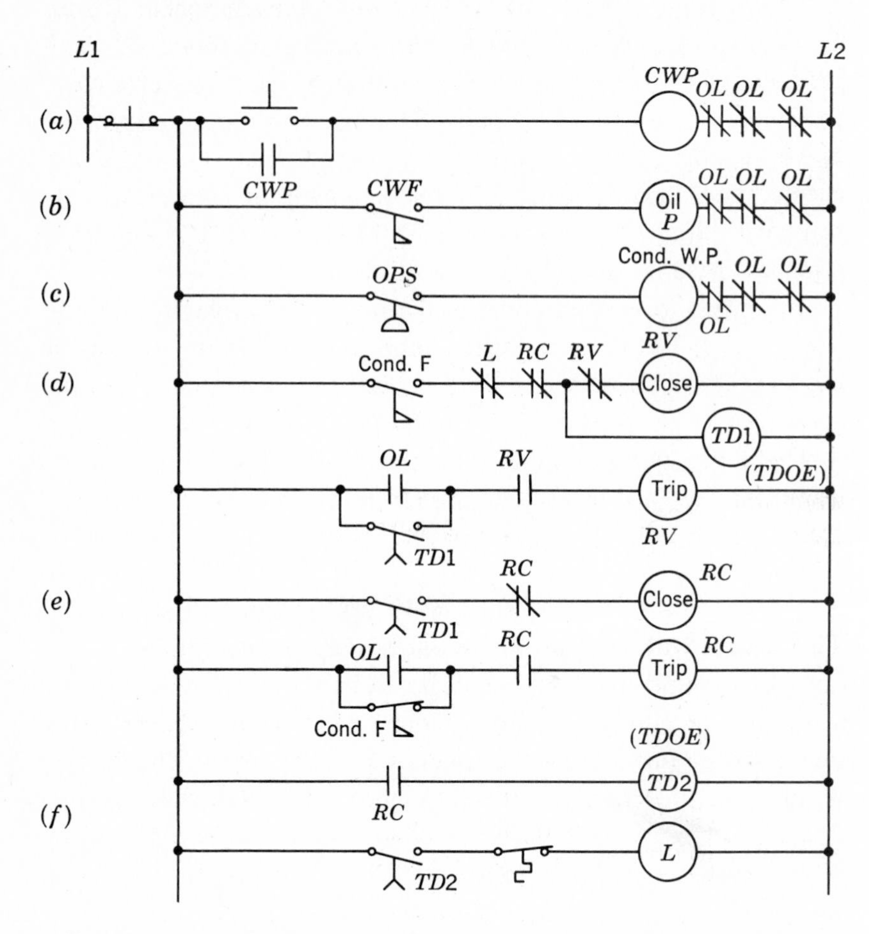

Fig. 6·14 Development of circuit 11. Chilled-water air-conditioning system.

When chilled-water flow is sensed by the flow switch, the oil pump on the compressor must be energized and correct pressure established (Fig. 6·14*b*). Once the oil-pressure switch senses sufficient pressure, it must energize the condenser-water pump (Fig. 6·14*c*) to provide cooling for the compressor and keep the head pressure from rising to a dangerous value.

The compressor has chilled-water flow, oil pressure, and condenser-water flow and is ready to be started under reduced voltage if the unloader assures that the load is less than 10 percent. There are many unloaders and vane positioners used. We shall indicate the unloaders as a contact *L* which is normally closed when the load is 10 percent or less. The unloader contact *L* should not allow the compressor to start if it is open. Figure 6·14*d* is the circuit for the reduced-voltage contactor closing coil (*RV*). Remember that this contactor is latched closed; therefore the coil must be energized only briefly.

The trip coil on the breaker must be energized if an overload occurs during reduced-voltage start and when the run contactor is energized by the time-delay relay. The trip coil must be energized only momentarily; therefore, a contact on *RV* must remove the coil from the line. The trip-coil circuit must be held open when the run contactor is closed; therefore, normally open (N.O.) contact *RV* is connected in series with this circuit.

When the reduced-voltage start timer *TD*1 times out, it closes its contacts to trip the reduced-voltage contactor and briefly energize the run-contactor closing coil (*RC*). The run contactor is now latched closed (Fig. 6·14*e*).

Protective control for the compressor while running must be paralleled normally open contacts in the trip-coil circuit. The chilled-water flow contact and the oil-pressure switch (*OPS*) contact will cause a shutdown of the condenser-water pump. The condenser-water flow contacts (Cond. F) in the start circuit will not shut down the compressor when they open; therefore, a set of contacts on this switch needs to be connected in the trip circuit of the run contactor for protection. These contacts would be normally closed when there is no flow, so that they will be open when flow is established. Normally open contact *RC* on the run contactor prevents the energizing of the trip coil unless the run contactor is closed.

All the circuit lacks at this point is the energizing of the loader device L after a short time delay to assure that the compressor is completely up to speed. The thermostat can then control the load of the compressor (Fig. 6·14*f*).

Summary

This selection of the book has been devoted to presenting a method of developing control circuits. While the few circuits developed here do not in any way approach the limitless number of possible circuits that the student of control will find in actual practice, they should provide the basic principles necessary for the development of any control circuit.

Develop the circuit one function at a time, adding only the components necessary to perform that function. Analyze the circuit after each addition to see that it has not interfered with any previous operation and that it actually does perform the function intended, before proceeding with any further additions to the circuit. If you follow these simple rules, you should have no trouble in developing a circuit to perform any desired function. The greatest danger in developing control circuits is to try to draw a complete circuit at one time. Therefore, it cannot be overstressed that step-by-step development will lead to fewer hours spent in trying to find out why a circuit did not work after it was wired.

It is highly recommended that the student practice developing circuits of various types and checking them to see that they actually should operate when wired. If a setup of control components is available to the student, he should at this stage of his study develop circuits of various types and then actually wire them and test them to see that they work. Should the circuit not operate as expected, a student should then troubleshoot this circuit and determine why it did not operate. If such a continuation of the study of the development of circuits is possible, a student will have a great advantage over others who have not had this practice when he tries to apply these principles on the job to actual circuits.

The principles involved in the development of control circuits have been clearly set down in this section of the book, and all that is required to perfect this technique to a satisfactory degree is practice by the student. Your own desires as to your degree of proficiency will dictate how much additional time you spend on practicing the development of control circuits.

Review Questions

Develop circuits for the following:

1. A motor controlled by a three-wire start-stop station.
2. Add to the above circuit a second push button for starting the motor from a different location.
3. Add to the above circuit a limit switch to stop the motor.
4. A motor controlled by a three-wire push-button station. When this motor stops, it starts a second motor which runs until stopped by pressing a STOP button.
5. Revise the circuit of question 4 so that the second motor runs for only 2 minutes and then stops automatically.
6. Three motors connected so that they are all started by one START button and interlocked so that if any one of them should fail to start, or should drop out, all will stop. The STOP button stops all motors.
7. Two pumps started and stopped one at a time by a pressure switch. Provide a manual switch to run the pumps alternately.
8. Add to the circuit of question 7 a second pressure switch to start the idle pump if the pressure continues to drop.
9. Replace the manual switch in the circuit of question 8 with a stepping relay to alternate the pumps automatically each time they start.
10. Replace the stepping relay in the circuit of question 9 with a time clock to alternate the pumps every 24 hours.
11. Four motors started in compelling sequence. Provide a 20-sec time delay between the starting of each motor.

12. Four motors started in selective sequence.
13. A three-speed motor with selective sequence starting. Provide control so that the speed may be reduced without pressing the STOP button. (HINT: This is similar to plugging without reversal of the motor.)
14. There are four exhaust-fan motors in a building. Each fan is also equipped with a thermostat known as a *firestat*. Should any one of the firestats, which have normally closed contacts, open from high heat, it will stop all fans.

7

ANALYSIS OF CONTROL CIRCUITS

Now that you have mastered the art of developing control circuits, you should be ready to analyze circuits developed by someone else. The first step in analysis of a circuit is to determine as much as possible about the operation of the machine or other equipment which the motor drives, so that the functions of the circuit can be more readily understood. To analyze any given circuit, it should be converted to a schematic diagram if one is not available. As stated earlier, if the schematic drawing is properly made, the sequence of operation of control should proceed more or less from the upper left of the drawing horizontally across the first line, and then each successive line of the circuit should proceed downward in the drawing. Not all schematics, however, are drawn with this sequence in mind, so do not expect that it will always apply.

7·1 BASIC PROCEDURE

The basic procedure for circuit analysis is very simple and should be readily understood if you have mastered the preceding chapter on circuit development. You merely consider the circuit one component at a time and decide what happens if a push button is pushed or a contact closes or opens, realizing that you must have a complete circuit from one line through the coil to the other line in order to energize any relay, contactor, or starter. If the circuit is open at any point, that particular coil will be deenergized and its contacts, wherever they may be found in the circuit, will be in their normal or deenergized position. When the circuit is complete to any particular coil, that contactor, relay, or starter is energized, and its contacts, wherever they may be, are opposite to their normal position. That is, if they are normally closed contacts, they are now open. If they are normally open contacts, they are now closed.

If a time-delay relay is used in the circuit, you must keep in mind whether its contacts are time opening or time closing to determine their normal position and their function in the circuit. When relays are used in the circuit, be sure that you consider every contact which is closed or opened by the relay whenever the coil is energized. Failure to consider one contact of a relay may lead to a misunderstanding of the whole circuit. When analyzing a circuit, be sure that you consider every component in its normal and energized positions so that you understand the whole operation of the complete circuit. Do not jump to conclusions when halfway through the analysis.

In the following section we shall analyze several circuits, using a step-by-step procedure which should give you the basic fundamentals of this operation so that you can apply it in actual job situations. The ability to analyze a circuit is a prerequisite to any efficient troubleshooting on motor control circuits.

7·2 ANALYSIS OF CIRCUIT 1

If we look at Fig. 7·1, it should be obvious that this is a control circuit for a forward and reverse starter. To analyze the operation of this circuit, we shall start with the upper left-hand side at $L1$. The first component is a STOP button which is normally closed. Therefore, the current may flow through as far as the normally open start button marked FORWARD. Also, the current may flow downward in the wire to the right of the STOP button to a single-pole switch, which is shown in the idle, or OFF, position, and also to the normally open push button marked REVERSE.

If now we press the FORWARD button, the current can flow through that button, through the normally closed contact of the REVERSE button, and through the normally closed contact identified with the letter $R2$ to the coil F, which is the contactor

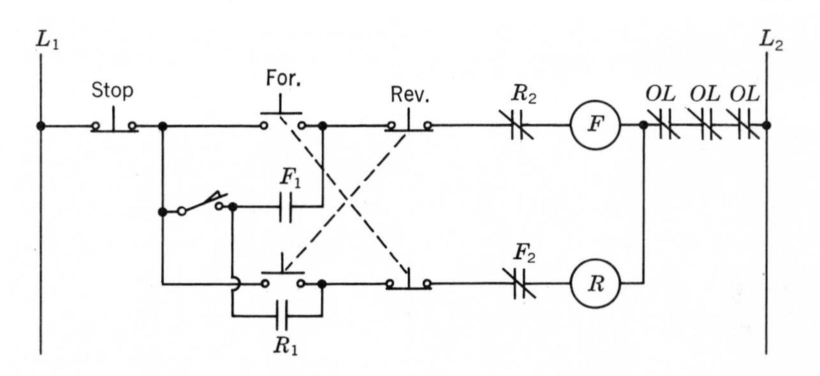

Fig. 7 · 1 Analysis of circuit 1. FORWARD and REVERSE control for a single motor.

coil for the forward direction. Thence it will proceed through the normally closed overload contacts identified as OL to $L2$. The circuit is therefore complete from line 1 through the forward-starter coil to $L2$, and coil F is now energized. The energizing of this coil will open the normally closed contact $F2$ and close the normally open contact $F1$. The opening of the normally closed contact has no immediate effect on the circuit, because the normally open push button for reverse has the circuit broken ahead of this contact. The closing of the normally open contact accomplished no immediate results, because the switch on the line side of this contact is open and there is no voltage present.

When the FORWARD button is released, the circuit from line

1 to coil *F* is broken at this point and, because there is no maintaining contact around this break in the circuit, the coil will drop out. Suppose now that we close the switch so that it connects line 1 to one side of the normally open contact *F*1 and again press the FORWARD button. The action of the circuit is the same as previously discussed, except that now when the normally open contact *F*1 is closed, it completes the circuit from line 1 around the normally open push-button contact. When this push button is released, the circuit is maintained through contact *F*1 and the motor will continue to run in the forward direction.

Suppose, now, that we press the REVERSE button. It will open its normally closed contact and close its normally open contact. The result of this action will break the circuit to coil *F* and make the circuit through the normally open contacts of the REVERSE button, through the normally closed contact *F*, through coil *R* to line 2, thus plugging the motor from forward to reverse. The running of the motor in reverse is maintained through the normally open contact *R*1, which is now closed. The forward starter is prevented from running by the opening of the normally closed contact *R*. If the switch is thrown to the OFF position and the REVERSE button is pressed, we have exactly the same operation as when we pushed the FORWARD button, except now the reverse starter is momentarily energized.

Now that we have analyzed the operation of the individual components of this circuit, we can sum them up by saying that this circuit provides forward and reverse run. It also provides plugging in either direction and by the position of the switch will also provide jogging in either direction at the will of the operator. The normally closed contacts *R*2 and *F*2 are electrical interlock between the forward and reverse starters. The switch shown in this diagram would be known as a *jog-run switch* because in one position it allows the motors to be jogged in either direction and in the other position permits the motor to run in either direction, as desired.

Looking at Fig. 7·2, we see only one contactor or starter coil, which would indicate that this is a circuit for the control of a single motor running in only one direction. Again applying our principle of analysis to the circuit to determine its operation, we shall see that the STOP button is normally closed so current can flow through it to either of two normally open push buttons and a normally open contact identified as *CR*.

Should we press the START button, it would complete the circuit through the coil identified as *CR* to line 2. If the designations used in this circuit are standard, it is safe to assume that

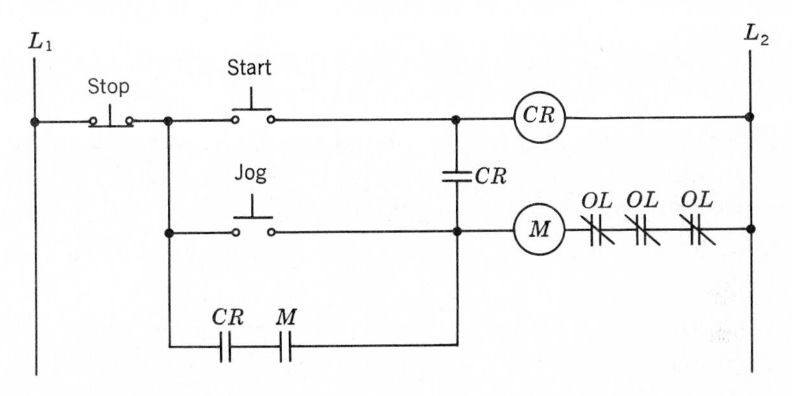

Fig. 7·2 Analysis of circuit 2. START, STOP, *and* JOG *service for a single motor using a jogging relay.*

this is a control relay which apparently has two normally open contacts used in this circuit. These normally open contacts identified with the letters *CR* will now be closed. The one energized from line 1 to the STOP button will allow current to flow only as far as the normally open contact labeled *M*.

The other control relay contact, which connects the wire following the START button down to the second horizontal line of the diagram, will permit current to flow through this normally open push button, through the now-closed relay contacts to coil *M*, and through the normally closed overload contacts to line 2, thus energizing coil *M* and the motor. The energizing of coil

M causes the normally open contact *M* to close, which will allow current to flow from line 1 through the normally closed STOP button, through the now-closed relay contact *CR*, through the now-closed contact *M*, and through coil *M*, maintaining the circuit to this coil and keeping the motor running even though we release the normally open START button. The motor can be stopped only by pressing the STOP button, which breaks the circuit from *L*1, allowing both the control relay and the starter coil to drop out.

Suppose now that we press the second or lower normally open push button, the JOG button. Current will flow directly from line 1 through the normally closed STOP button, through the button we have pressed, to coil *M*, then through the overloads to line 2, and the motor will be energized. Energizing coil *M* again closes its normally open contact; but this will not maintain the circuit when the push button is released, because the normally open contact *CR* is open and has the circuit broken from line 1. When we release this push button, the motor is dropped from the line.

This circuit provides jogging with the additional safety protection of a relay which definitely prevents the starter from locking in during jogging service. When the START button is pressed, both the control relay and the starter are energized, and the starter is locked in through the relay contacts. When the JOG button is pressed, only the starter is energized, and it is definitely prevented from locking in by the normally open relay contacts.

7·4 ANALYSIS OF CIRCUIT 3

A careful study of Fig. 7·3 shows us that the top three horizontal lines contain the line contacts of the starter identified by the letter *M*, the overload heater elements, and the three motor terminals identified as *T*1, *T*2, and *T*3. The next two horizontal lines contain first the contacts *DB* and then the primary of a transformer identified as *PT*. The secondary of this transformer is connected to a full-wave bridge rectifier with the d-c terminals marked with a plus and a minus sign. The output of this rectifier is applied to terminals $T1$ and $T3$ of the motor through contacts *DB*. The part of the circuit so far considered is part of the internal wiring of the controller, and the remaining section of the circuit contains the external start-stop control for the controller.

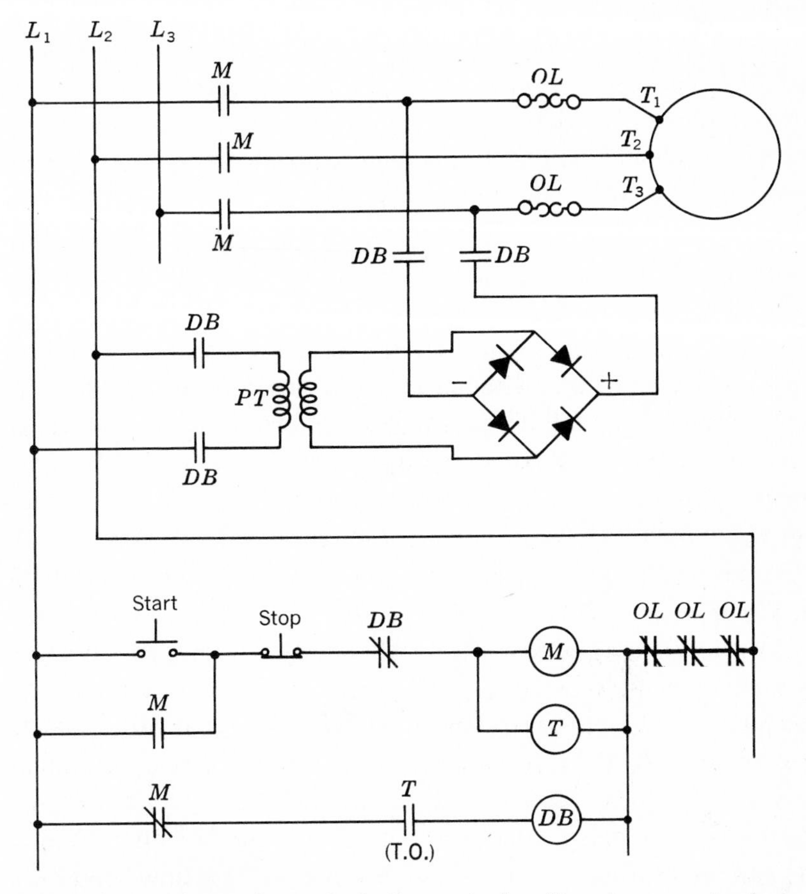

Fig. 7 · 3 Analysis of circuit 3. Dynamic breaking for a squirrel-cage motor. (Cutler-Hammer, Inc.)

Considering the balance of this circuit, if we apply our analysis technique, we will find that pressing the START button will energize coil *M* because all the other components in this circuit are normally closed. The energizing of coil *M* will close all its contacts, and, if we consider this in the drawing, the motor will be energized through the closing of the three line contacts. The auxiliary contact in parallel with the START button will close, thus sealing in the circuit and maintaining the motor in the running position. The opening of the normally closed contact *M*, located in the bottom line of the drawing, will prevent coil *DB* from being energized.

Simultaneously with the energizing of coil *M*, coil *T* is energized. This seems to be a time-delay relay because its contact *T* is indicated to be time opening. If we now press the STOP button, coil *M* is dropped out along with all its contacts, which will return to their normal positions. The opening of the line contacts *M* breaks the circuit to the motor and stops the flow of current.

The auxiliary contact in parallel with the START button opens, which has no effect on the circuit at this time. The returning of the normally closed contact *M* to its closed position, however, will energize coil *DB* because contact *T* is still closed. We know that this contact is closed because it is designated to be time opening, and even though its coil is now deenergized, the timer would maintain this contact in a closed position.

With coil *DB* energized, all contacts indicated by the letters *DB* will now be in their operating position. The normally closed contact in series with coil *M* will be open, thus preventing a reenergizing of this coil until the time-delay relay has opened contact *T*. The closing of the four normally open *DB* contacts associated with the transformer and rectifier will, in effect, apply d-c voltage to *T*1 and *T*3 and hold this voltage on the motor until the time-delay relay has timed out, thus opening contact *T*, which returns the circuit to normal at rest condition.

What is the purpose of applying d-c voltage to a motor when you press the STOP button? The application of d-c voltage to a rotating squirrel-cage motor has the effect of a smooth but positive braking action and will bring the motor to a rapid but very smooth stop. You may wonder why the time-delay relay is necessary in this circuit. If we did not remove the d-c voltage from the motor at almost zero speed, the low d-c resistance of the motor winding would allow excessive current to flow, thus overheating and possibly damaging the motor windings. This time-delay relay should be adjusted so that it will apply the d-c voltage to the motor windings practically down to zero speed and remove it so that the motor comes to an immediate stop.

This circuit seems to provide a normal across-the-line start for a squirrel-cage motor, but in addition provides a rapid, smooth braking effect on its stop. This circuit may well be applied to any piece of equipment where a smooth, fast stop is required or where it is desired to have the motor shaft free for manual rotation when the power is disconnected. It also provides a stop without any tendency to reverse, such as is encountered with a plugging stop. This type of braking is also an advantage where the braking effect must be applied frequently. It requires less maintenance than a mechanical brake, thus reducing maintenance cost. It also provides less shock to the drive system than a mechanical brake and less heating than with a plugging stop. This type of braking is known as *dynamic braking*.

7·5 ANALYSIS OF CIRCUIT 4

Looking at Fig. 7·4, we see a double set of line contacts identified as 1*M* and 2*M* which connect lines 1, 2, and 3 to the motor terminals. Also in this part of the circuit we have contacts identified as *S*, which seem to connect some of the motor windings. In the lower control section of the diagram, we have a START button, a STOP button, and a coil *S*, which seems to be some sort of auxiliary contactor. Also, we have coil 1*M*, which apparently is a line contactor for the motor. Coil *TR* appears to be a time-delay relay. Coil 2*M* appears to be a second line contactor for the motor.

In analyzing this circuit, suppose we press the START button, which will energize coil *S*, since all the contacts and push buttons in this circuit are closed. The energizing of this coil will operate all its contacts, which will energize coil 1*M* and also will prevent the energizing of coil 2*M* by the opening of its normally closed

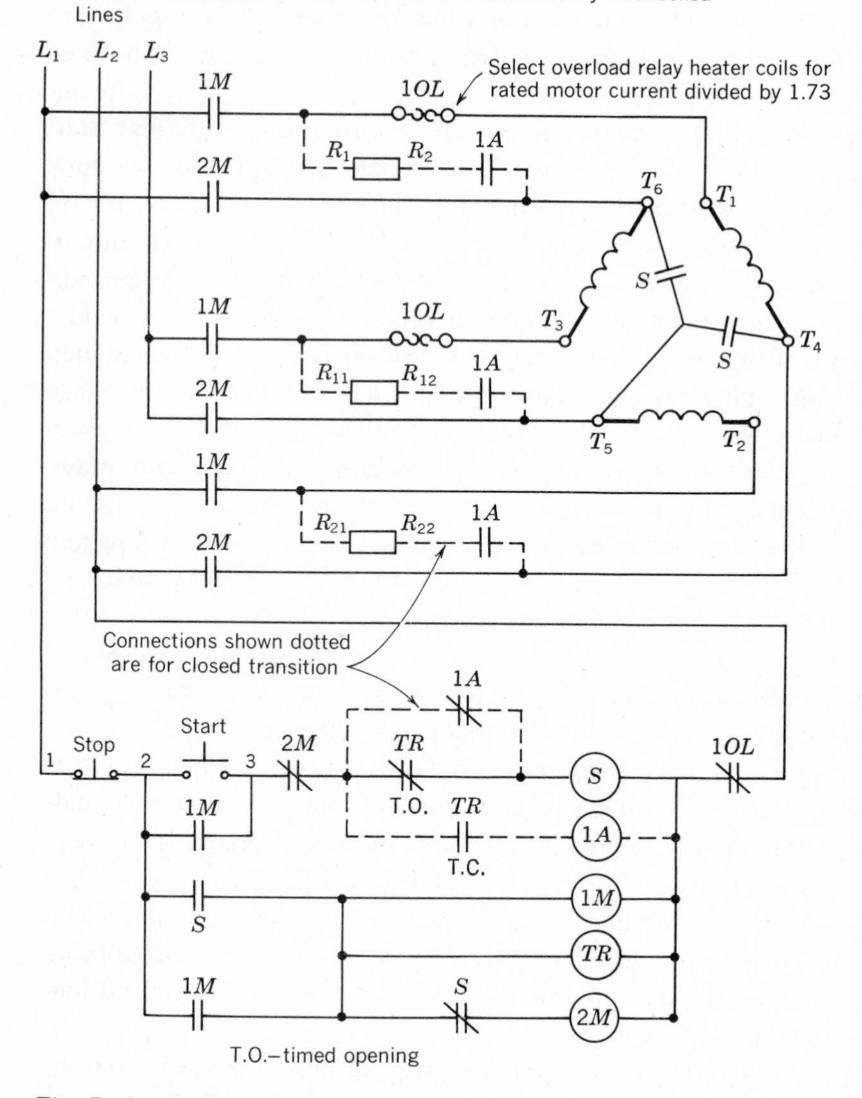

Fig. 7·4 Analysis of circuit 4. A wye-delta controller for a squirrel-cage motor. (Cutler-Hammer, Inc.)

contact. The two normally open contacts *S*, which connect the three motor terminals, will now be closed, forming a wye connection for the motor coils. The energizing of coil 1*M* closes all its contacts, three of which are line contacts for the motor, thus energizing the motor and starting it running. One such contact is in parallel with the START button and acts to maintain the control circuit. Another contact is in series with coil 2*M*, but at this time it has no effect on the circuit, because the normally closed contact *S* is now open.

At this point, we have a wye-connected squirrel-cage motor running across the line. At the time coil 1*M* was energized, coil *TR* was energized, and the timing action of its normally closed contact *TR* was started. When this contact times out and opens, it breaks the circuit to coil *S* and returns all its contacts to their normal position. An opening of the two contacts connecting the motor windings breaks the wye connection for the motor windings. The opening of the contact in series with coil 1*M* has no effect on the circuit, because this circuit is completed through contact 1*M* in parallel with it. The closing of the normally closed contact in series with coil 2*M* now completes the circuit to this coil and causes its contacts to close, connecting the motor terminals directly to the line, forming a delta connection for the motor.

If you have any trouble visualizing these motor connections, you should draw them separately on a sheet of paper to see that the first connection was a wye and the second a delta connection of the three motor windings. Of course, pressing the STOP button deenergizes all the coils and returns the circuit to its normal at-rest condition. This circuit shows three resistors and three contacts to connect them, along with a coil and other associated contacts which would be necessary to establish a closed transition for the starting of this motor.

Our analysis of this circuit shows that it is a wye-delta-type motor controller used for the purpose of giving a reduced-voltage effect to the starting of this motor as discussed in Chap. 3. In applications where a closed transition is necessary or desirable, the additional connections are shown for adding resistance to bridge the motor connections during the transfer from wye to delta. This is a rather common circuit and deserves some concentrated study as to principle of operation. Again, however, a warning is in order not to memorize this circuit as being the only possible way to give wye-delta starting to squirrel-cage motors. The use of the time-delay relay with its time opening contacts gives us a definite-time type of acceleration for this control circuit. This controller, as you may have noticed, involves a two-pole and two three-pole magnetic contactors along with the necessary mechanical interlock to assure a sequence of operation and to prevent two connections at the same time, which would cause a short circuit.

7·6 ANALYSIS OF CIRCUIT 5

Considering the circuit of Fig. 7·5, we find that the resistance in series with the motor leads would seem to indicate that this is a primary-resistance reduced-voltage starter. Looking at the control section of the diagram, we have what seems to be an ordinary three-wire control circuit to energize coils 1*CR* and *TR*. If we press the START button, current may flow through the normally closed STOP button, the START button and contact *R*2, and coils 1*CR* and *TR* will be energized. The energizing of coil 1*CR* will cause all its contacts to close. Contact 1*CR*1 is in parallel with the START button and will perform the function of maintaining the circuit to the coil. Contact 1*CR*2 will close and energize coil *S*. The energizing of this coil will cause the line contacts *S* to close and will energize the motor through the series resistances. The presence of resistance in series with the motor leads will cause a reduced voltage to be applied to the motor, thus reducing the inrush current to the motor and providing reduced-voltage starting.

The motor is now accelerating under reduced voltage, and the time-delay relay *TR* is timing out. When relay *TR* times out, it will close contact *TR*. When this contact closes, it will energize coil 2*CR* because contact *S*1 is closed by coil *S*. The energizing of coil 2*CR* will cause contact 2*CR*1 to close. This contact is in parallel with the START button and forms an additional maintaining circuit for the coil. The closing of contact 2*CR*2 causes coil *R* to be energized, closing the line contacts identified as *R*. These contacts are in parallel with the resistors

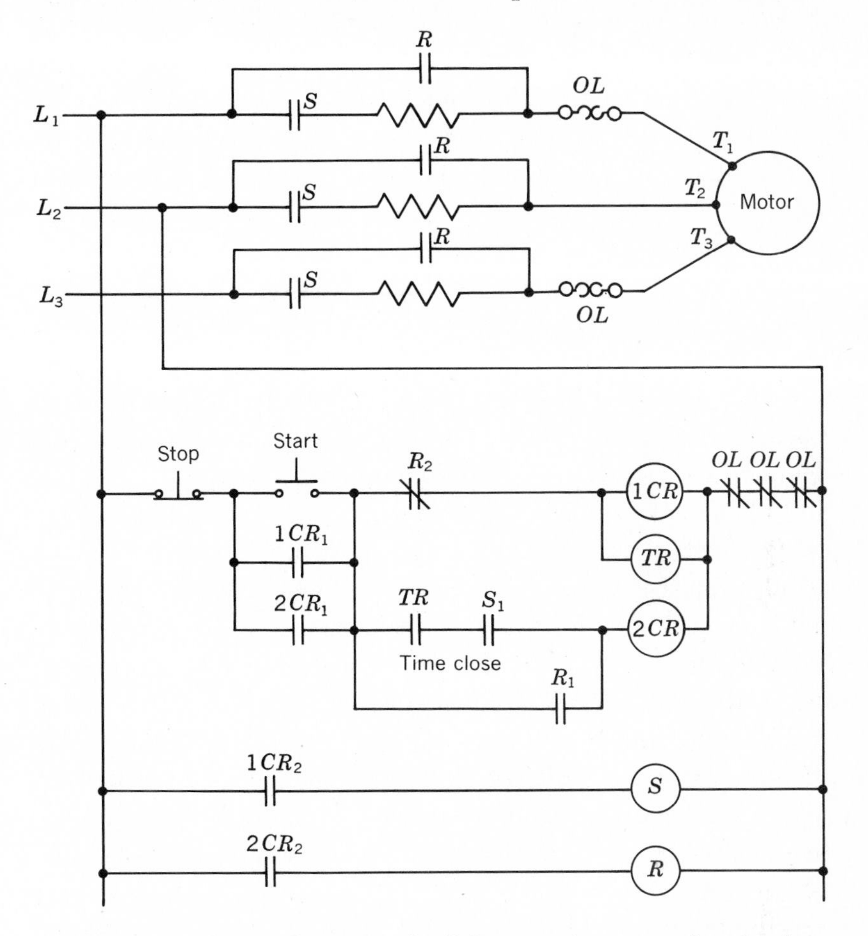

Fig. 7·5 Analysis of circuit 5. Primary-resistance reduced-voltage starter.

and effectively short them out of the circuit, thus applying full line voltage to the motor, which will enable it to accelerate to its full speed and run across the line.

The energizing of coil *R* also closes contact *R*1, which is in parallel with contacts *TR* and *S*1. The opening of contact

*R*2 will cause the dropping out of coils 1*CR* and *TR*. The contacts associated with these two coils will now return to their normal position, and we shall have the motors running through the control circuit, which consists of the STOP button, contact 2*CR*1, contact *R*1, and coil 2*CR*. This circuit maintains the circuit to the run coil through contact 2*CR*2. Should the STOP button now be pressed, all contacts would return to their normal position and all coils would be deenergized, thus opening the line contacts to the motor, and the motor would come to a stop.

This circuit obviously is one for a primary-resistance reduced-voltage starter. Again it must be pointed out that this is only one of the many arrangements of coils and contacts which could be used to achieve the same results. Different manufacturers will use variations of a similar circuit in the control of their starters, but the basic principle of operation is the same if a definite-time sequence of starting is employed.

This circuit could be expanded to give several more stages of acceleration by the addition of more units of resistance in series with the motor, with a control relay and a time-delay relay for each step or stage of acceleration. This is a two-stage, or two-step, starter, since it provides two steps of acceleration, one under reduced voltage and one under full voltage.

The only critical adjustment in this circuit will be found in the time-delay relay *TR*, which must not be allowed to maintain the motor under reduced voltage longer than the time it takes to accelerate to its maximum speed under reduced-voltage conditions. Prolonged operation of the motor under reduced voltage may quite possibly overheat and damage the motor windings and cause the resistance elements to be seriously damaged or burn out.

This controller consists of a start contactor *S*, which must be three-pole, and a run contactor *R*, which also must be three-pole. In addition to the two contactors, there are two control relays and one time-delay relay. This equipment would be found generally mounted in one enclosure with the START-STOP station either mounted on the door of this enclosure or separately wired to any convenient location in the building.

At this stage in your study of controls and analysis of control circuits, you should consider a circuit from the standpoint of what would happen if a particular coil burned out or a particular contact failed to open or close, as the case may be. For instance, suppose that the time-delay relay *TR* were to have a burnt-out coil. What would be the effect on this circuit? The circuit would function up through the closing of the start contactor *S,* and the motor would be energized under reduced-voltage conditions. If contact *TR* does not close, however, then the second control relay cannot be energized and the run contactor cannot be energized. Thus, the motor would continue to run under reduced-voltage conditions. The current under this condition is such that it will open the overload relay contacts and will drop out coil 1*CR*, thus stopping the motor and returning it to its normal position. These overload units should be manual-reset units so that the operator will have to reset them and determine the cause of trouble before restarting the motor. This control circuit will give overload protection and incomplete sequence protection (Sec. 2·15).

7·7 ANALYSIS OF CIRCUIT 6

The circuit of Fig. 7·6 is a partial circuit used to illustrate a lockout circuit, which is extensively used where malfunction of some part of the equipment must require attention by the operator before the equipment is restarted. Looking at the circuit of Fig. 7·6, you will see dotted a contact which represents the normal control components such as START and STOP buttons, limit switches, or other devices which normally start and stop the machine. This circuit concerns itself only with the lockout components. The normally closed contacts represented as *PS*1, 2, and 3 are pressure switches which will open only when the required pressure is not maintained in their particular part of the machine or process. Coils *A, B,* and *C* are relay coils paralleled by pilot lights.

To start the operation of this equipment, it is necessary to press the RESET button, which will close the three associated and mechanically interlocked switches. The three relays will be energized, thus closing their normally open contacts. Contacts $A1$, $B1$, and $C1$ are used to maintain the coil circuit. Contacts $A2$, $B2$, and $C2$ are maintained in their closed position as long as the operation of the pressure switches is normal, thus enabling the normal control components to energize coil M at will. If

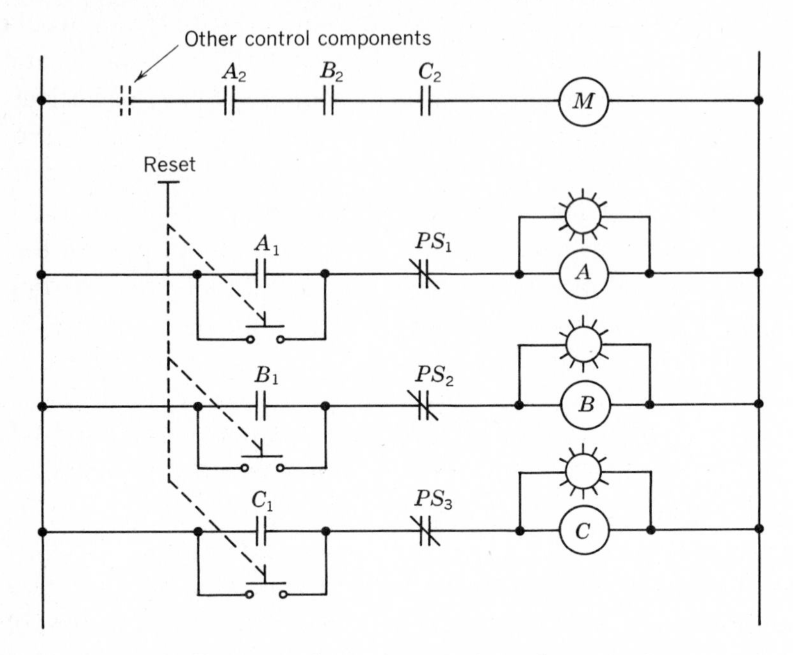

Fig. 7 · 6 Analysis of circuit 6. Lockout circuit.

the pressure drops or rises, as the case may be, from its normal value at any one of the three places where pressure switches are located, it will open one of the normally closed contacts. For instance, if $PS1$ opens, coil A will be deenergized, which will open contact $A2$ and drop out the motor. At the same time, contact $A1$ will open. If the proper pressure returns to pressure switch 1, its contact will close but coil A will not be reenergized, because the circuit is broken at contact $A1$. The pilot light in parallel with this coil will be out and will indicate which of the protective circuits is not functioning. The operator will know that pressure switch 1 was the cause of the shutdown of the equipment. In order to restore the machine to operation, the pressure sensed by pressure switch 1 must be restored to normal. Then the RESET button must be pressed in order to energize coil *A*, thus closing its contacts and allowing the normal operation of the control circuit. Of course, the same procedure will apply to the second and third pressure switches and associated contacts and coils.

This type of circuit is generally applied to fully automatic equipment, where the machine or process is allowed to start and stop by itself under the control of pilot devices which sense the condition of the process or material as the machine does its work. When machinery operates under these conditions, it is generally desirable to have some means to stop the process whenever a malfunction occurs and to prevent it from restarting until it has received the attention of an operator.

Summary

The preceding circuits and their analysis should form a basis upon which you can build a skill of analyzing circuits as found in everyday use in industry and on the job. While these circuits do not in any way represent all or even a major part of the possibilities in motor control, the procedure and method of analyzing their operation, if properly understood, may be applied to any and all control circuits and followed through to a complete understanding of the operation of the equipment and control components associated with it.

The student who wishes to obtain proficiency in control will apply these basic principles to other circuits which are at his disposal until he is satisfied that he can, with reasonable speed, interpret and analyze control circuits of all types.

The danger in circuit analysis lies in the tendency to jump to conclusions, that is, to decide what the circuit does and how it operates when you have analyzed only a fraction of all of its possibilities. Learn to study a circuit contact by contact and coil by coil until you have completely traced its operation through its normal sequence from beginning to end, and you will save many headaches in the future.

There are no review questions at the conclusion of this section. The procedure you should follow is to obtain additional circuits and practice analyzing them until you obtain proficiency. The mark of distinction between a good troubleshooter and a poor troubleshooter generally lies in the ability to analyze the control circuit and determine quickly which of the many components could cause the malfunction of the machine which he is trying to correct.

8
MAINTAINING CONTROL EQUIPMENT

If there is a single rule which applies to all maintenance procedure in all plants and under all conditions, it is *be careful.* Carelessness and failure to observe safety precautions are two things that the maintenance man cannot afford.

8·1 GENERAL PROCEDURE

The first procedure in any organized maintenance of equipment should be periodic inspection to prevent serious trouble from arising. This inspection should include not only electrical equipment but the machine as well, should point up the wear and tear on the electrical equipment, and should provide a basis on which replacement of parts and correction of danger spots can be taken care of before they cause serious trouble.

One of the greatest causes of failure of control systems is the presence of dust, grease, oil, and dirt, which must be removed periodically in order that the equipment may function properly. The removal of dust and dirt may be accomplished by dusting or wiping with rags, but this is not always effective with oil and grease. These substances generally should be removed by the use of a solvent such as carbon tetrachloride. Care should be exercised whenever these solvents are used, because the inhaling of any appreciable quantity of their fumes is quite likely to be very harmful. Therefore, adequate ventilation should always be provided.

Periodic inspection should always include a check for overheating of electrical equipment and mechanical parts, because excess heat is always an indication of trouble to come. The value of checking for excess heat depends upon your knowledge of the proper operating temperature of bearings, coils, contacts, transformers, and the many other pieces of equipment associated with machinery, motors, and control.

Bearings of motors and mechanical equipment should be checked for proper lubrication. It is very seldom, however, that bearings of electrical equipment such as starters and switches are oiled. They are designed to operate dry, and, generally speaking, oiling the bearings will eventually cause a gum to form, causing the equipment to malfunction.

Another frequent cause of failure of control equipment is loose bolts and electrical connections. Each connection should be periodically checked for tightness, and the inspection should include the checking of possible loose bolts and nuts on the equipment.

Short circuits and grounds in the electrical wiring may be prevented by proper inspection of insulation and by using a Megger insulation tester on motors and cables in associated equipment.

If you are to maintain the same equipment over a period of time, the first law to follow is to be familiar with your equipment. Know your equipment mechanically and electrically so that you will sense trouble before it develops.

The second law is to be observant. Whenever you pass a

piece of equipment for which you are responsible, listen and look. Quite often this is all that is necessary to tell you that trouble is on its way. Good maintenance procedure can be summed up in a very few words: Keep it tight, keep it clean, keep it lubricated, and inspect it frequently.

8·2 MAINTENANCE OF MOTOR STARTERS

The most frequent trouble encountered with motor starters is contact trouble. Contacts should be inspected for excessive burning or pitting and for proper alignment. If they are pitted, copper contacts may be filed, but care must be exercised not to remove too much contact surface or to change their shape appreciably. Copper contacts are subject to heat and oxygen on closing and opening of the circuit, and copper oxide may be formed on the surface of the contact. This oxide is an insulator which must be removed if it covers a large part of the contact surface. Most contacts made of copper are arranged to be of the wiping type, which allows the mechanical closing of the contacts to remove this oxide as it forms. If the contacts are silverplated, the silver oxide is a good conductor and need not be removed; in fact, silver contacts should never be filed.

The contacts should be inspected not only for pitting but for proper alignment and for proper contact pressure. Improper alignment or lack of contact pressure will cause excessive arcing and pitting of the contacts.

8·3 CAUSES OF TROUBLE

One of the most frequent causes of failure of automatic equipment is improper adjustment of contacts and time-delay circuits. Generally, the manufacturer of contollers for automatic equipment will supply the proper contact-clearance distances and other information necessary for the proper timing of the circuit. This information should be readily available to the maintenance man so that he can periodically correct these adjustments. A check of these adjustments should be part of the regular inspection of this type of equipment.

The second most prevalent cause of trouble in motor starters and contactors is coil burnout. Coils on modern starters are well built and well insulated, which has eliminated considerable trouble due to vibration and moisture. Coils are still subject to burnout, however, chiefly because of one of two things. The most frequent cause of coil burnout is failure of the contactor magnetic circuit to close, causing a gap in this circuit which increases the normal current through the coil to dangerous levels. The normal current to start the movement of a magnetic pole piece may be as high as 40 or 45 amp, but as the magnetic circuit is closed, this current usually drops to a very low value of 1 to 1½ amp, which is all that is required to maintain the magnetic circuit. If this circuit does not close, the coil will maintain a current somewhere in between these two values, which can very easily cause it to overheat and burn out its windings.

The second most frequent cause of coil burnout is improper voltage. If the voltage applied to the coil is exceedingly high, the current through the coil can reach dangerous levels and cause it to burn out. If the voltage applied to the coil drops so low that the magnetic circuit cannot be completed, we have a gap which will cause exceedingly high currents and cause coil burnout. In view of the above-mentioned causes, the proper procedure when it is found that a coil has burned out on a starter is to check the mechanical linkage to see that the contactor can close completely and to check the voltage applied to the coil under load to see that it is sufficient but not excessive. Check for spring tension to see that the springs themselves are not causing the magnetic circuit to remain partially open.

Should the contactor be equipped with flexible leads, they should be checked for fraying and broken strands and should be replaced if these conditions exist. Should the starter be equipped with arc shields, they should be inspected for proper alignment around the contacts. They should be checked for accumulations of dust and dirt, and if carbon deposits have built up on the inside of these shields, these deposits should be carefully removed, since carbon reduces the arc path and can be the cause of serious arc-overs, particularly under high-voltage conditions.

Spring tension for proper contact pressure is very important in a starter and should be checked against manufacturer's standards if they are available. They should at least be checked to see that each contact has approximately the same spring tension so that the contact pressure will be equal on each contact. Improper or unequal spring tension is one of the most common causes of contact chatter and starter hum, so be sure that when these conditions exist, you check the spring tension on every contact to determine if it is sufficient and that they are all equal.

8·4 MAINTENANCE OF RELAYS

Generally speaking, the maintenance of voltage relays is the same as that for motor starters and contactors with only the additional precaution that, in general, relays operate on lower currents and thus are provided with less power. This lower power demands a smoother, easier operating mechanical linkage and mechanism and thus requires more careful attention to the matters concerning it.

Current relays must be checked to see that they are receiving the proper amount of current for closing their contacts and that the spring tension and contact spacing are correct to give the proper pull-in and drop-out currents. Wear of contact surface and change in spring tension can cause a geat deal of variance in these values of pull-in, drop-out, and differential currents, which may make the circuit operate in a manner detrimental to the equipment.

Overload relays are devices which normally do not operate for long periods of time; therefore, they are subject to accumulations of corrosion, dust, and dirt, which must be removed after periodic checking to see that they can operate when needed. If proper equipment is available, overload relays should be tripped by current periodically to see that they do function. Excessive tripping of overload relays is generally not an indication of relay failure so much as it is of overload on the circuit. The maintenance man should first determine the current value at which the overload unit actually trips and compare this with the allowable current to determine whether the fault lies with the overload unit or with the circuit itself.

Time-delay relays, whether of the pneumatic type or the dashpot type, require periodic adjustment to compensate for normal changes in their operating characteristics. The dashpot relay should be checked for dust and other foreign matter in the oil reservoir, since any impurities in the oil will affect the accuracy of the timing of the dashpot.

Quite frequently, relay contacts may be of the make-before-break or break-before-make type, and here again spring tension and contact spacing become very important and require a check to determine that they are functioning as they were intended to.

8·5 MAINTENANCE OF PILOT DEVICES

Generally speaking, pilot devices require very little maintenance other than a check of their mechanical operation and their contact condition. Where the pilot device is a form of pressure switch or vacuum switch, the range of its operation should be checked occasionally to see that the contacts open and close at the pressure they were set up to operate on. The contact surfaces should be examined to see that they have not accumulated a coating of copper oxide, dust, or oil. They should be operated through their pressure range several times to check for consistency of operation.

Float switches are subject to troubles because of float rods being bent or because of water in the float due to a leak. A check of the proper operation of the float, the float rod, and the mechanical linkage to the float switch itself will determine the amount of wear and can generally indicate a replacement of parts before any serious trouble can develop. Of course, contact condition is a must on checking this as well as other pilot devices.

When limit switches are an integral part of a control system, they are a very likely source of trouble because they perform many thousands of operations per day on an active piece of equipment. They are prone to mechanical failure because of worn bearings and cam surfaces as well as contact surfaces and spring tension. The only solution to limit-switch failure is frequent and accurate inspection to determine their mechanical and electrical condition. When the mechanical condition of a limit switch becomes questionable, it should be replaced or repaired before it causes serious trouble with the other equipment.

8·6 MAINTENANCE OF BRAKES AND CLUTCHES

The chief cause of failure of brakes is, of course, worn brake lining or brake disks, as the case may be. This is an inexcusable cause of failure, however, if periodic inspection is employed. Never allow brake lining to reach the dangerous condition of wear.

The second most prevalent cause of brake failure is excessive wear and improper adjustment of the linkage from the electric solenoid or other operating device to the brake shoe or brake disk. These must be maintained in their proper mechanical alignment and condition. Improper linkage adjustment is a frequent cause of coil burnout on brake solenoids, since it may not permit the proper closing of the magnetic circuit, which in turn causes excessive current to flow in the solenoid coil.

Solenoid-operated clutches are subject to the same types of trouble as solenoid-operated brakes. Therefore, the inspection and maintenance procedure for these units should be the same as that for brakes.

Summary

While this chapter has attempted to point up some of the basic principles of good maintenance, the actual maintenance of a specific piece of equipment must be determined by its operating cycle, the complexity of its equipment, and the amount of maintenance time available. The chief difficulty in most maintenance situations is a lack of the understanding of the word "maintenance," which means to *maintain* the equipment in operation as compared with repairing the equipment after it has broken down. Again, inspect it, keep it clean, and keep it tight, and you will be doing maintenance, not repair.

Review Questions

1. What is the chief cause of coil burnout on starters, contactors, and relays?
2. Can copper oxide be the cause of trouble on contacts?
3. Should silver contacts be filed frequently?
4. What are some of the results of improper spring tension on starters?
5. What is a probable cause of contact chatter and starter hum on motor starters?
6. What would you expect to be the result of low voltage applied to the coil of a magnetic starter?
7. What is the proper procedure to determine the cause of too frequent tripping of overload relays?
8. What is a likely result of having a float half full of water when if operates a float switch?
9. What is likely to happen if an accumulation of carbon is allowed to form in the arc shields of a starter or contactor?
10. What is the best method of removing oil and grease from contacts and other surfaces where it might be harmful?
11. When using cleaning materials, what precautions should be taken?
12. What will be a likely result of poor adjustment of the linkage on a brake?
13. What causes a change in the timing on dashpot-type time-delay relays?
14. What two adjustments are likely to change the setting of the pull-in, drop-out, and differential currents of a current relay?
15. What is the difference between maintenance and repair as applied to control circuits and components?

9 TROUBLESHOOTING CONTROL CIRCUITS

Troubleshooting is a field of control work which generally separates the men from the boys. Many a man who can do a beautiful job of wiring a new control circuit from a circuit diagram is lost if the circuit fails to function as expected. Your chief asset in this field is an analytical mind trained in all of the aspects of control functions, components, circuits, and circuit analysis. The secret to efficient and accurate troubleshooting lies in determining the section of the control circuit that contains the trouble component and then selecting the proper component to be checked. This can only be accomplished by efficient and accurate circuit analysis, not by trial and error, long, extended wire tracing, or indiscriminate checking of components at random.

9·1 GENERAL PROCEDURE

First let us consider a new circuit which has just been wired but fails to function as expected. Here there is a possibility that the wiring has been misconnected or even that the circuit was not properly designed. If we check all of the connections in all of the wiring, however, it becomes a trial-and-error process and generally involves a considerable waste of time.

The first procedure should be to analyze the circuit to determine that it has been properly designed and should work as expected if the wiring was done properly. The next step is to follow the operation of the equipment through the expected sequence until we find the section of the circuit which is not properly operating. When you have located the section of the circuit which is giving trouble, it should be simple procedure to check the wiring and operation of the components involved in this section of the circuit and clear whatever the trouble might be.

In this process you have already made use of your knowledge of analysis of circuits and your knowledge of components and their proper functions to determine whether or not they are operating as they should. Any lack of knowledge on your part of control functions, control components, control circuits, or circuit analysis will cause undue delay and wasted time in this process. When you have located the trouble in this section of the control circuit, the sequence should be started over and run through until either it has operated successfully or another section of the control circuit has been determined to be malfunctioning.

When considering troubleshooting an existing circuit, we can generally eliminate the possibility of improper connections. If the circuit had been improperly wired, it would not have operated originally. It is surprising, however, how many men will begin their troubleshooting procedure by checking out the wiring, connection by connection, to determine if it was properly made. This procedure is an injustice to the plant owner and the operator of the machine, who are interested in speedy and efficient repair rather than time-consuming experimentation.

The first step in troubleshooting an existing circuit which has developed trouble is to understand that circuit and to understand the operation of the machine it controls. On complex circuits time generally does not allow the service man or troubleshooter to digest the complete circuit. With the help of the operator, however, you can determine how much of that circuit is operating. Follow the machine through its cycles until it reaches the point where it does not function properly. Having determined this point, you can analyze the circuit, starting with the section that does not operate. A careful check of this circuit and a location of the components involved in this section of the circuit will generally lead you to the source of the trouble you are seeking. The malfunction of some control component must be the cause of the circuit failure.

In the rare case where insulation breakdown is the cause of the trouble, it should be evident from a visual inspection of the components and the wiring. Quite frequently, however, a grounding of a wire in the control circuit may escape detection in a visual inspection, and if it is suspected that a ground is the cause of the trouble, careful checks should be made with the power off. With an ohmmeter determine the resistance to ground of the wires in this particular section of the control circuit.

Let us assume that you have now located the section of the control circuit which seems to be causing the trouble. The first step is to locate the components involved in this part of the circuit. There must be a coil of a relay, a contactor, or some other device which is energized by this section of the control, and the machine should be run through its sequence to determine if this coil does receive energy.

If the contactor or relay does not close as it should, the circuit should be disconnected and the wires removed from the coil of the relay or contactor so that a voltage check can be taken. Apply a voltmeter across the wires which were connected to the coil and again energize the circuit operating the control sequence up to this point. If the voltmeter indicates a proper voltage applied, then the trouble most likely is in the windings of the coil itself. Do not attempt to check the voltage or resistance of the coil while it is connected in the circuit, since false readings are likely to result from feedback and parallel paths in the control circuit.

If it is suspected that the coil is at fault, disconnect the power from the circuit and with an ohmmeter check the resistance of the coil, which should be very low on a d-c resistance check. If the coil is burnt out, you will receive a high resistance reading or a reading of infinity on the ohmmeter, indicating that the coil needs to be replaced. Do not depend on the coil smelling burnt or showing any visible evidence of being burnt out, since this is not always the case.

Suppose that our voltage check showed that the voltage did not reach the coil when it should have in the sequence of operation of the control circuit. This indicates that some contact is not closing when it should, thus deenergizing the circuit to the coil. A careful study of this section of the control circuit following the principles outlined in Chap. 7 should easily show what contacts should close in order to energize this coil. You must now locate the components which contain these contacts and again operate the machine through its sequence, observing the operation of the relay, limit switch, float switch, pressure switch, or other device that contains these contacts, to determine whether it operates mechanically as it should. If this component does operate mechanically, it indicates two possibilities. The first and most likely is that the contacts involved are not properly closing or are coated with copper oxide or other insulating material which prevents them passing current to the coil as they should. The other possibility is an open circuit due to a broken or burnt wire. This generally, however, is the least likely cause of trouble. Having checked the contacts and eliminated the trouble, which probably will be found there, again operate the control circuit with all coils connected, and if it does complete its sequence, then proceed to apply the above procedure to the next section of the control which does not function.

The above procedure is based on years of experience and an understanding of the fact that control circuits are made up basically of only two things: contacts, which make and break the circuit, and coils, which operate these contacts. If the contacts close and open as they should, then the proper voltages will be applied or disconnected from the coils as they should. If this is true, then the malfunction must lie in the coil itself. If the contacts do not operate properly, however, then the trouble must be in the contacts or in the associated wire which carries this current from the contact to the coil.

The most important rule in troubleshooting is to change only one thing at a time. If you find a set of contacts that you suspect is not properly functioning, correct this trouble and try the circuit again before changing anything else. If you find a coil you suspect to be burnt or otherwise causing trouble, repair or replace it and try the circuit again before attempting any other changes. One of the most common mistakes of troubleshooters is to change or correct several supposed troubles at one time before trying the circuit for operation. Quite frequently several changes made at one time may introduce more trouble than you had originally. This should be made a cardinal law in your work as a troubleshooter and will put you far ahead of the field in efficiency of your work. It is very seldom that several parts of a machine would wear out at the same instant. Therefore, even though the overall condition of the control components may be poor, it still remains probable that only one component has failed completely.

If the machine that you are troubleshooting is not thoroughly familiar to you, do not underestimate the value of the operator in your process of determining the cause of trouble. His knowledge of the normal operation of this piece of equipment can be put to work to eliminate a lot of wasted time on your part in determining how the machine should operate. Depend on him to help you locate components which may be hidden by parts of the machine, since he probably knows where they are. In short, make use of every available source of information to shorten the time necessary for you to arrive at the spot of trouble.

All failures of electric control circuits are not necessarily caused by electrical troubles. Quite frequently, the mechanical malfunction of some component may be the sole source of trouble, so remember to examine suspected components not only for electrical trouble but also for mechanical trouble.

It must also be pointed out that a man who is attempting control troubleshooting who is not equipped with a voltmeter, an ammeter, and an ohmmeter is wasting valuable time and money. He must also be trained and competent in the proper use of these instruments and the proper interpretation of the readings that they give him. Even though you may know many men who do not make use of all these instruments in their troubleshooting work, it is an indisputable fact that their efficiency could be greatly increased by a proper understanding and application of these instruments to their work.

9·2 TROUBLESHOOTING CONTROL COMPONENTS

All that need be said of the individual problems involved in the various components of control has been covered in Chap. 8. The trouble spots recommended in this chapter for checking under maintenance procedure are identical with those trouble spots which will have to be detected in the process of troubleshooting and repairing the circuit after it has failed to perform as it should.

Again, the best equipment for efficient and proper troubleshooting of individual components is a complete knowledge of their proper operation and a familiarity with as many manufacturers' versions of each component as possible. Much of this knowledge will of necessity have to be gained through experience. The student may obtain a sizable portion of this required knowledge by a study of manufacturers' literature and by making a concerted effort to familiarize himself with the various components he comes in contact with in his daily work.

9·3 STEP–BY–STEP PROCEDURE

In order to make the procedure outlined in Sec. 9·1 clearer to you, we shall now consider a circuit and determine the probable cause of some troubles which we shall assume to have occurred in this circuit.

The circuit of Fig. 9·1 is that of a chilled-water air-conditioning compressor. The components as shown on the diagram are as follows: The coil *CR* is a control relay. The coil *M*1 is the starter for the chilled-water pump. The coil *M*2 is the starter for the condenser-water pump. The coil *M*3 is the starter for the oil pump on the compressor itself. The coil *M*4 is the compressor-motor starter. The contact identified as *T* is a thermostat

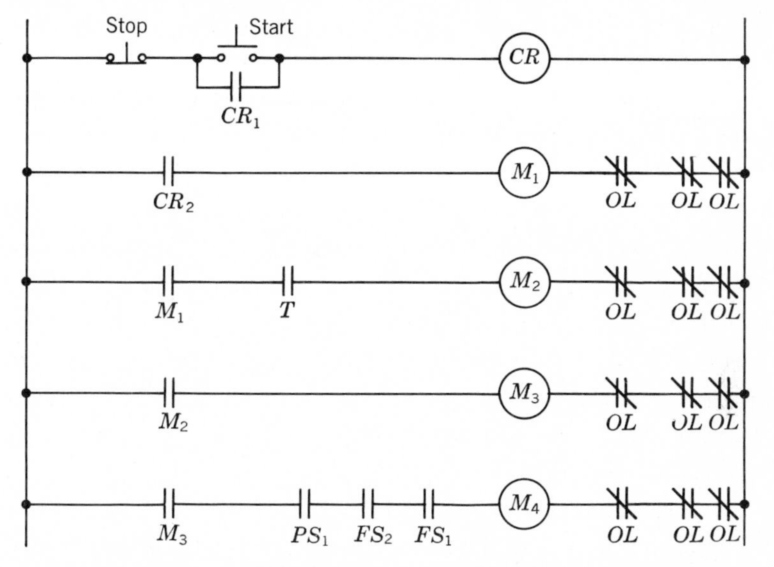

Fig. 9·1 Circuit for chilled-water air-conditioning compressor control.

which senses the temperature of the chilled-water return. Its function is to start the condenser-water pump when this temperature reaches a predetermined high level. The contact identified as *PS*1 is an oil-pressure switch whose function it is to stop the compressor should the oil pump fail and also to prevent its starting before the proper oil pressure has been obtained. The contact identified as *FS*1 is a flow switch in the chilled-water piping system. Its function is to prevent the compressor from running unless there is sufficient flow of chilled water. The contact identified as FS_2 is a flow switch in the condenser water piping system.

Suppose now that you are called in to troubleshoot this circuit. The first step should be to determine from the owner or operator what trouble he is having with his circuit. Suppose that he tells you that the condenser-water pump does not start as it should. Then from a study of a diagram we can assume that the section of the circuit for the control relay is functioning properly, that contact *CR*2 closes, and that the chilled-water pump runs as it should. Something must be wrong in the third line of our schematic diagram.

The first procedure is probably to check the overload relays and determine that they were not tripped. Having done this, we next check the thermostat to see that its contact is closed as it should be. Here it must be pointed out that determining the setting of this thermostat and the actual water temperature will indicate whether it should be open or closed. We are assuming that through the shutdown of the machine the water temperature has increased to a point that demands that these contacts be closed. Let us assume that the thermostat contacts are closed; then inspection of the starter for the chilled-water pump is indicated to determine if contact *M*1 is closing when this contactor is energized.

If our inspection of this starter shows that this contact seems to be closing properly, then the next procedure is to disconnect the wires from coil *M*2 and apply an ohmmeter to the coil to determine whether it was open or not. From the preceding analysis, it is almost certain that this coil will be found open, and for the sake of this illustration we will assume that it is. Before you replace this coil, the starter should be examined for proper mechanical operation. Determine that the contact arm which raises and lowers or swings to move the contacts is free from bind and that the spring tension is not excessive. Also examine the faces of the magnetic pole pieces to see that they have not been abused and possibly damaged by someone forcing them or even through the many operations of closing of the contactor. When all mechanical problems have been eliminated, install a new coil in the starter.

It would be good practice to check the voltage at the ends of the wires which feed this coil before putting it back into service. This can be done by connecting a voltmeter between the ends of these wires and operating the control circuit up to this point. If the voltage is excessively low or excessively high, then the cause of this trouble must be determined and eliminated. Otherwise, the new coil will also burn out.

Suppose that this circuit did not malfunction in this way, but instead the report was that everything seemed to work except the compressor itself. Then our operation would be to energize the circuit and watch its sequence to determine for ourselves where it failed. We would see that the control relay operates, the chilled-water pump starts, the condenser-water pump starts, and then that the oil pump on the compressor starts.

Here we shall assume that our sequence stopped and the compressor did not come on the line as it should. Again examining our circuit, we find that we have a contact on the oil-pump starter which could cause trouble. We have a pressure switch and two flow switches which might be the source of trouble. So again we must determine which of these components is not properly functioning. If these components are readily accessible, a physical examination of each of them may immediately disclose the trouble. If they are inaccessible, however, a good procedure to follow is to disconnect the wires from the starter coil and operate the control circuit to determine if voltage is reaching the coil, thus eliminating the possibility of trouble being in the coil itself.

Let us assume that the contact *M*3 is properly functioning and we have checked it. The two flow switches have been determined to be properly functioning and their contacts closed. Then the examination of the pressure switch is the only remaining possibility. It may even be necessary in some cases to recalibrate pressure switches with known pressures to see that they are operating at the settings which show on their indicating dials. Again, however, the procedure is to inspect physically and determine the actual cause for the part not functioning properly.

Summary

While this procedure may seem oversimplified, as you are guided through the diagram on a supposed troubleshooting job, it is the basis upon which good troubleshooting practice is laid. No matter how complex the control circuit is, it can be separated into simple branches such as we have illustrated here and in other sections of this book. The efficient troubleshooter will narrow his trouble down to one of these simple branches of even a very complex circuit, so that the actual process of locating the troublesome component will be as simple as outlined here.

Review Questions

1. When is it necessary to check completely the connections of a whole control circuit?
2. Why must the wires be disconnected from a coil in order to determine accurately whether the coil winding is damaged or not?
3. Is all control-circuit trouble necessarily electrical trouble?
4. Does the fact that contacts appear to be touching indicate that the electric circuit is complete through them?
5. Why should the troubleshooter operate a machine through part of its sequence before starting to look for the trouble?
6. What are the two possible causes for repeated tripping of overload relays?
7. Should the troubleshooter try his circuit after repairing one fault, or should he attempt to fix everything that seems as if it might be causing trouble before trying the circuit?
8. What are the most frequent sources of trouble in motor starters?
9. When troubleshooting a circuit which has been operating, is it wise first to check to see if the wiring was done properly?
10. What is the chief source of failure of pilot devices such as float switches or limit switches?

11. In Fig. 9·1 what would be the most likely cause of the circuit operating only as long as the START button were held down?
12. What is the most likely source of trouble if, when we press the START button, the control relay remains energized but coil $M1$ does not pull in?
13. What would be the results if the overload relays on the circuit for coil $M2$ were to open while the compressor was running?
14. Which is generally the most difficult, finding the source of trouble or repairing the trouble after it is located?
15. Which of the above requires the most skill?

10

BASIC CONCEPTS OF STATIC CONTROL

Just as the magnetic starter liberated the machine from the line shaft, static control is liberating the machine and operator from servitude to the slow-acting, ever-failing, short-lived magnetic relay and contactor. The advent of static control opens a vast new field of possibility for rapid, fully automated machines and processes.

The previous chapters of this book have been concerned with the language of magnetic control, that is to say, control by means of moving contacts and magnetic cores. This chapter will present a new language of control. We may define *static* as "pertaining to or characterized by a fixed or stationary condition." This definition gives us the key to the meaning of *static control,* or control by means of devices without moving parts.

The ever-present problem with magnetic control has always been the failure of components. The magnetic switching devices, such as relays and contactors, have coils which require relatively large currents to operate the mechanical linkage attached to the contacts. These coils tend to burn out, and the linkage parts are constantly subject to wear. The contacts themselves are often victims of dirt, grease, and other foreign matter which cause arcing and burning or pitting of the contact surfaces themselves. For any installation of a single motor with relatively simple control functions and where a few million operations provide a satisfactory life factor, the magnetic control circuit is and will continue to be the most practical and economical solution to the control problem. However, when the demands of the circuit require a significant number of control functions, when the rapidity of switching becomes a significant factor, and when long life in terms of number of operations is essential, static switching through the use of logic circuitry becomes not only economically feasible but almost mandatory. One other factor which must always be considered when selecting a system for control of a machine or process is the requirement of space for the control components. The use of magnetic control for a complex system puts serious demands upon the available space, in contrast with a fraction of that space required by static switching devices. Atmospheric environment may also be sufficient reason for using static control.

Static switches operate at low d-c voltages, usually 10 to 20 volts, and very low current. They have no moving parts which would be subject to wear or require adjustment. There are also no contacts to burn or to collect dirt and other foreign matter; therefore no cleaning of contacts is required.

Static control offers several advantages over magnetic control. The first and very important advantage is the increased reliability of the circuit. A static system has a much greater ability to produce a signal output when and only when an output is called for. The long life of static switches, which is completely independent of the number of operations performed, makes them almost indispensable for automated control systems. Static switching provides a much higher speed of operation which is often required by modern machines and processes. Many control functions must be performed in adverse environments where magnetic control devices would be destroyed or at least limited to short life by the chemicals or other atmospheric content. This is generally not a major factor when static switching devices are used. Static switching also provides a much simpler circuit design than magnetic control. Circuit simplification in processes which must sense and evaluate many factors is provided through the basic concept that a static switch is a multiple-input and single-output device, as contrasted with the relay or contactor which is inherently a single-input multiple-output device. The single output of the static switch may be used to provide inputs to many other static switches; this phenomenon is referred to as "fan out."

This chapter will be concerned with digital control or, in more familiar terms, switching-type control. Semiconductors or solid-state devices are also used in industrial control circuits to perform analog functions, but these are not a subject for this chapter, which will discuss logic circuitry.

10·1 ESSENTIALS OF STATIC CONTROL

The language of static control consists of only a few words, five to be exact. These five words are: AND, OR, NOT, MEMORY, and DELAY. There are also a few derivatives and combinations of the basic words such as NOR, which is really a combination of OR and NOT and is sometimes called an OR-NOT.

If you feel slightly confused at this point, you have fallen victim to the limitations of a five-word language. The very simplicity of the language makes it sound like double-talk if you are not careful. The student who does not let himself get confused by the simplicity will have no real trouble with this new but astonishingly useful language called "logic."

Consider the possibilities of a control system in which even the most fantastically complicated specifications can be met by the use of a handful of basic building blocks in the proper combinations; this is static control. Each of the words in static control represents a basic building block called a *logic function* or *logic element*. Each logic function has a symbol used in what is known as a *logic diagram*.

The first word in the static language is AND. To understand the meaning of this word it is necessary to remember that all logic elements have multiple inputs and only one standard output. This, of course, is just the opposite of a relay which can have only one input, the coil, and may have multiple outputs, the contacts.

Consider the logic symbol of Fig. 10·1. This is the common form of the AND symbol. The requirement of an AND element is that all inputs must be present in order to have an output.

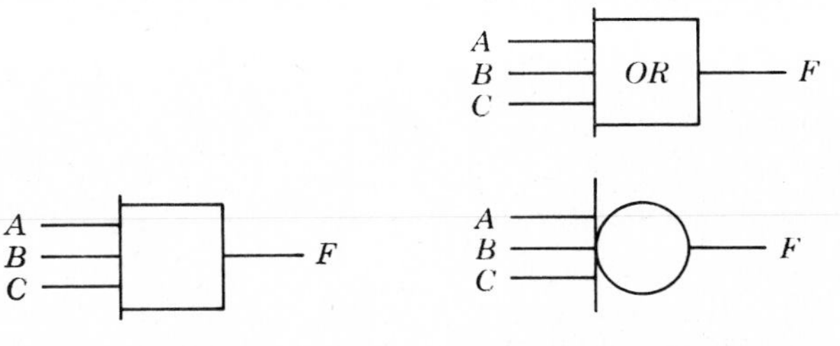

Fig. 10·1 The AND *symbol.*

Fig. 10·2 Two forms of the OR *symbol.*

In the case of Fig. 10·1, inputs *A* and *B* and *C* must be present in order to have an output. The output is at *F*. It should be remembered that the loss of any one input will turn off the output.

At this point do not be concerned about what an input or an output is, or how the logic element performs its functions; just learn to read the language in symbol form.

The second word in the static language is OR. The two common forms of the OR logic symbol are shown in Fig. 10·2. The requirement of an OR logic element is that it will have an output when any one or more of its inputs are present. The symbol of Fig. 10·2 says that there will be an output at *F* if input *A* or *B* or *C* or any combination of these inputs is present.

The third word in the static language is NOT or NOR. The NOT symbols are shown in Fig. 10·3. The requirement of a NOT is that it will produce an output when and only when its input is not present. The NOR is merely a multiple-input NOT and will have an output only when all its inputs are not present. When any one or more inputs to the NOR element are present, the output is turned off.

A third form of a negated-input logic element similar to the basic NOT or NOR is the NAND (Fig. 10·4). The requirement

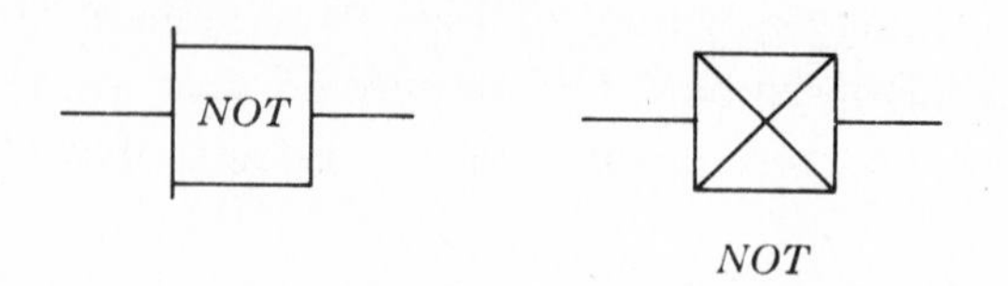

Fig. 10·3 The NOT *symbol.*

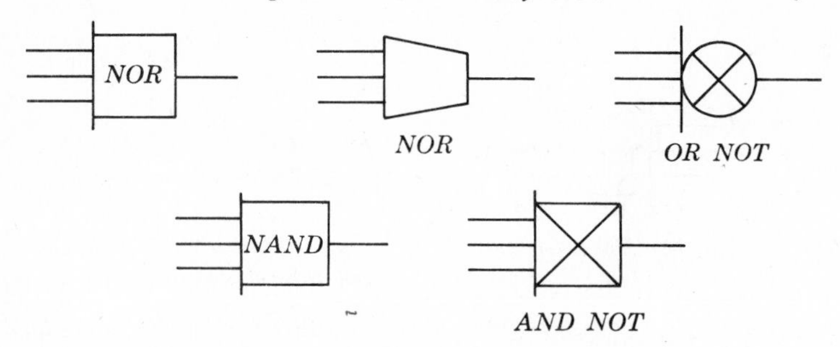

Fig. 10·4 Multiple-input NOT *symbols.*

of a NAND is that it will have an output unless all its inputs are present. When any one or more of the inputs to the NAND element are not present, it will have an output.

The fourth word in the static language is MEMORY. Elementary symbols are shown in Fig. 10·5 for the basic form of MEMORY elements. The dotted line indicates the NOT output of the MEMORY. The MEMORY element remembers the condition of its output as long as the power remains on. The *retentive* MEMORY element remembers the state of its output even after the power is turned off. This resembles the action of a manual switch which mechanically remembers which way it was last thrown.

The *off-return* MEMORY remembers the state of its output until the power is turned off and then always goes to the OFF condition. This is similar to the action of the magnetic starter and a three-wire control circuit.

The fifth word in the static language is DELAY. Elementary symbols are shown in Fig. 10·6. The function of the DELAY element is to provide an output after a specific delay following the application of an input. The above function would be known

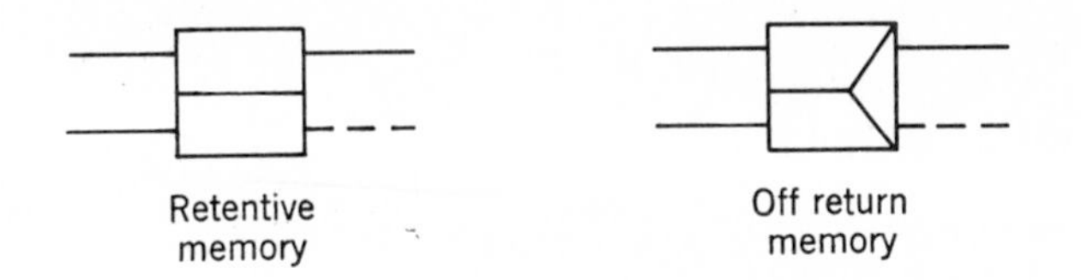

Fig. 10·5 The MEMORY *symbols.*

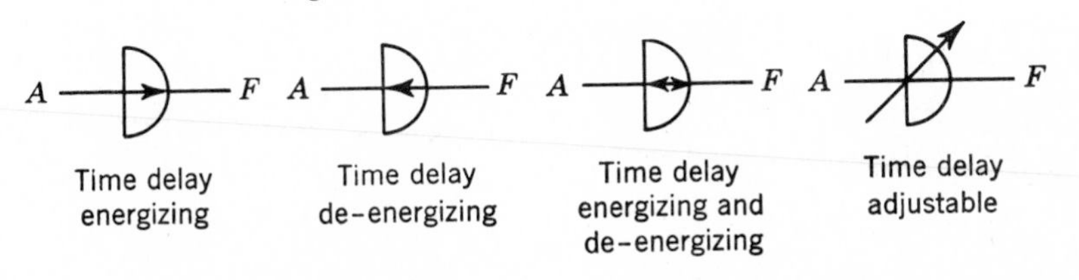

Fig. 10·6 Four forms of the time DELAY *symbol.*

as time-delay-on-energizing. DELAY elements can be built to provide delay upon deënergizing as the needs of the circuit require. The symbols of Fig. 10·6 illustrate the four common forms of DELAY logic elements.

At this point it would be well to test your understanding of this new language. Consider a wall with three push buttons mounted below a lamp. You walk up and press one button at a time; nothing happens. You then press the buttons in pairs; nothing happens. You then press all three buttons at once. The lamp lights. What type of logic element are the push button and lamp connected to? Of course, the logic element is an AND.

What type of logic element would necessarily have been used if the lamp came on when you pressed any one or any combination of buttons. Of course, it must be an OR element.

You supplied the input by pressing the buttons, and the lamp indicated an output by turning on. The other logic elements can be as easily understood by applying the same reasoning to their specifications. If the lamp had been on and was turned off only by pressing all three buttons at once, the logic element would have been a NAND.

All logic-function elements operate at very low power levels; therefore they must be followed by an amplifier in order to bring the level of power high enough to operate the device which is to be controlled. Figure 10·7 shows the symbol for an ampli-

Fig. 10·7 Two forms of the amplifier symbol.

fier. It is used regardless of the physical or electrical makeup of that amplifier.

You should also be familiar with the USASI standard logic symbols, although they are not used in this book:

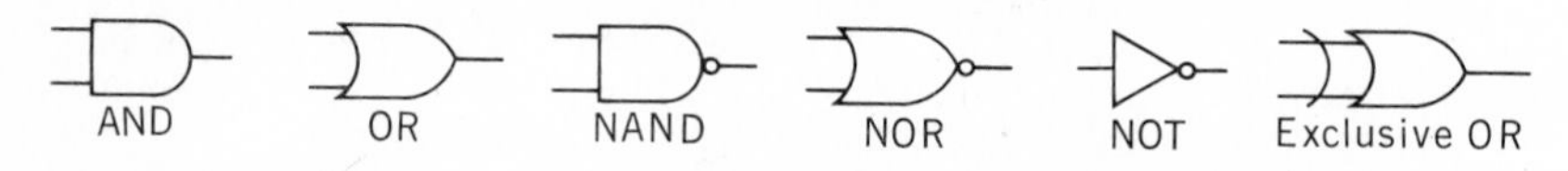

10·2 DEVELOPMENT OF LOGIC CIRCUITS

Any explanation of the control of a machine can be expressed in terms of the logic relationships of each function. Therefore, a control system breaks down into basic logic functions. The designer of conventional control circuits probably is not conscious of this, but examination of the progressive steps in designing a circuit will illustrate that the logic-function technique was actually the method used to determine the circuit design.

Consider a circuit whose requirement is that a coil *M* be energized when either a pressure switch *PS*1 or a limit switch *LS*1 is closed. If this circuit were developed for magnetic control, it would be as shown in Fig. 10·8. The logic statement of this circuit is: The coil will be energized when either *PS*1 or *LS*1 is closed. Therefore the logic element required would be a two-input OR, as indicated in Fig. 10·9. Note the amplifier which must be inserted between the OR element and coil M to raise the power level sufficiently to energize M.

Possibly the greatest difficulty when studying static control for the first time stems from the difference in the circuits of Fig. 10·8 and Fig. 10·9. The conventional, or electromagnetic, circuit provides a means of tracing the flow of current from line 1 through the control devices through the coil back to line 2 and is easy to understand. However, the logic diagram does not indicate the power circuit as such; rather, it is a block diagram of the control function of the circuit and at this stage may tend to leave something to be desired from the student's point of

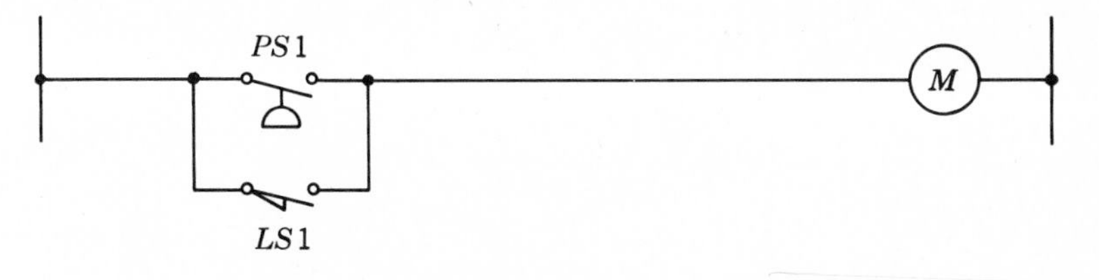

Fig. 10·8 Magnetic control circuit.

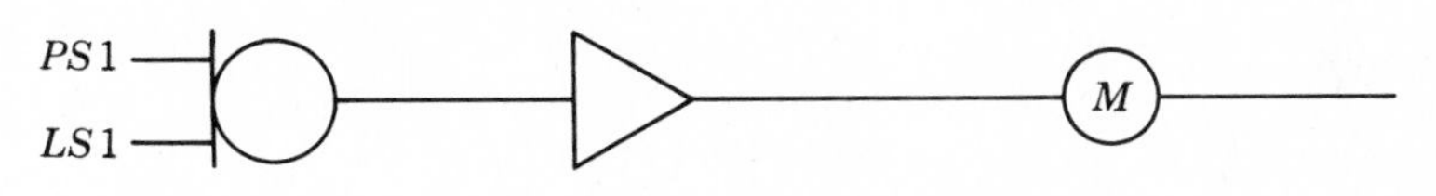

Fig. 10·9 Basic logic control circuit.

view. As logic circuits become more complex, the desirability of omitting the actual wiring and using symbolic representation of control functions will become more understandable. The OR element of Fig. 10·9 could be made up of vacuum tubes, saturable reactors, or, more probably, solid-state devices such as transistors. The amplifier of Fig. 10·9 might well be a relay, or a vacuum-tube amplifier, or a saturable reactor, or a silicon-controlled rectifier used in a switching mode to provide the necessary power amplification. The logic diagram does not in any way indicate the actual circuitry or components used within the logic element. This information would be found in the circuit for each element and will vary with the manufacturer involved. The student who feels it necessary to visualize a completed circuit with the logic diagram might be aided by considering that there is a common, or ground, bus which is not shown and all input and output voltages are taken from this common bus. Normal practice dictates that the common bus not be shown in order that the diagram will not be unnecessarily cluttered, since its presence adds nothing to the information given by the logic diagram.

Suppose that we add to the circuit of Fig. 10·9 a further specification that coil *M* will be energized by *PS*1 or *LS*1 only when contacts *T*1 and *T*2 are closed. Our circuit would need to be modified to that of Fig. 10·10. The logic function we

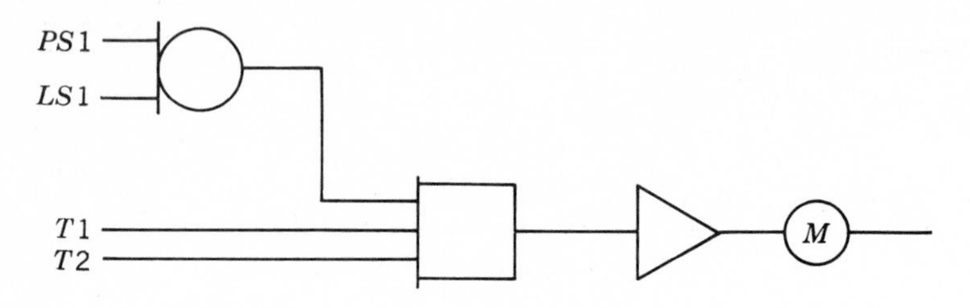

Fig. 10·10 First addition to logic circuit.

have used could be stated: The outputs of the OR and *T*1 and *T*2 must all be providing an input or ON condition before *M* is energized. This naturally indicates the use of a three-input AND unit, as shown in Fig. 10·10.

A further development of the logic circuit of Fig. 10·10 might include the following as the total logic statement: Coil *M* is to be energized when *PS*1 or *LS*1 and *T*1 and *T*2 are closed, and only if there is no input from *T*3 or *T*4. The logic requirements assigned to *T*3 and *T*4 would indicate the use of an OR-NOT element, since the presence of an output must be accompanied by the lack of an input from *T*3 and *T*4. The circuit for our new specifications is that given in Fig. 10·11. The two-input AND could be eliminated by using one four-input AND.

To read the diagram of Fig. 10·11 when no specifications are provided is relatively simple. The logic elements supply the key words to indicate the operation of the circuit. If we start at the top of the left-hand part of our circuit, the diagram will read as follows. Whenever *PS*1 or *LS*1 provides an input to the OR element, there will be an output. When there is an output from the OR, it provides one of the three inputs to the first AND element, which requires an input from *T*1 and *T*2 in order that it have an output. When the above conditions are met, the first AND unit provides an input to the second AND unit, which must also have an input from the OR-NOT element. The OR-NOT will have an output only when there is no input from *T*3 or *T*4. When this condition is met, both inputs are present at the second AND element, thus providing an output.

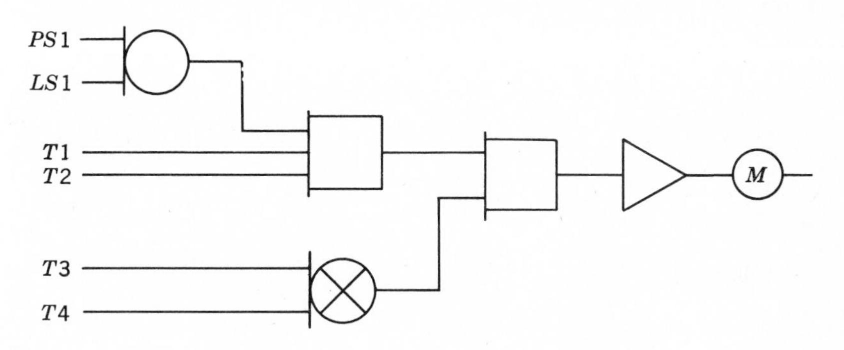

Fig. 10·11 Second addition to logic circuit.

The output of the second AND element provides an input to the amplifier which supplies the power amplification necessary to energize coil *M*.

At this point the student may well wonder what *PS*1, *LS*1, *T*1, *T*2, *T*3, and *T*4 are. The details of the field wiring or sensing devices are normally found on a separate diagram and not shown on the logic diagram. These might well be contacts of pressure switches, limit switches, and thermostats. A later portion of this book will be devoted to bringing the various parts of the overall control system together in order that the student may fully understand the interrelationships of the sensing devices as shown on the field-wiring prints; the action part of the control circuit is represented in the logic diagram, and the power or utilization circuit normally is shown in a third diagram.

Consider the circuit of Fig. 10·11 from an operational standpoint. If *PS*1 is closed but *LS*1 is open, there will still be an output from the OR element. If *T*1 and *T*2 are closed, they will complete the necessary inputs to the AND element, thus providing one input to the second AND element. If *T*3 and *T*4 are open, there will be an output from the NOR element. The second AND element now has an output, and the coil is energized. Consider what would happen if *T*3 closed, thus providing one input to the NOR element. The result of *T*3 providing an input to the

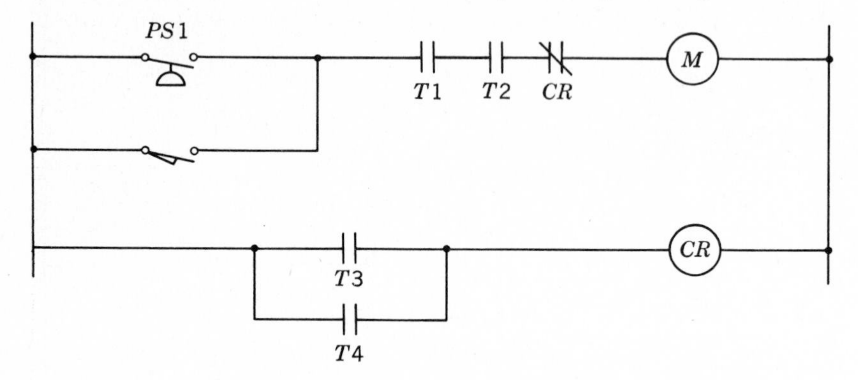

Fig. 10 · 12 Electromagnetic circuit equivalent of final logic circuit.

NOR element is that it will lose its output. Since the second AND element now has only one input present, it no longer has an output. The amplifier has no input; therefore it has no output, and coil *M* is no longer energized. Consider that instead of *T*3 closing and providing an input to the NOR element, *T*2 opens and thus eliminates one input to the first AND element. The net result would be the same in both cases—coil *M* would not be energized.

What would the circuit of Fig. 10·11 look like if it were an electromagnetic circuit and *T*1, *T*2, *T*3 and *T*4 were thermostat contacts, and if we assumed that *PS*1 is a pressure switch and *LS*1 is a limit switch? The circuit is shown in Fig. 10·12. This may seem to the student to be a much simpler representation of our circuit than the logic diagram of Fig. 10·11. But suppose now we add a further requirement to the circuit, that it provide complete isolation of coil *M* from the sensing devices and their contacts. Now the circuit becomes more complex and might possibly be wired as shown in Fig. 10·13. Neither the logic diagram nor the schematic diagram is a particularly complex circuit, but you should be able to visualize the simplicity of the logic diagram when applied to the complete automation of a machine or production line.

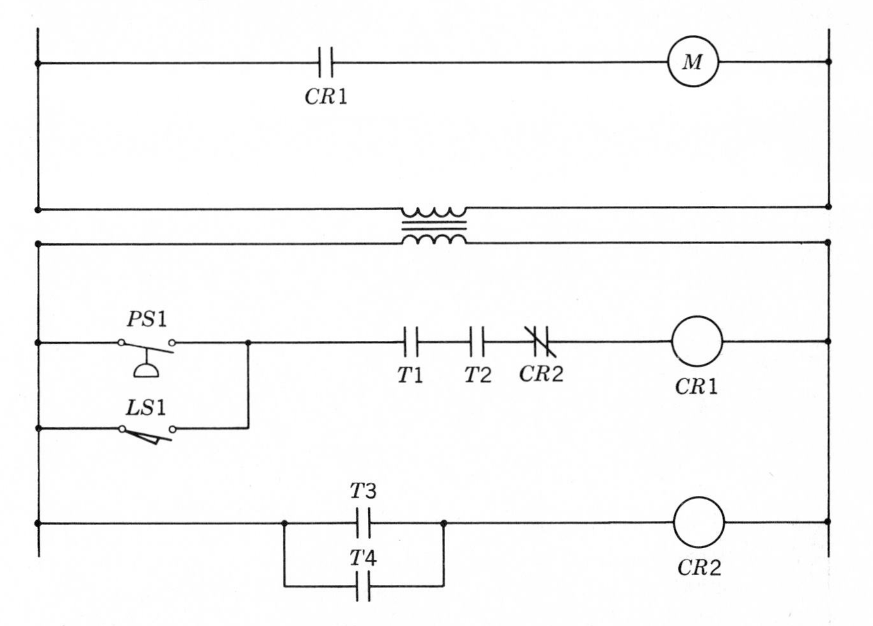

Fig. 10·13 Electromagnetic circuit equivalent of final logic circuit providing isolation of pilot devices.

The three-wire control circuit which was used so frequently in the study of electromagnetic control circuits can be represented and accomplished in the logic circuit by the use of feedback. The schematic for an electromagnetic circuit and the equivalent logic diagram are shown in Fig. 10·14. To understand the operation of the logic diagram, consider that one input is always provided to the AND element by the STOP button. When the START button is pressed, the second and final input to the

AND element is provided, and an output is produced. Once an output is produced, the internal feedback loop maintains an input even though the START button is released. This is equivalent to the action of the maintaining contacts parallel with the START button in the electromagnetic circuit. If the STOP button is pressed, the AND element loses its output, and there is no feedback to provide an input to substitute for the START button. The circuit is now in its original off state, and even when the STOP button is returned to the closed position, the START button must be pressed before there will again be an output from the AND element.

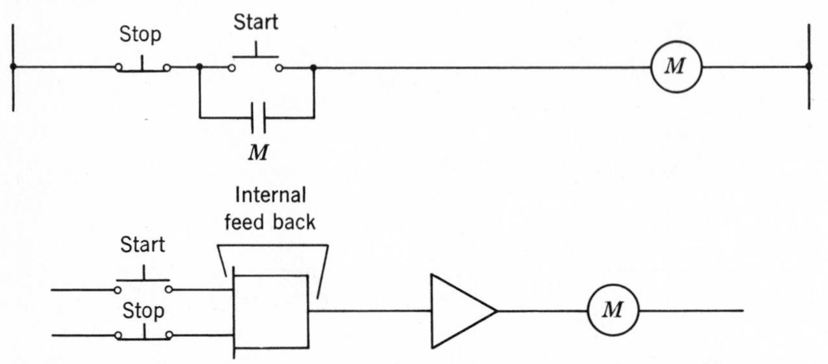

Fig. 10 · 14 Electromagnetic three-wire control circuit and logic equivalent.

10·3 APPLICATION OF STATIC ELEMENTS

Any control system can be divided into three major parts (Fig. 10·15). The sensing devices such as push buttons, limit switches, and thermostats constitute the information-gathering section. As the information is gathered, it must be acted upon in order that a decision can be made about what the system should do. In electromagnetic circuits this section contains mostly relays. Having arrived at a decision based upon the information gathered, the system must take appropriate action. The action section of the control system consists of the final output device or devices, such as motor starters, indicating lamps, and solenoids.

Static control in its digital form, logic circuitry, generally is applied to the decision section of the control system. Logic elements are low-voltage, low-power devices. They therefore require signal converters, sometimes called original inputs, to reduce the necessary high voltage of the sensing section to proper logic-signal values. They also require amplifiers to convert their low output power to the level of power required by the action section of the system.

The information section of the control circuit can be made entirely static by the use of static sensors such as the proximity limit switch. This section is much more likely to contain the

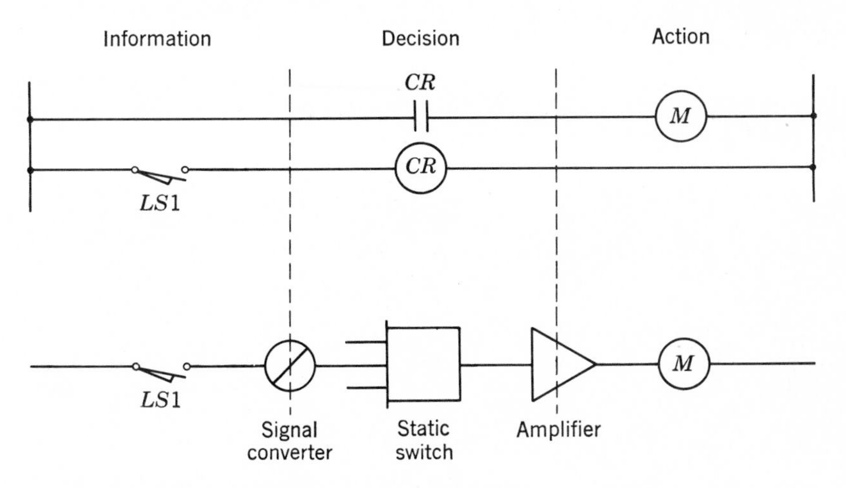

Fig. 10 · 15 The three major divisions of control circuits.

familiar contact-type devices found in electromagnetic control. The opening and closing of contacts can be used as direct input to the logic elements, provided the proper value of low-voltage direct-current is used through the contacts. When this is done, the contacts may become unreliable because the voltage is not high enough to overcome the resistance of dirty contact surfaces. The resistance of long leads can produce excessive voltage drop in the sensing circuits, which also will produce unreliable operation. Experience has shown that reliable operation of static logic fed from contact sensors requires a high-voltage contact circuit, usually 48- to 125-volt d-c or a-c if desired. This value of voltage used on contacts which have good wiping action will provide reliable input to the logic elements. The voltage used in the sensing section must be reduced and sometimes converted from alternating to direct current by means of an original-input signal converter (Fig. 10·16). Generally speaking, the signal converter should be mounted as close to the logic elements as possible to reduce electromagnetic interference (EMI) problems, commonly referred to as "noise."

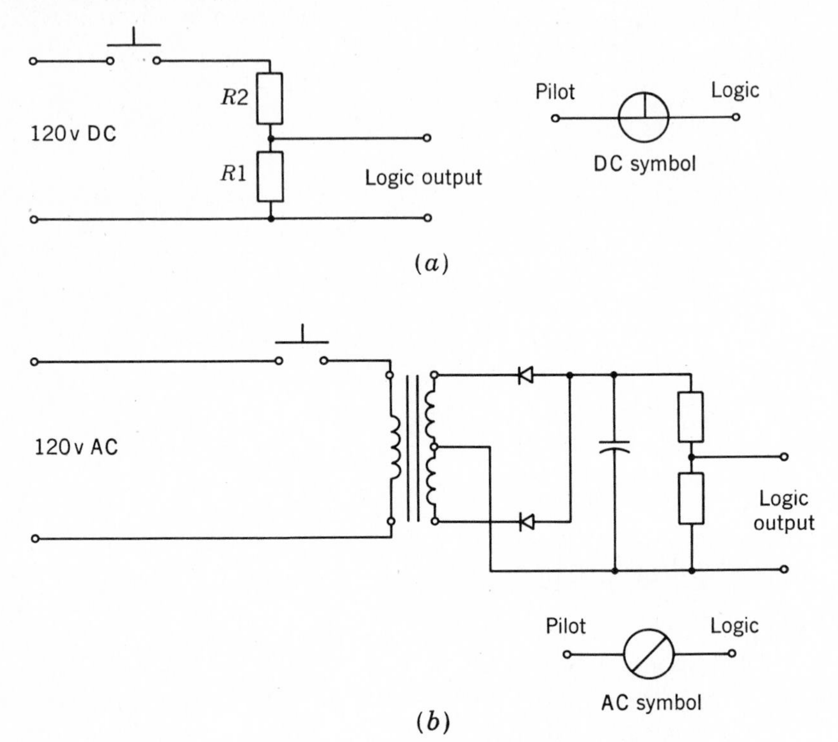

Fig. 10·16 Signal converters and their symbols.

The d-c signal converter of Fig. 10·16*a* is a simple voltage divider used to lower the d-c input to the proper value. This type of converter is the least expensive and probably the most used. This simple circuit can be improved by adding a transistor circuit to the voltage-divider network.

The circuit of Fig. 10·16*b* is a simple transformer and rectifier-supply a-c signal converter which would be greatly improved by the addition of a transistor circuit to its output. Most manufactured signal converters are supplied with a pilot lamp to indicate the condition of the pilot device, on or off. Good practice with static switching is to wire each pilot device to its own original-input signal converter, even though it might be connected in parallel or series with other devices (Fig. 10·17). The purpose of the lamp is to provide instant trouble indication at the logic panel.

Manufacturers' specifications should be followed very carefully in regard to noise suppression and the location of original inputs. The logic element operates at such low power levels that noise spikes can produce false switching if not properly suppressed.

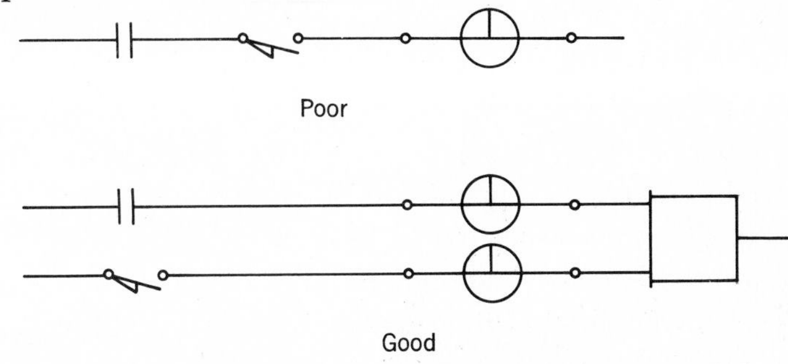

Fig. 10 · 17 Connections for signal converters.

The action section of the control circuit is where the work is done, that is, where power is consumed or switched. When the action required is to light a pilot lamp, the power may be a watt or two. When a large solenoid valve or starter must be actuated, the power requirements are much greater. The logic element is made to be a milliwatt device and therefore cannot switch such loads. Static amplifiers are used to directly switch loads up to several hundred watts or to activate relays and starters for larger loads. It is quite possible that future development in semiconductor devices will make it practical for all action devices to be static. Amplifiers are available using power transistors, reed relays, and silicon controlled rectifiers (SCR). Manufacturers' specifications should be carefully followed, and a unit should be selected which fills the requirements of the load.

The decision section of the control circuit is the chief application of logic elements and therefore has been left until last. Remember, the input to the logic has been adjusted by the signal converter to optimum value, and the amplifier will convert the logic output to match the load. This gives the designer of the logic circuit freedom of choice of components limited only by the English logic of the decisions to be made. He does not even need to know how each element works or its internal circuitry.

10·4 DEVELOPMENT OF LOGIC DIAGRAMS

The first step in the development of a logic diagram is to convert the specifications into simple English logic statements. Each logic statement will be the equivalent of one line on the schematic diagram studied in Chap. 6.

The second step is to draw each sequence or statement in logic-symbol form.

The third step is to integrate and interconnect the individual sequences where necessary to provide a complete control circuit.

The fourth step is to examine the total circuit to see if some functions can be combined to reduce the number of logic elements required.

The fifth step is to review the circuit for possible conflicts between sequences and to check for correct overall circuit satisfaction of the specifications.

10·5 DEVELOPMENT OF CIRCUIT 1

The specifications for this circuit are as follows: a solenoid valve (*SOL*) is to be energized whenever a normally open push button *PB*1 is pressed, regardless of other inputs, or whenever pressure switch *PS*1 and thermostat *T*1 are closed and extreme-limit pressure switch *PS*2 and extreme-temperature thermostat *T*2 are both open.

The first step is to convert to logic statements. The first English logic statement is: *SOL* will be energized when *PB*1 or *PS*1 and *T*1 are closed. The logic diagram for this statement is shown in Fig. 10·18*a*.

The second English logic statement is: Statement 1 will be

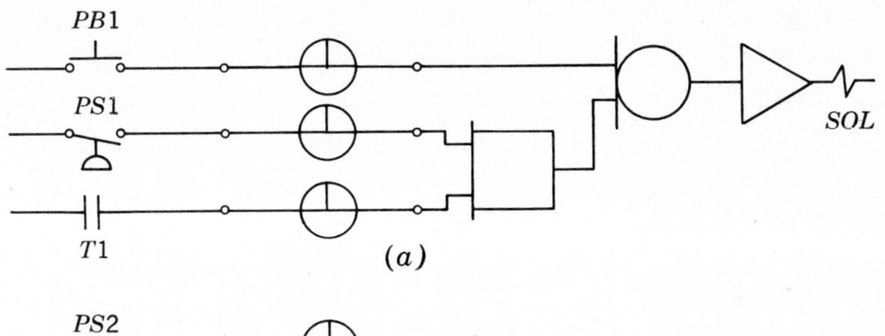

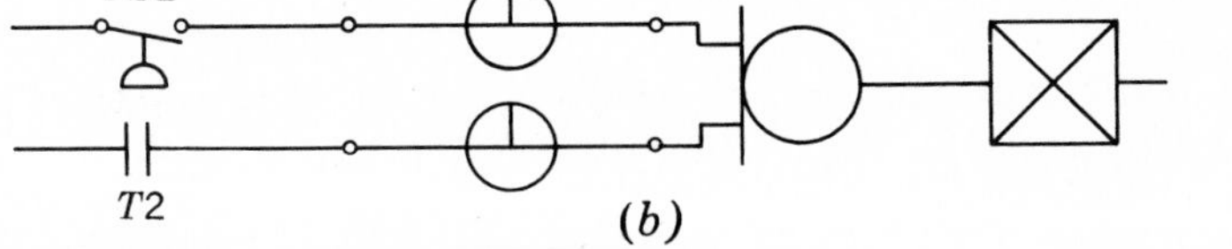

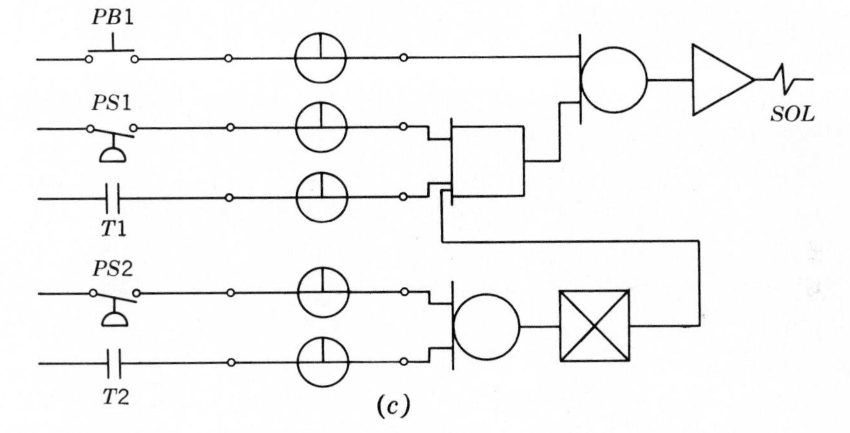

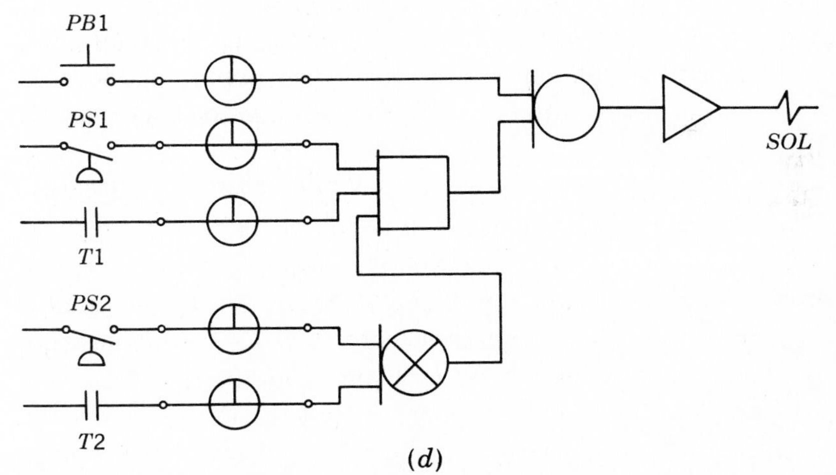

Fig. 10 · 18 Development of circuit 1.

true only if *PS*2 and *T*2 are not energized. This statement calls for the NOT output from an OR function as shown in Fig. 10·18*b*. The first two steps in development are now complete. The third step is to combine the statements in the logic diagram as in Fig. 10·18*c*.

The fourth step is to combine logic elements where possible and simplify the circuit if this is practical, which can be done in this case by using an OR unit with a built-in NOT output as shown in Fig. 10·18*d*.

The fifth step is to analyze the circuit to be sure it will perform the functions specified. We see from 10·18*d* that when *PB*1 is closed it provides an input to the OR through its signal converter. The OR requires only one input to have an output; therefore there is an input to the amplifier. The amplifier has been chosen to provide proper output for the solenoid when it is provided with an input; therefore *SOL* is now energized and satisfies the specifications for *PB*1.

If *PS*1 is closed but *T*1 is open, there will be only one input to the AND; therefore there will not be an output. If *T*1 closes while *PS*1 is closed, there are two inputs provided to the AND, but this will not produce an output unless there is an output from the NOR element. We must now examine *PS*2 and *T*2. If these two devices are open, there is no input to the NOR element, and therefore it will have an output and supply the third input to the AND element. When all three inputs are present at the AND, it will have an output. The output of the AND provides an input to the OR and energizes the solenoid through the amplifier.

10·6 DEVELOPMENT OF CIRCUIT 2

This circuit is for a three-stage air-conditioning system. Machine 1 is the smaller of two machines and should run whenever the chilled-water flow switch *FS*1 and the main-control thermostat *T*1 are on and the second-level thermostat *T*2 is not on. This provides normal operation for most conditions.

When machine 1 cannot keep up with the load, the second-level thermostat *T*2 must first shut down machine 1 and then

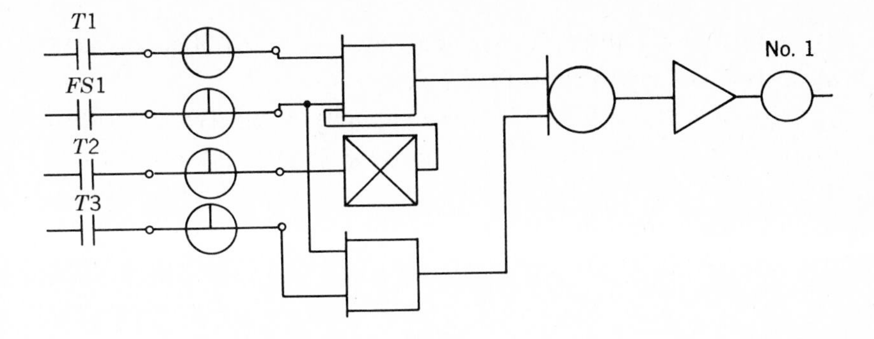

(a)

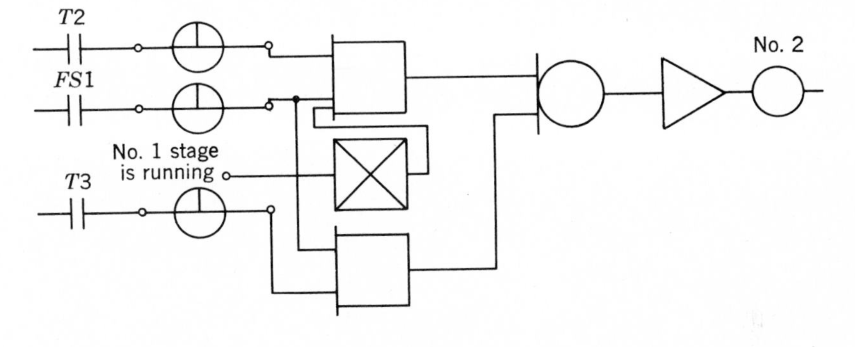

(b)

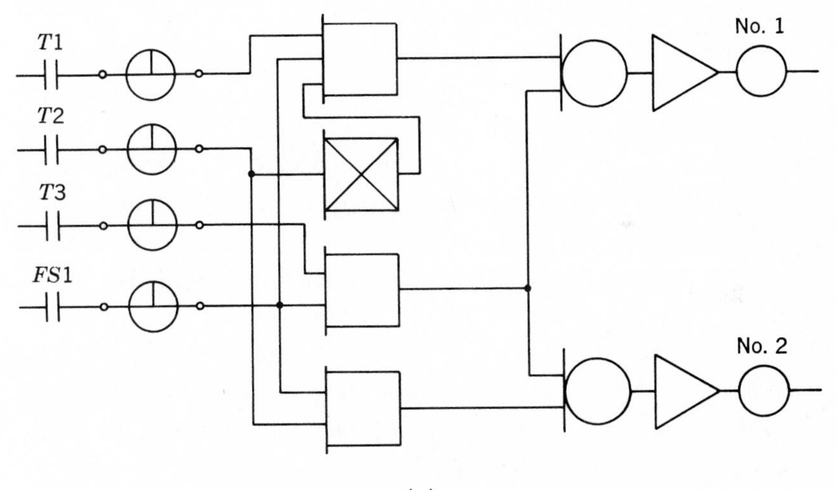

(c)

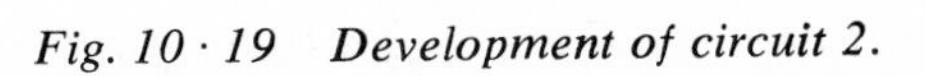
Fig. 10 · 19 Development of circuit 2.

cause machine 2 to start, provided *FS*1 is still on. This provides a greater capacity to carry the load.

Whenever the load exceeds the capacity of machine 2, a third thermostat set at a higher temperature than those for *T*1 and *T*2 closes. This will keep machine 2 running and start machine 1. This provides a third capacity to carry large heat loads; in this case, both machines are used.

The first step in the development will be to reduce the specifications to logic statements. Then a logic diagram will be drawn for each.

The statement for machine 1 for all three stages of operation is: Run when *T*1 and *FS*1 are on and *T*2 is not on or when *T*3 and *FS*1 are on. The logic diagram for machine 1 is shown in Fig. 10·19*a*.

The statement for machine 2 for all three stages of operation is: Run when *T*2 and *FS*1 are on and machine 1 is not on or when *T*3 and *FS*1 are on. The logic diagram for machine 2 is shown in Fig. 10·19*b*.

The third step in the development is to combine the two logic diagrams into a complete circuit using only those components actually needed. This complete circuit is shown in Fig. 10·19*c*. Careful analysis of the final circuit will show that one of the AND units has been eliminated because one unit meets the need of the circuit. One NOT unit has also been eliminated for the same reason.

10·7 DEVELOPMENT OF CIRCUIT 3

This is to be a circuit for three conveyors. There are two START buttons, one located at each end of the conveyor system. There are three STOP buttons, one located at each conveyor. Each conveyor is to be protected by a limit switch.

Pushing either START button will start all conveyors in sequence. Operation of any STOP button or limit switch will stop its conveyor and the conveyor immediately preceding it in the sequence.

The logic statement for conveyor 1 is: Run when either START button 1 or START button 2 is pressed, provided STOP button 1 and *LS*1 are closed and push button 2 and *LS*2 are closed.

The logic statement for conveyor 2 is: Run when conveyor 1 is running and *PB*2 and *LS*2 are closed and *PB*3 and *LS*3 are closed.

The logic statement for conveyor 3 is: Run when conveyor 1 and conveyor 2 are running and *PB*3 and *LS*3 are closed. The complete circuit is shown in Fig. 10·20.

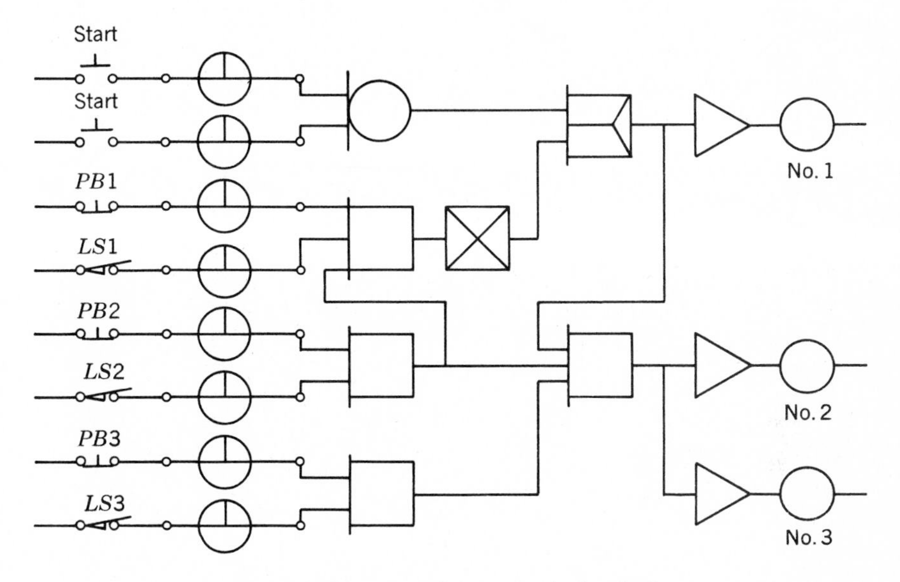

Fig. 10·20 Development of circuit 3.

10·8 DEVELOPMENT OF CIRCUIT 4

This circuit is to provide sequence speed control with definite-time delay. The motor must be started in its first speed by START button 1 and can then be raised to its second speed by means of START button 2, after a time delay to allow for acceleration. The motor may then be raised to its third speed by START button 3, after a time delay. The STOP button stops the motor regardless of the speed at which it is running.

The logic statement for the first speed is: Run when the START button is closed and the STOP button is closed. A MEMORY will be needed because of the momentary contact of the START button.

The logic statement for the second speed is: Run when the START button is closed and the time delay has lapsed since the first speed was energized.

The logic statement for the third speed is: Run when the START button is closed and a time delay has lapsed since the second speed was energized. The completed circuit is shown

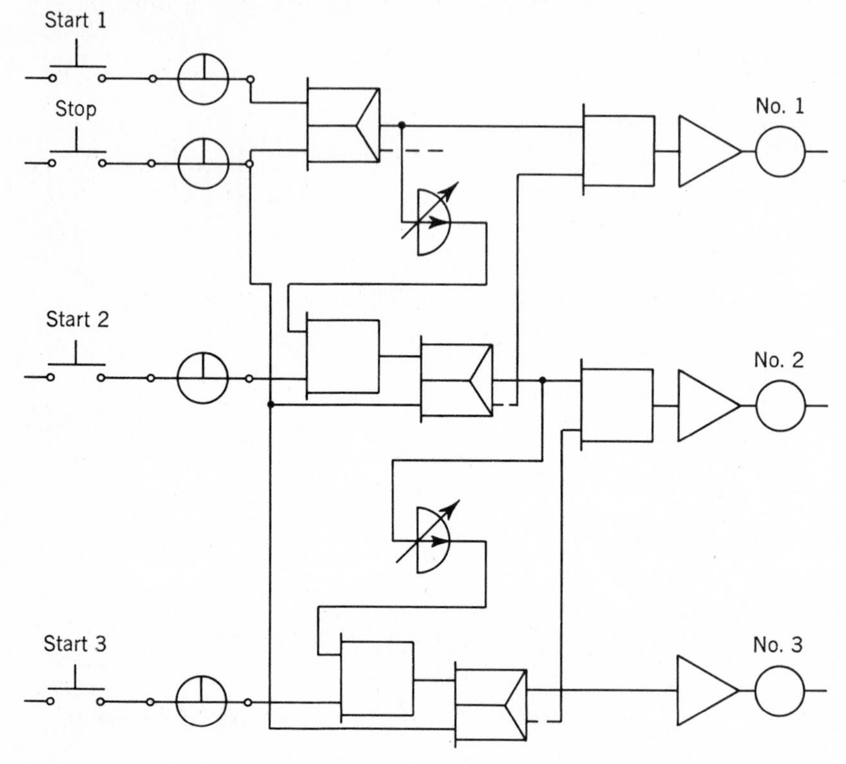

Fig. 10·21 Development of circuit 4.

in Fig. 20·21. The NOT outputs of the MEMORYs are used to shut down each lower speed when necessary.

10·9 DEVELOPMENT OF CIRCUIT 5

There are two solenoids on the machine. The first solenoid, *SOL* 1, is to turn on when momentary-contact limit switch *LS*1 is closed. The solenoid must remain on until shut off by the closing of momentary-contact limit switch *LS*2. The closing of *LS*2 must also turn on a second solenoid, *SOL* 2, which will remain on until shut off by the momentary closing of *LS*1 on the next cycle of the machine.

Since the contacts of *LS*1 and *LS*2 are closed only momentarily and the solenoid must remember whether they were turned off or turned on, the circuit will require the use of *retentive*-MEMORY elements.

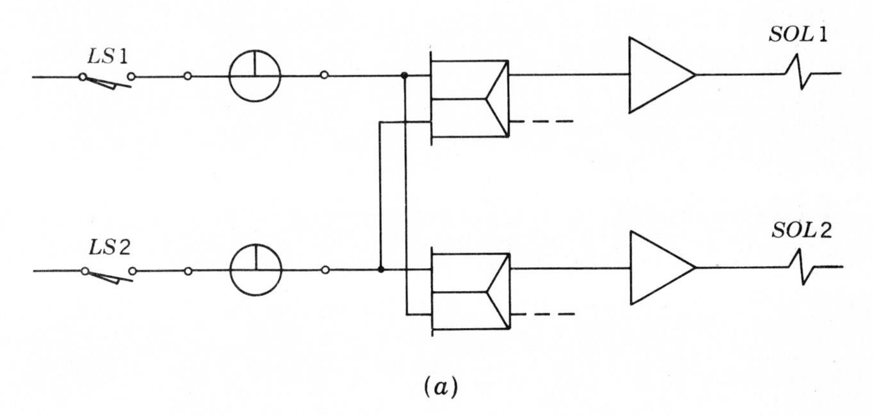

(*a*)

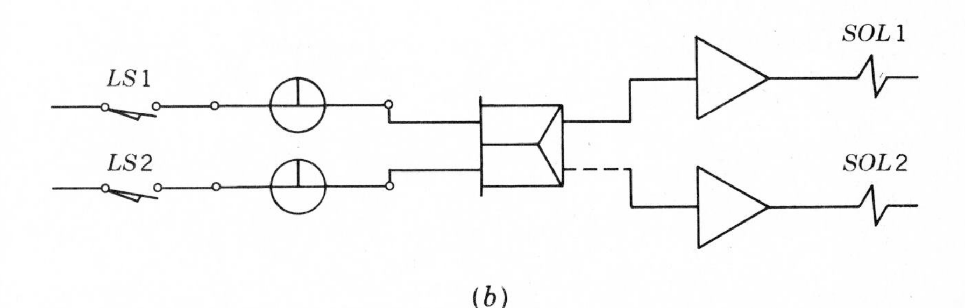

(*b*)

Fig. 10·22 Development of circuit 5.

When *LS*1 is connected to the ON input of one MEMORY unit and the OFF input of a second MEMORY unit (Fig. 10·22*a*), half of the requirements of the circuit will be satisfied.

When *LS*2 is connected to the OFF input of the first MEMORY and the ON input of the second MEMORY, the required circuit will be complete. This circuit can be simplified by utilizing the inverted, or NOT, output of only one MEMORY as shown in Fig. 10·22*b*.

Any number of relay contacts connected in series (Fig. 10·23*a*) can be represented by equivalent AND circuitry (Fig. 10·23*b*). Sometimes it may be necessary to use more than one AND element to provide the required number of inputs. Figure 10·23*c* and *d* illustrates how nine inputs can be provided when only three-input AND elements are available.

The logic equivalent of paralleled normally open relay contacts is the OR (Fig. 10·24).

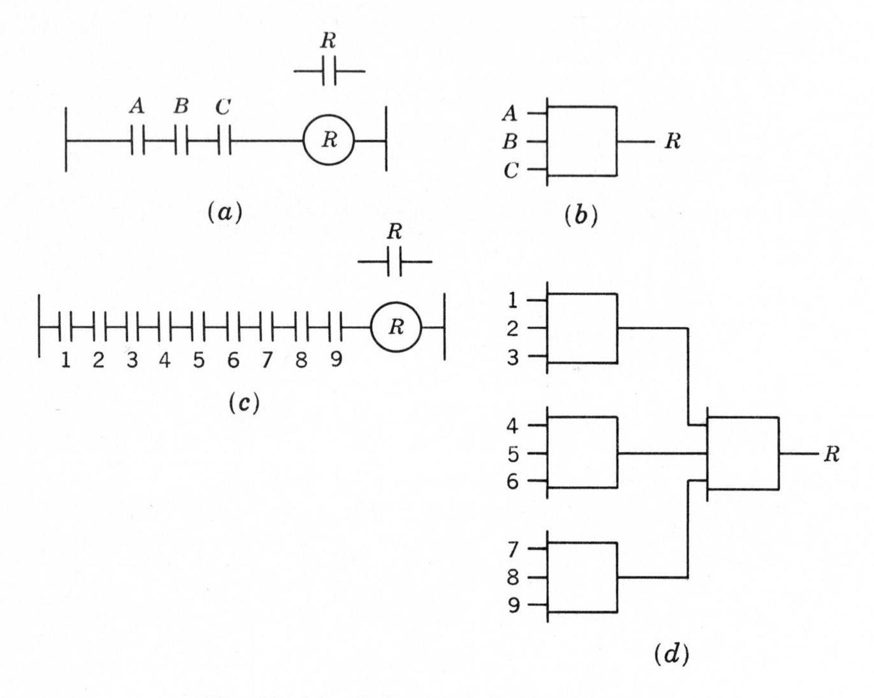

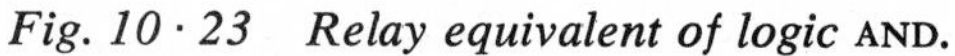
Fig. 10·23 Relay equivalent of logic AND.

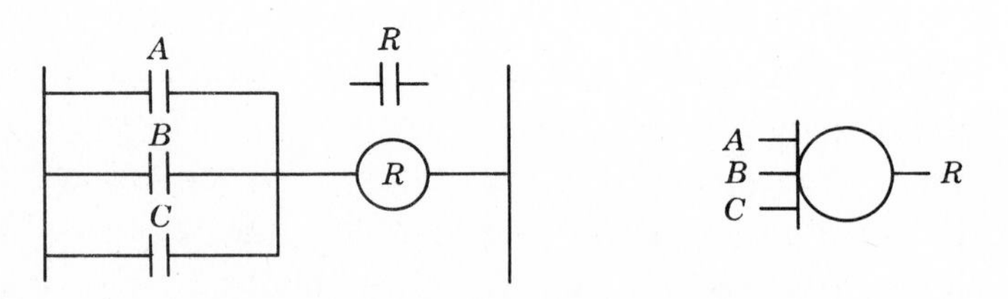

Fig. 10·24 Relay equivalent of logic OR.

Relays with normally closed contacts become a NOT (Fig. 10·25*a*), a NOR (Fig. 10·25*b*), or a NAND (Fig. 10·25*c*), depending upon the number of contacts and how they are connected.

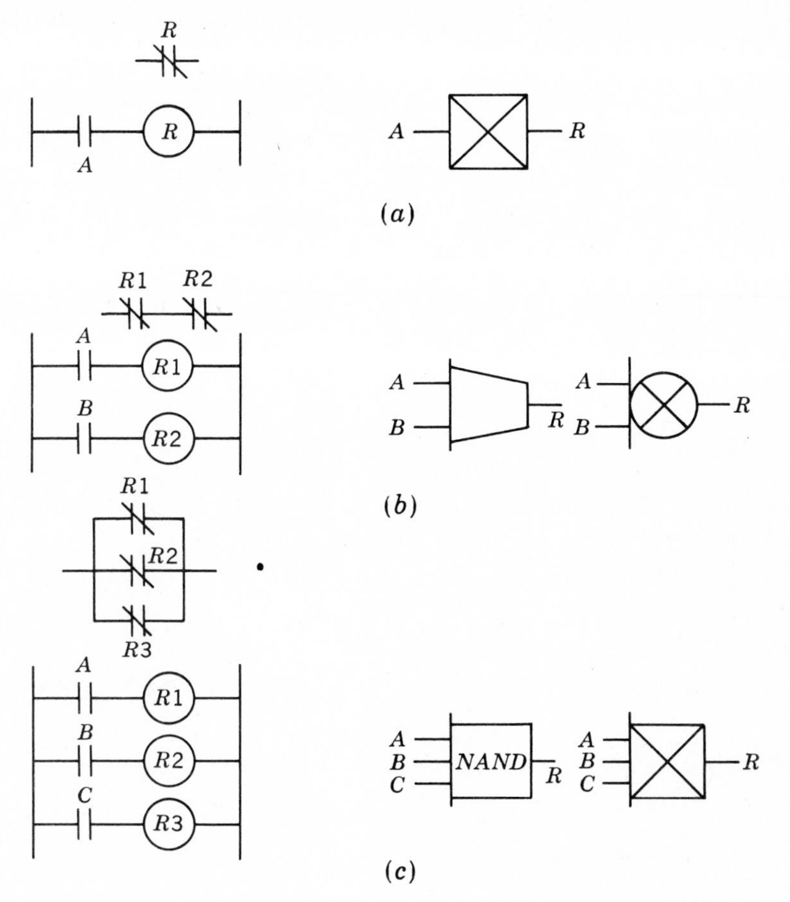

Fig. 10·25 Relay equivalent of logic NOT, NOR, *and* NAND.

The popular off-return of the magnetic control using three-wire control (Fig. 10·26*a*) can be duplicated in logic circuits by using an *off-return* MEMORY (Fig. 10·26*b*). The static circuit can also have the NOT, or inverted, output of the intended function shown by dotted lines on the symbol or by a NOT symbol within the MEMORY symbol. The circuit of Fig. 10·26*c* provides the same basic control using a *sealed* (feedback) AND and allows other inputs to be used.

Interlocking, which is so important in machine and process control, is easily achieved in logic circuitry by using the output of a logic function in one part of the circuit as one of the inputs to another logic element in a different part of the circuit (Fig. 10·27).

10·11 THE TRANSISTOR AS A STATIC SWITCH

The logic diagram gives complete information on the operation of the control system but tells nothing about the circuit of the

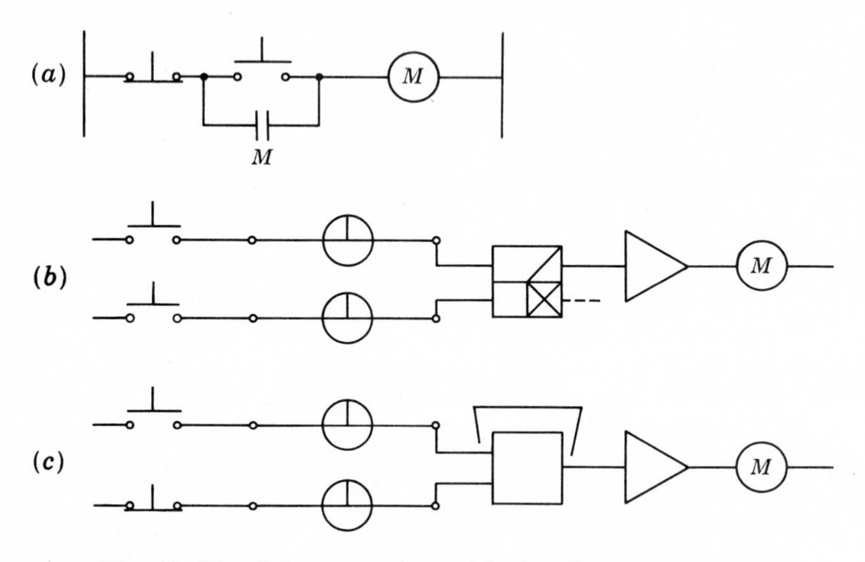

Fig. 10 · 26 Relay equivalent of logic off-return MEMORY.

logic element itself. When a system is installed it must use the logic elements made by only one manufacturer, because two systems are not generally compatible. The voltage and current requirements for the logic elements differ between manufacturers.

Once a particular manufacturer's system is chosen, the installer can be confident that the individual logic elements will work when properly connected in accordance with the logic diagram. Power supplies are designed to supply all the correct voltages. Signal converters are designed to supply the correct input for the logic elements. The output of each logic element is designed to provide the proper input for other logic elements or amplifiers. Since the interconnection requirements are built into the logic elements, it is not absolutely necessary for the installer or serviceman to fully understand the actual circuit of the logic element itself. General practice is to replace a defective element or return it to the factory if repair is to be attempted. The installer or serviceman will need specific installation information on the individual system he is working with, and will be much better pre-

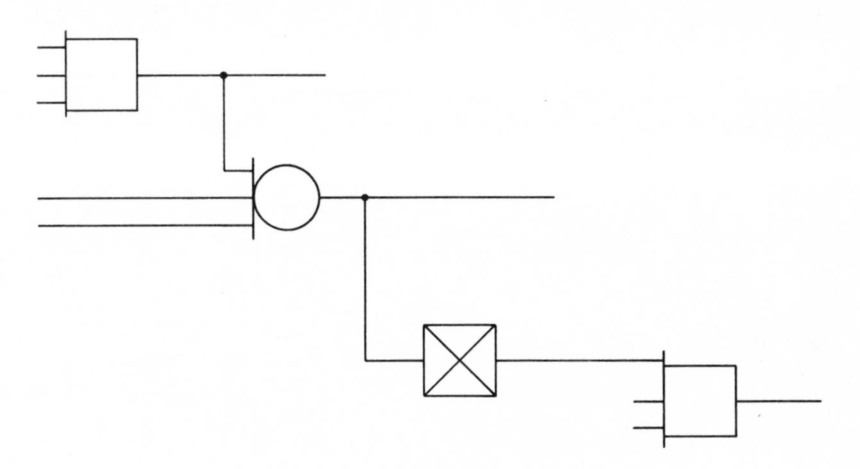

Fig. 10·27 Interlocking in logic circuits.

pared to clear or prevent trouble if he understands how the logic elements work.

All commercially available logic components use the transistor as the basic switching device. During the late 1950s, logic elements were designed around a saturable reactor; however, these are no longer manufactured except for replacement in existing systems.

To understand transistorized logic components, you must first understand how a transistor responds to voltage and current. Figure 10·28 shows the symbols for NPN and PNP transistors with the necessary polarity for full conduction. The three external leads are identified by their correct names: *C* for collector,

B for base, and *E* for emitter. Conventional current flow is always used in transistor circuits and is indicated by arrows on the symbols.

The transistor is a semiconductor device which has two circuits: emitter to base and emitter to collector. The emitter-to-base circuit is generally referred to as the *base circuit*. The emitter-to-collector circuit is referred to as the *power* or *output circuit*.

Consider the circuit of Fig. 10·29. When contact *B* is closed, the base of the transistor is at the same potential as the emitter,

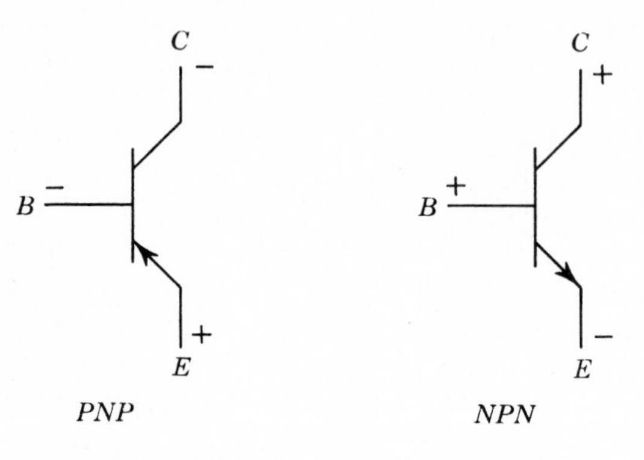

Fig. 10·28 Transistor symbols.

and the transistor will not conduct from emitter to collector. For general circuit analysis, consider the emitter-to-collector circuit to be an open switch whenever the base of a PNP transistor is at zero voltage difference from the emitter. The same result will be achieved whenever the base is more positive than the emitter.

Whenever the base of a PNP transistor is made sufficiently more negative than the emitter, the transistor will conduct emitter to collector. The emitter-to-collector circuit acts as a closed switch, with practically no voltage drop across the transistor. When contact *A* (Fig. 10·29) is closed and contact *B* is open, the base is at −12 volts. The transistor will conduct from emitter to base and, by transistor action, will also conduct from emitter to collector. The current is determined by the resistance of R. If the output voltage is taken from the output lead to the negative bus, the output voltage will be 12 volts. This is, of course, due to the voltage drop across R, which must equal the line voltage.

The previous explanation should make the switching action clear. When the base of the transistor (Fig. 10·29) is positive, with contact B closed and contact A open, the output is at zero volts because there is no voltage drop across R. When the base is negative, with contact A closed and contact B open, the output is at 12 volts because of the voltage drop across R.

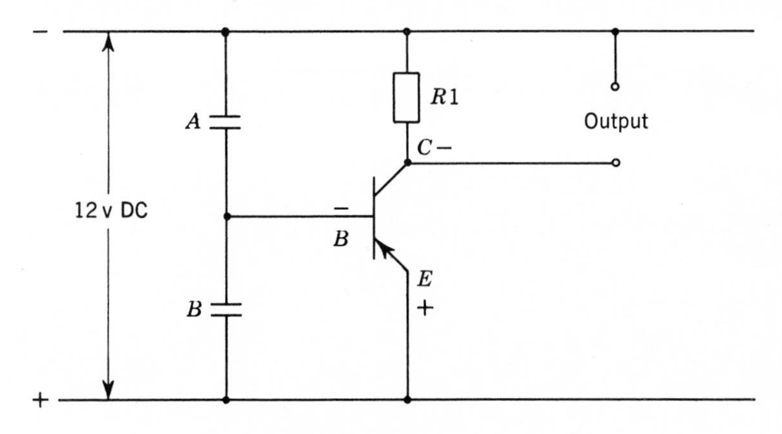

Fig. 10·29 Basic transistor switch.

Because of various design considerations, it is desirable that the switch be conducting when no signal is applied and stop conducting when a signal is applied. In a slight modification of the circuit (Fig. 10·30), contact A is replaced by resistor $R2$, and resistor $R4$ is added. Resistors $R2$ and $R4$ form a voltage divider across the line voltage. Proper selection of $R2$ and $R4$ will provide the ideal value of negative potential at the base in order to provide full conduction in the emitter-to-collector circuit. The transistor is now conducting and has an output voltage of 12 volts. The closing of contact B will connect the base lead to the positive line and cause the transistor to cut off, thus reducing the output to zero volts. Consider contact B to be the pilot device used to control the switch. When the contact is closed, there is an input to the switch, but there is no output. When the contact is not closed, there is no input to the switch, but it does have an output. This action is known as a NOT or inverted logic. This is the building block upon which all transistorized logic elements are built. Different manufacturers will use different circuit arrangements to achieve the NOT operation of the transistor. However, the analysis given above will suffice to explain the operation of the transistor switch itself.

When the NOT action of the transistor switch (inverted logic) is not desirable, a second transistor is added to the basic circuit.

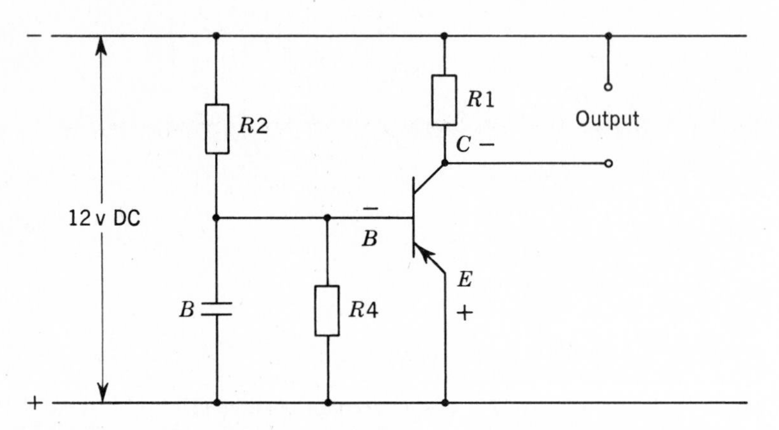

Fig. 10·30 Transistor switch as a NOT.

Figure 10·31 shows the new circuit. When contact *B* is closed, *T*1 will not conduct. The base of *T*2 is then at a negative value in relation to its emitter, because of the circuit from emitter to base through *R*1 to the negative line. The negative potential on the base of *T*2 will cause it to conduct, producing a current through *R*3 and an output of 12 volts.

When contact *B* is open, *T*1 conducts and brings the base of *T*2 to the same potential as the positive bus; therefore *T*2 is cut off, and there is no output. The transistor switch is now a combination of two NOT basic circuits and produces an output when there is an input. When there is no input, there is no output.

The basic transistor switch can be made to perform the AND function by proper input circuitry. The actual circuits will be considered in later chapters, but a symbolic representation here should improve the student's understanding.

Figure 10·32 shows the basic switch connected through contacts *A*, *B*, and *C* in series. The switch cannot have an output

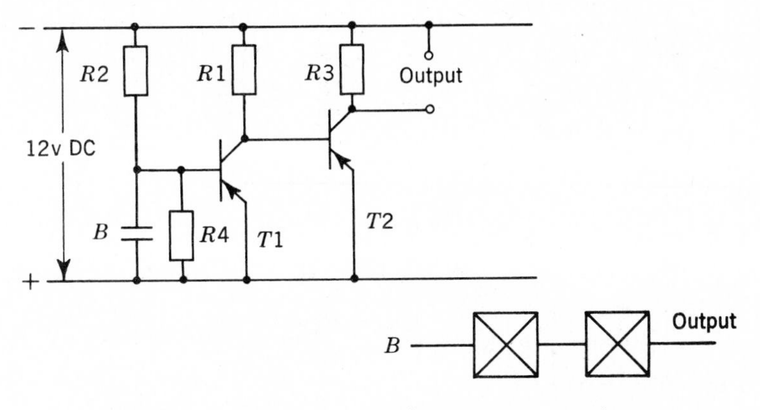

Fig. 10·31 Transistor switch using two transistors.

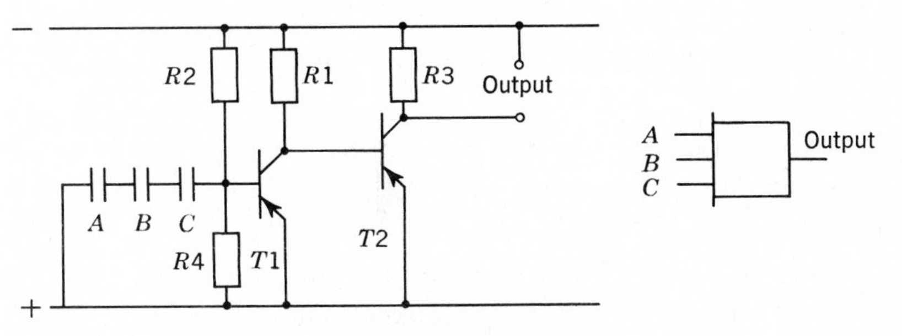

Fig. 10·32 Transistor switch as an AND.

until the base of *T*1 is connected to the positive bus and cuts off *T*1. Resistance voltage dividers are generally used for a practical input circuit. The circuit is arranged so that proper input voltage must be applied to all inputs before the base voltage will become sufficiently positive to cut off *T*1.

If the second transistor, *T*2, had been left off of the basic transistor switch, the result would have been an AND-NOT element.

10·13 THE TRANSISTOR SWITCH AS AN OR ELEMENT

Figure 10·33 shows the basic transistor switch connected to three contacts in parallel. If any one of these contacts, *A* or *B* or *C*, were closed, the base of *T*1 would be connected to the positive line and *T*1 would be cut off.

The actual input circuit for the transistor switch is made up of resistors and/or diodes. When they are properly connected, the OR input circuit must make the base of *T*1 positive when any one or a combination of its inputs has the proper input voltage applied.

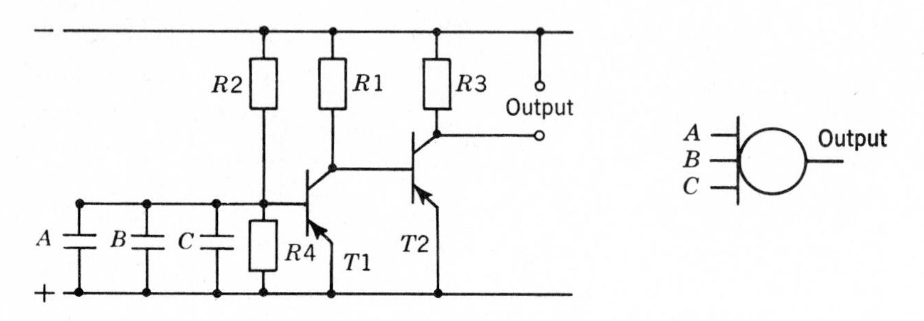

Fig. 10·33 Transistor switch as an OR.

If the second transistor *T*2 had not been used in the transistor switch circuit, the result would have been an OR-NOT or NOR element.

Chapters 11 to 13 will take up the actual commercial circuits in detail. Each manufacturing concern has its own design to accomplish what the symbolic circuits in this chapter illustrate.

Summary

Static control systems are built by properly interconnecting five basic building blocks: AND, OR, NOT, MEMORY, and *time* DELAY.

The information section of the control circuit usually consists of conventional contact-type sensing devices. Signals of relatively high voltage from the information section of the system are modified by signal converters to the proper low-voltage direct current required by the logic elements of the decision section. The action section of the system converts the low-voltage, low-power signal from the decision section into the power required by the device to be controlled.

All available logic elements are built around the basic transistor switch but differ in the application of specific circuitry.

Development of static control circuits has been approached as an extension of electromagnetic control development as covered in Chap. 6. The principles are rather simple but require practice. The serious student should practice developing many circuits to perfect his approach. When it is possible to actually wire the circuits after development, a higher degree of proficiency should result.

The five steps in logic circuit development are:

1. Convert the specifications into English logic statements.
2. Draw each statement in logic-symbol form.
3. Integrate and interconnect the individual statements.
4. Combine functions where possible to simplify the circuit.
5. Review the overall circuit and check it against the specifications.

Review Questions

1. What are the requirements of an AND element?
2. What are the requirements of an OR element?
3. What are the requirements of a NOT element? What are the requirements of a NOR element?
4. Describe the action of the *off-return* MEMORY element when the power fails and returns.
5. Describe the action of the *retentive* MEMORY element when the power fails and returns.

6. Draw the proper symbol for each of the following: AND, OR, NOT, NOR, NAND, *retentive* MEMORY, *Off-return* MEMORY, and the four types of *time* DELAY.
7. Develop a logic circuit that will energize a solenoid when a normally open momentary-contact START button is pressed. The solenoid must remain on until a second normally open momentary-contact STOP button is pressed.
8. Develop a logic circuit for the following specifications. A coil must be energized when normally open limit switch *LS*1 is closed and pressure switch *PS*1 is closed and thermostat *T*1 is closed or when normally open push button *PB*1 is closed and thermostat *T*2 is closed.
9. Develop a logic circuit for the following specifications. A signal light is to turn on whenever normally open contacts *A*, *B*, *C*, and *D* are all closed, provided normally closed contacts *E* and *F* are not open.
10. What is the function of the signal converter?
11. What is the function of the output amplifier?

11
GENERAL ELECTRIC COMPANY SOLID-STATE LOGIC

Each company has its own approach to the problem of circuit design and application for practical transistorized static control. This chapter will take up the details of the General Electric Company system. The material for this chapter was furnished by the General Electric Company.[1] Other systems will be covered in Chaps. 12 and 13.

11·1 THEORY OF OPERATION

Basic transistor circuits are used throughout the system. Figure 11·1 shows a circuit with —12 volts d-c connected to the system bus, and with three resistors connected between the bus. Current *i*1 will flow through the resistors as shown, since a potential difference exists across each resistor. Because of the particular resistors used in this circuit, the input connection point will be at —4 volts.

[1] SOURCE: General Electric Company publication GET 3551.

If the input terminal is now connected to the zero-volt bus (see Fig. 11·2), current will flow from the zero-volt bus through resistor $R1$ to the —12-volt bus. Since the connection between resistors $R1$ and $R2$ is at zero volts, no current will pass through resistors $R2$ and $R3$ because no potential difference exists across them.

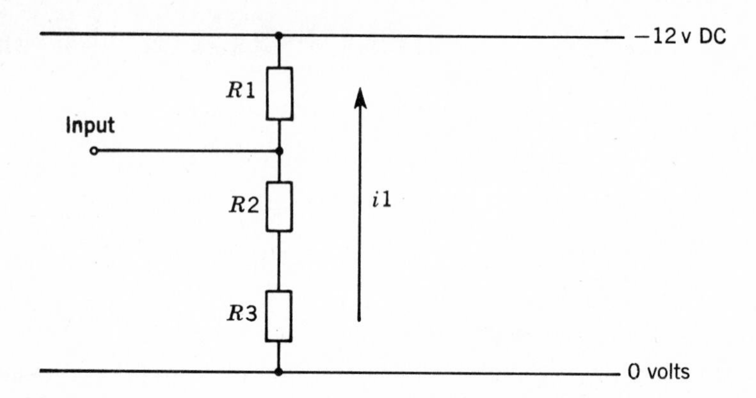

Fig. 11·1 Basic input circuit, input open. (General Electric Company)

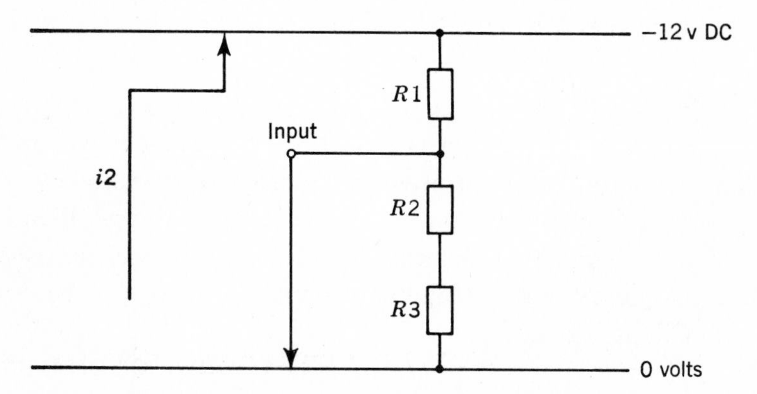

Fig. 11·2 Basic input circuit, input closed. (General Electric Company)

If a transistor and appropriate load resistors $R4$, $R5$, and $R6$ (with the same values as $R1$, $R2$, and $R3$, respectively) are connected between the two buses (see Fig. 11·3), the output connection will be at —4 volts unless there is conduction through the transistor.

The silicon-type PNP transistor shown in Fig. 11·3 has

three connection points called the emitter, the base, and the collector. If there is adequate current flowing from emitter to base, a much larger current will flow from emitter to collector. With no current flowing from the emitter to the base, no current will flow from the emitter to the collector. The transistor is satu-

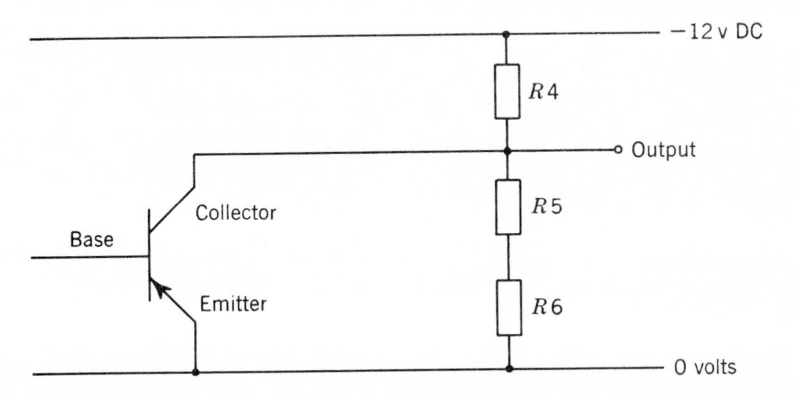

Fig. 11·3 Basic output circuit, preceded by a transistor. (General Electric Company)

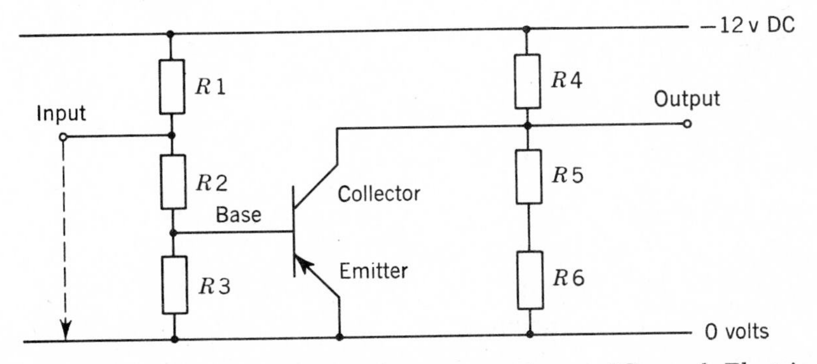

Fig. 11·4 Basic NOT *circuit with output resistors. (General Electric Company)*

rated when additional emitter-to-base current flow causes no increase in emitter-to-collector current flow. As the transistor is driven to saturation, its emitter-to-collector resistance changes from a very high value to a very low value, essentially acting as a switch in the circuit.

The circuit shown in Fig. 11·4 is obtained by combining Figs. 11·1 and 11·3. With the input connection at −4 volts, current will flow from emitter to base of the PNP transistor, resulting in current flow from the emitter to the collector through load resistor *R*4. This action results in the output connection becoming essentially zero volts because of the very low resistance of the transistor during saturation.

In contrast, if the input terminal were connected to the zero-volt bus (dotted line), no emitter-to-base current would flow because the emitter and base would be the same voltage—zero volts. This would result in no emitter-to-collector current flow because of the very high emitter-to-collector resistance, since the transistor is not in saturation. The voltage at the output

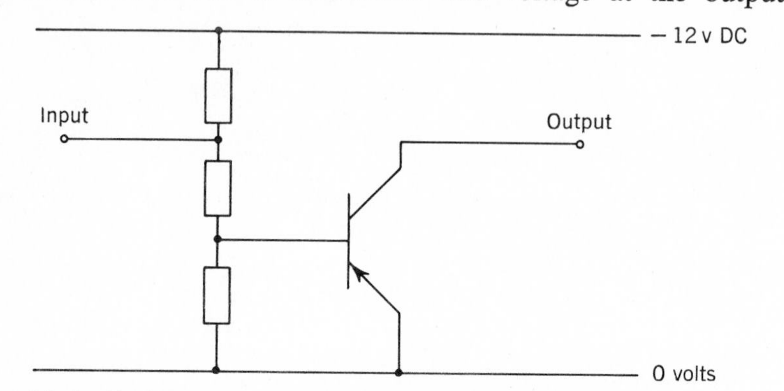

Fig. 11·5 Basic NOT *as manufactured.* (*General Electric Company*)

terminal is therefore —4 volts, as determined by resistors *R*4, *R*5, and *R*6. In this static control system, a signal of 0 volts d-c is an ON signal, and a signal —4 volts d-c is an OFF signal. The term "0 volts" means a load short, not the absence of voltage.

Figure 11·4 is a basic NOT circuit. An ON input produces an OFF output. Conversely, an OFF input produces an ON output.

In Fig. 11·5 and succeeding circuits, the final output of the device does not have resistors *R*4, *R*5, and *R*6, since the output will be connected to the inputs it drives of logic elements such as AND, OR, etc. The input circuits act the same as resistors *R*4, *R*5, and *R*6.

The NOT circuit is shown in Fig. 11·6. This is identical to Fig. 11·5, except —12 volts d-c is indicated by *P*—, and zero volts d-c is indicated by *P*+. The input and output are represented by typical pin connections, numbered 1 and 4 or 5 and 8, respectively. The output terminal can drive up to 12 inputs.

If additional parallel input circuits are added (Fig. 11·7), input terminals 1 *and* 2 *and* 3 must *all* be connected to zero volts (an ON signal) in order *not* to have current flow from

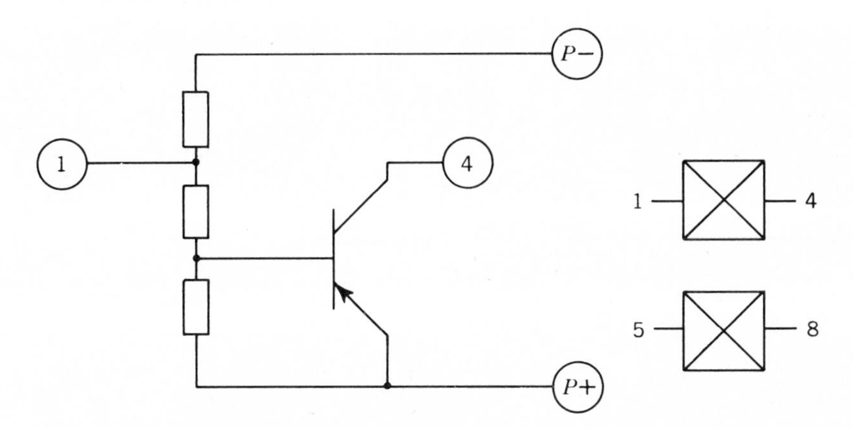

Fig. 11·6 General Electric Company NOT *with pin connections shown. (General Electric Company)*

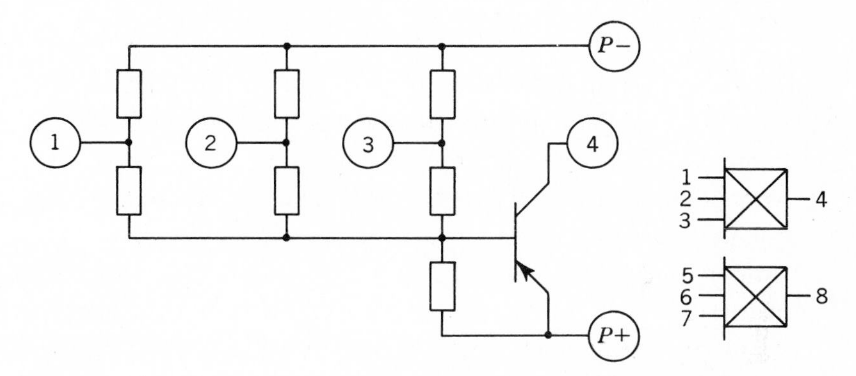

Fig. 11·7 Three-input AND-NOT. *(General Electric Company)*

emitter to base to the *P*— bus. No emitter-to-base current flow will cause output terminal 4 to be at —4 volts or OFF, as a result of its connection to the next element's input resistors. This device is functionally an AND-NOT. Terminals 1 *and* 2 *and* 3 must all have an ON signal present *not* to have an output.

Whenever it is stated that "there is an input" or "an output exists," the meaning is that a zero-volt ON signal is present at the terminal point being discussed.

A seven-input AND-NOT element is also available which requires an ON signal to exist at terminals 1 to 7 in order to cause the logic function not to have an ON signal at terminal 8. The output terminal of the AND-NOT can drive up to 12 other inputs. This transistorized static control system is sometimes called an AND-NOT system, since the basic internal circuits utilized actually perform that function. A single-input AND-NOT is more simply expressed as a NOT, since no "AND situation" exists with a single-input device.

If a NOT is followed in a circuit by another NOT (Fig. 11·8), the input signal (OFF or ON) to the first NOT function is the

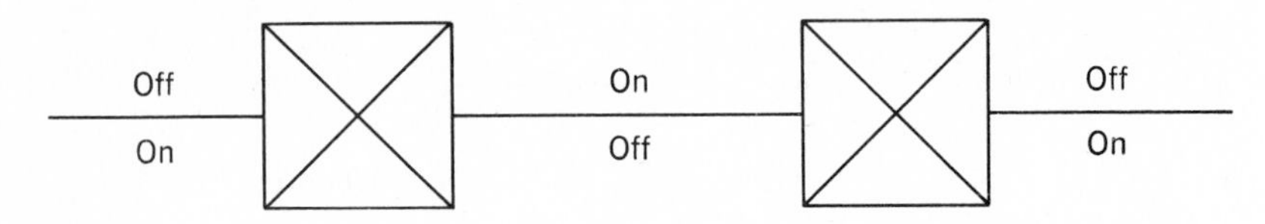

Fig. 11·8 Signal inversion through two NOTs *in series. (General Electric Company)*

same as the output of the second NOT function. For example, if the input signal to the first NOT is OFF, its output is in the ON condition; therefore, the input to the second NOT is ON and the output is OFF. To obtain the AND function, a NOT circuit is added to the output of an AND-NOT (Fig. 11·9).

If all three inputs to the AND-NOT are present, there will not be an output; that is, an OFF signal is present. With this OFF signal to the input of the NOT, an ON signal will be produced as an output. Thus, a three-input AND is developed and is manufactured as a complete logic unit (Fig. 11·10).

An additional output, the built-in NOT of the standard output, is available in this AND-NOT system and is a very useful feature. It is an inversion of the standard output and exists at the input connection to the last portion of the function (refer to Fig. 11·9). Figure 11·11 shows that one input connection is elimi-

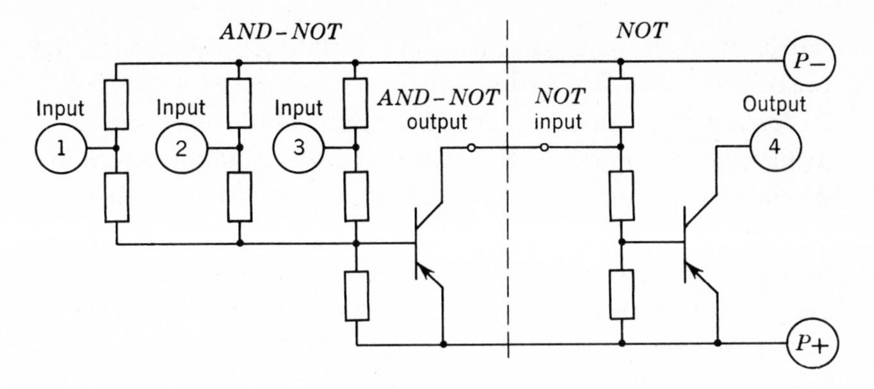

Fig. 11·9 Basic AND-NOT *followed by a* NOT. *(General Electric Company)*

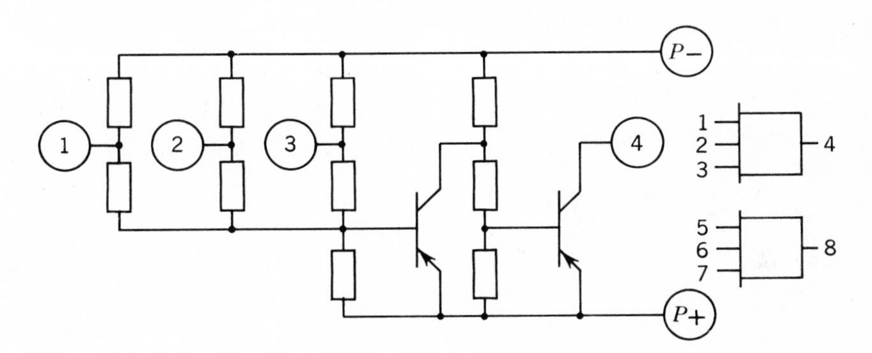

Fig. 11·10 Three-input AND *with pin connections shown. (General Electric Company)*

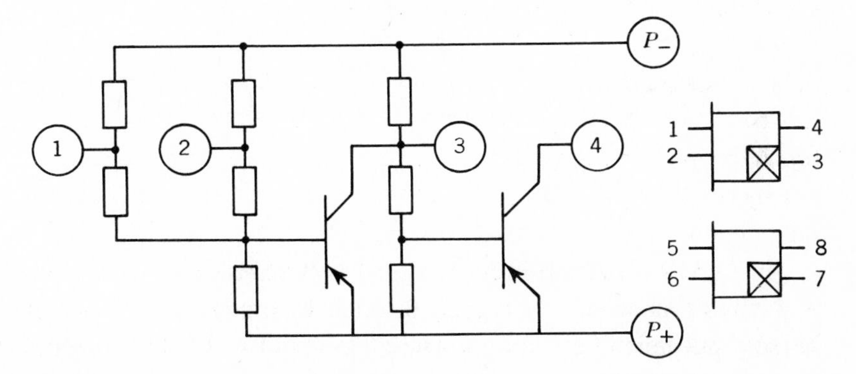

Fig. 11·11 Two-input AND *with* NOT *output. (General Electric Company)*

nated from the basic AND (Fig. 11·10) to gain a terminal for the additional built-in NOT output.

A six-input AND with an additional built-in NOT output is available with input terminals 1 to 6, standard output at terminal 8, and NOT output at terminal 7. Also a seven-input AND, with input terminals 1 to 7 and output at terminal 8, is available to complete the family of AND logic functions. Each output from an AND unit can drive up to 12 other input terminals.

The OR logic function will produce an ON signal at its output terminal with an ON signal present at any one of its input terminals. The OR element circuitry differs from the AND function in that each input circuit is brought in through an isolating diode to prevent interaction between input signals (see Fig. 11·12). Two additional diodes, *D*4 and *D*5, are also added to offset the forward voltage drop across *D*1, *D*2, or *D*3 when one or more input signals are present.

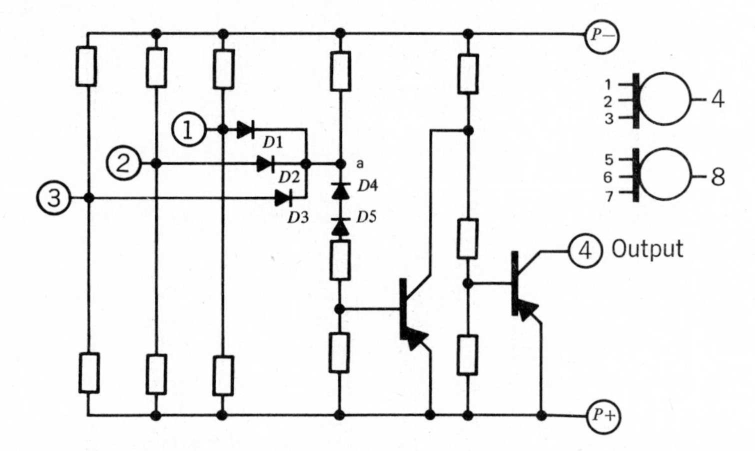

Fig. 11 · 12 Three-input OR *with pin connections. (General Electric Company)*

In Fig. 11·12, each input is at —4 volts, and point *a* is also at —4 volts. When an ON signal, 0 volts, is applied to any of the three input terminals, point *a* swings to about —0.6 volts, and no current flows from emitter to base of the first transistor. Therefore the collector of the first transistor is at —4 volts, and the second transistor goes into saturation, producing an ON signal (0 volts) at the output terminal. Additional ON signals applied to the input terminals continue an ON signal at the output terminal. Removal of all ON input signals swings point *a* back to —4 volts, which puts the first transistor into saturation and turns off the second transistor. The output terminal then reverts to its previous OFF signal, —4 volts (as supplied by the resistor input configuration of the following logic element to which it is connected). The output of the OR can drive up to 12 subsequent units. The OR function also comes in a two-input form with NOT output and a six-input form with NOT output.

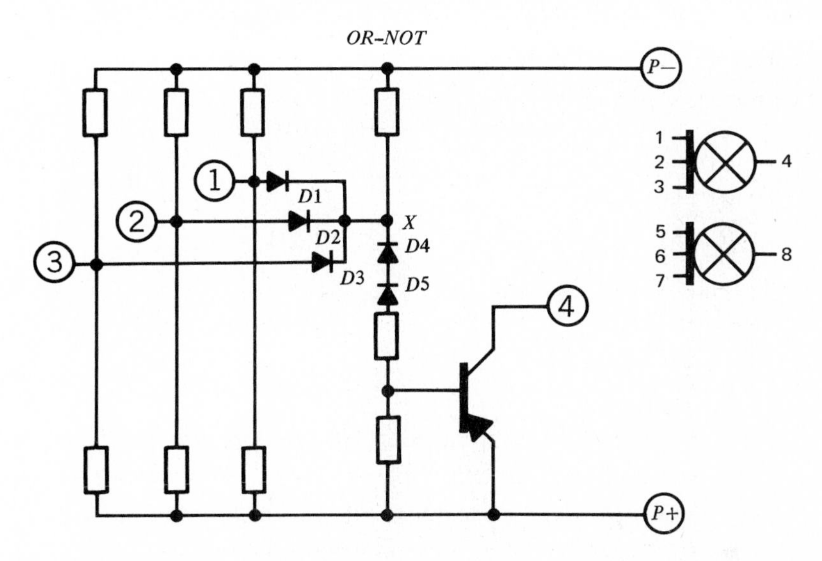

Fig. 11·13 Three-input OR-NOT. (*General Electric Company*)

The OR-NOT logic function (Fig. 11·13) will not have an ON signal at its output terminal with an ON signal at any of its input terminals. With no ON signal at the input terminals, both sides of the isolating diodes will be at —4 volts. The PNP transistor can then have emitter-to-base current flow, causing it to be saturated and to have an ON signal at the output terminal.

If an ON signal is applied to any input terminal, approximately 0 volts will appear at the cathode side of the isolating diodes. Since at least 1 volt is required to break down the forward drop of the two diodes in series, no current will flow through them with a voltage near zero volts applied at point X. This causes the PNP transistor to drop out of saturation, and an OFF signal appears at the output terminal. Removing the ON signal to the

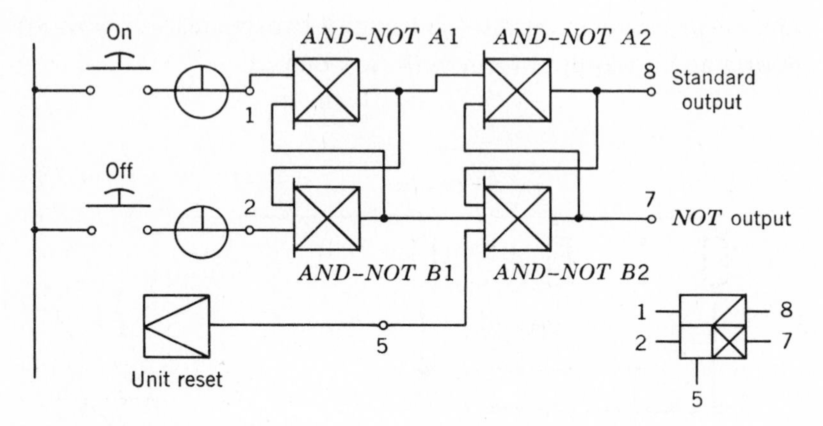

Fig. 11·14 Block diagram of the off-return MEMORY. (*General Electric Company*)

input terminals will again cause conduction of the PNP, as the voltage at point X will again be -4 volts. The output of the OR-NOT can provide a maximum of 12 other input signals to subsequent logic functions in the logic control.

The *off-return* MEMORY operation can best be understood by breaking its logic down to its simplest terms as shown in Fig. 11·14.

Assume initially that power has just been applied, and that neither push button is actuated. The unit reset, which is discussed later, provides a delayed ON signal.

As soon as power is applied, the $A1$ and $B1$ AND-NOT elements produce an output immediately, but the unit reset delays its output long enough to assure that AND-NOT $B2$ turns on. With AND-NOT $B2$ and AND-NOT $A1$ both driving AND-NOT $A2$ with ON signals, AND-NOT $A2$ cannot turn on, and therefore the standard output is left in the OFF condition.

When the ON push button is actuated, AND-NOT $A1$ loses its output, causing AND-NOT $A2$ to turn on and put the third ON input to AND-NOT $B2$, which then loses its output. The loss of this output removes the other input from AND-NOT $A2$, assuring an output at terminal 8 regardless of the condition of AND-NOT $A1$, which might again have an output because of the releasing of the ON push button. Actuating the OFF push button will turn off AND-NOT $B1$, resulting in the turning on of AND-NOT $B2$, which will turn off AND-NOT $A2$.

The first push button actuated will take precedence as long as it is being actuated. For example, actuating the ON push button removes the output from AND-NOT $A1$, which in turn removes an input to AND-NOT $B1$. Once the input from AND-NOT $B1$ is removed, actuating the OFF push button has no effect. The first input signal locks out the effect of the second while the first is being maintained. Each output of the *off-return* MEMORY can drive 12 other inputs.

The *retentive* MEMORY logic function is a modification of the *off-return* MEMORY, with the difference being what occurs upon restoration of system power. The *retentive* MEMORY will resume its previous output condition, whereas the *off-return* MEMORY returns to the OFF condition. As evident from Fig. 11·15, the output condition the flip-flop assumes depends upon where the unit reset is connected.

With the unit reset connected to AND-NOT $B2$, the *retentive* MEMORY will return to the OFF condition, and an ON signal will be present at terminal 7 when system power is established. With the unit reset connected to AND-NOT $A2$, the MEMORY will return to the ON condition, and an output will exist at terminal 8 when system power is established.

A completely sealed single-pole double-throw reed switch is used to perform this switching function. Referring to Fig. 11·15, we see that permanent-type bias magnets hold the movable pole to its respective side of the switch, depending on which way the coil around the switch caused the movable reed to position itself. The coil, connected between terminals 7 and 8, creates sufficient flux to override the bias magnets. Current will flow in either direction through the coil, depending upon which output terminal is at zero volts and which is at −4 volts.

The reed switch has essentially an infinite mechanical life, and in this application switches no current when the movable reed moves. Current is carried, but not switched, during a brief

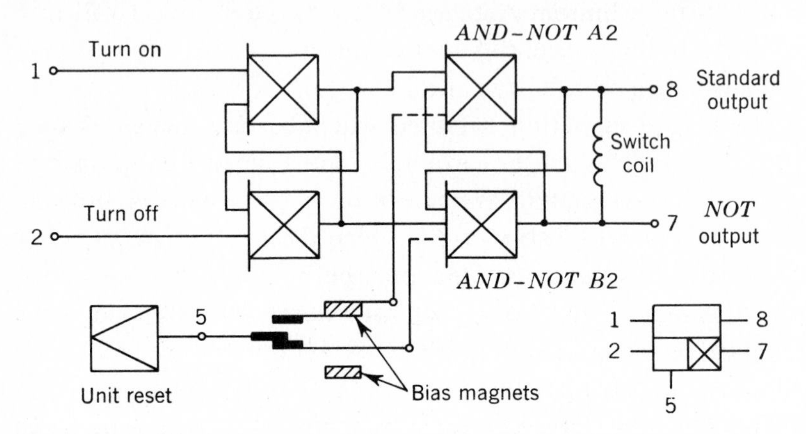

Fig. 11·15 Block diagram of the retentive MEMORY. (*General Electric Company*)

period that system power is applied, that is, during the period when the unit reset delays its continuous ON signal. Thus the mechanical and electrical life of the reed switch in the *retentive* MEMORY is very compatible with the reliability of the transistorized static system.

The speed of operation of the *retentive* MEMORY is not dependent upon the speed of the reed switch. The flip-flop can switch at its maximum rate, and the reed switch can follow at a slower rate, as the last position the reed switch is in when power is lost determines its retentive position. The slow decay of 12-volt d-c power upon loss of 115-volt a-c system power is quite adequate to switch the reed to its proper position.

Because of the loading of the output terminals by the coil, the driving capability of the *retentive* MEMORY is 7 units, compared with the normal rating of 12 units of load.

A unique three-input AND can be formed by combining three AND-NOTS as in Fig. 11·16. This element, called a *sealed* AND, requires inputs to be present at terminals 1 *and* 2 *and* 3 to cause an output at terminal 8, and then inputs at terminals 1 and 2 can be removed without affecting the output condition. The NOT of the standard output is available at terminal 7, and an auxiliary connection is required at terminal 5 to a unit reset

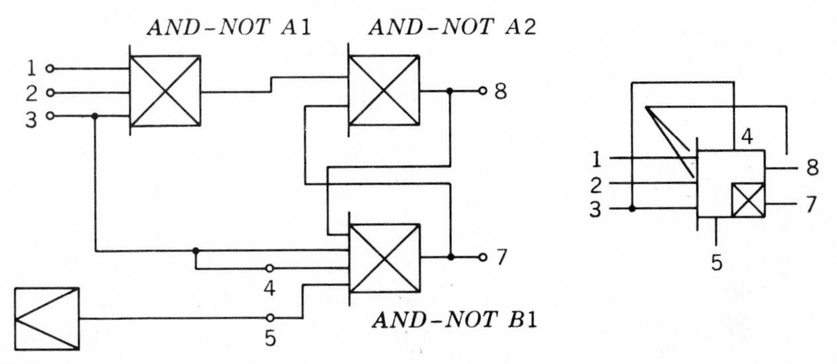

Fig. 11·16 Block diagram of the sealed AND, *inputs 1 and 2 sealed. (General Electric Company)*

which supplies a delayed ON signal upon application of system power.

To analyze the internal operation of the *sealed* AND (Fig. 11·16), assume power has just been applied to the 12-volt bus and no inputs exist at terminals, 1, 2, or 3. Terminal 4 has been connected to terminal 3, which will cause identical effects in AND-NOT *B*1. AND-NOT *A*1 produces an output immediately upon applying system power, since NOT inputs are present to this AND-NOT. The unit reset always delays its output to assure that AND-NOT *B*1 will momentarily be lacking one input, which results in an output from AND-NOT *B*1. With AND-NOTS *A*1 and *B*1 providing an input to AND-NOT *A*2 with both inputs, no output will exist at terminal 8; this causes another input to be missing to AND-NOT $B1$. This continues an output from AND-NOT $B1$ when the unit reset gains an output, and thereafter remains in the ON condition.

When an ON signal is applied to terminal 3, no change in output occurs in AND-NOTS $A1$ or $B1$, since each continues not to have all inputs present. Inputs must simultaneously be present at input terminals 1 *and* 2 *and* 3 to turn off the output of AND-NOT $A1$. This removes an input from AND-NOT $A2$, which results in an output at terminal 8 and also a loss of an output at terminal 7, since all inputs are now present to AND-NOT $B1$. With AND-NOT $B1$ not providing an input to AND-NOT $A2$, inputs to terminals 1 and/or 2 can be removed without effecting AND-NOT $A2$. Only removing the input to terminal 3 will cause a change; AND-NOT $B1$ gains an output, which removes the output from terminal 8 because all inputs are again present to AND-NOT $A2$.

The combination of AND-NOTS $A2$ and $B1$ is called a *flip-flop,* and with terminal 4 connected to terminal 3, inputs at terminal 1 and 2 become "sealed" after all three inputs are simultaneously present and the flip-flop yields an output at terminal 8. The line between terminals 8 and 1 and 2 *is not a wire;* it is a symbol representing the sealing action. But the line from terminal 4 to terminal 3 *is a wire*. With terminal 4 connected to terminal 2, *only* terminal 1 is sealed (refer to Fig. 11·17). This arrangement in the same element still requires all three inputs to be present to gain an output at terminal 8, but now loss of input at terminals 2 or 3 will cause AND-NOT $B1$ to have an output at terminal 7 and subsequent loss of output at terminal 8. The *sealed* AND is then *not* "picked up," an ON signal being present at the built-in NOT, at terminal 7. Each output terminal of the *sealed* AND can drive 12 other outputs.

The DELAY element provides time delay upon energizing with a variety of individual timing ranges. Timing occurs when an RC network is being charged and results in the firing of a unijunction transistor. The RC time constant is variable by an adjustable linear potentiometer. A unijunction transistor is a unique semiconductor having three leads—an emitter and two bases. Fig. 11·18 depicts a unijunction transistor in a typical timing circuit. With the switch closed (no emitter voltage), there is no conduction between base 1 and base 2, and $Vb1b2$ is close to 12 volts ($R1$ and $R2$ have low values). When the switch is opened, the capacitor starts to charge. When the emitter voltage reaches approximately 75 percent of $Vb1b2$, conduction occurs. At this point, the impedance between base 1 to base 2 drops, and a significant current flows through $R2$, producing an output signal. The timing period is initiated when charging current is permitted to flow into the RC network, and the length of time is determined by adjustment of $R3$.

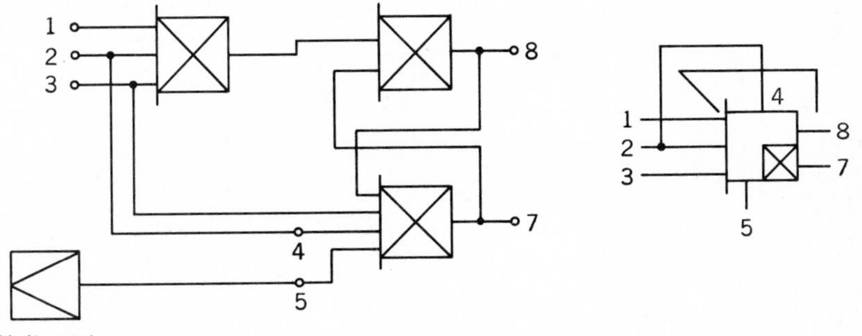

Fig. 11·17 Sealed AND, *only input 1 sealed.* (*General Electric Company*)

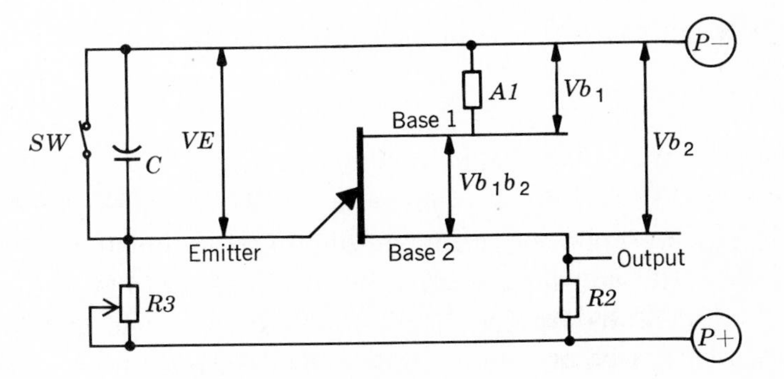

Fig. 11 · 18 Unijunction transistor. (General Electric Company)

When an ON signal is applied to terminal 1 (Fig. 11·19), the instantaneous NOT output at terminal 2 turns off, and the transistor in the following NOT goes into conduction. This conduction establishes a base-to-base voltage difference at the unijunction

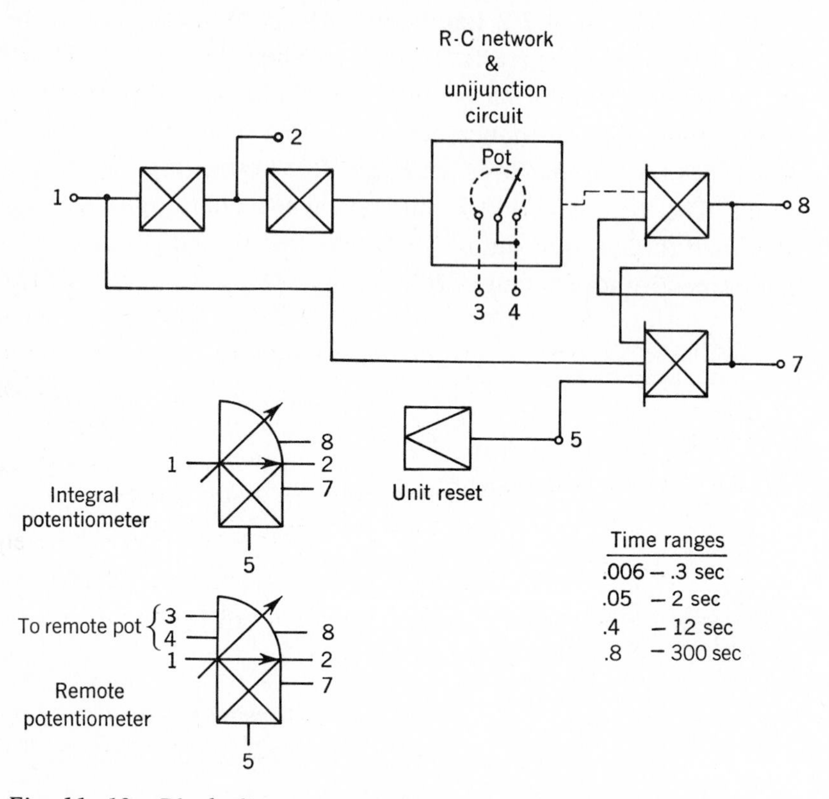

Fig. 11 · 19 Block diagram of the DELAY *unit. (General Electric Company)*

transistor and simultaneously permits charging current to flow into the RC network. As voltage in the RC network builds up, a time delay upon energizing results. The unijunction triggers when the RC network's voltage rises sufficiently, and a pulse is introduced in the flip-flop. Terminal 8 turns on, and this causes terminal 7 to turn off, setting the flip-flop. Only loss of an input to terminal 1 will reset the flip-flop and completely discharge the

RC network by removing the base-to-base voltage difference on the unijunction transistor.

Terminal 7 is the NOT of terminal 8 and can be considered equivalent to the normally closed time opening contact on a time-delay-on-energizing pneumatic timer. Terminal 8 is equivalent to the normally open time closing contact; terminal 2, the NOT of the DELAY input signal at terminal 1, is equivalent to the instantaneous normally closed contact; and terminal 1 is equivalent to the instantaneous normally open contact when compared to a conventional pneumatic timing relay. The remote time-adjusting potentiometer, an optional form of the basic DELAY, would be connected to terminals 3 and 4.

The repeatability over the operating range of the DELAY is ± 5 percent of the time setting of the device, or ± 1 percent if ambient temperature is held constant. The recommended minimum reset time for the above repeatability is 0.5 percent of the maximum time setting of the individual DELAY.

A *TDOD* (time-delay-on-deenergization) form of the DELAY is also available (see Fig. 11·20). This comes in ranges of 0.4 to 12 seconds and 8 to 300 seconds. The internal circuitry effectively is that shown in Fig. 11·20. An ON signal at the input causes an immediate output at pin 8. Loss of the ON signal at pin 1 results in loss of the ON signal at pin 8 x seconds later.

11 · 2 GENERAL ELECTRIC ORIGINAL INPUTS

The pilot-device signals which supply the information to the logic control must each be connected to a device that accepts the signal and converts it to a logic-level ON or OFF signal of the proper magnitude. Electrical contacts are the most frequent inputs and are normally operated at 125 volts d-c with static control. One hundred and fifteen volts a-c is also used as an input to static control.

Electrical contacts do not exhibit reliable circuit continuity or contact fidelity at the low voltages used as direct inputs to the logic elements. At higher voltages such as 115 volts a-c or 125 volts d-c, the voltage will burn away an oxidation film and assure good contact fidelity.

The 125-volt d-c original-input operation is as follows. The LEPS provides a source for 125 volts d-c to be used with this input. This −125 volts d-c is connected to one side of the contacts of pilot devices such as limit switches, push buttons, and

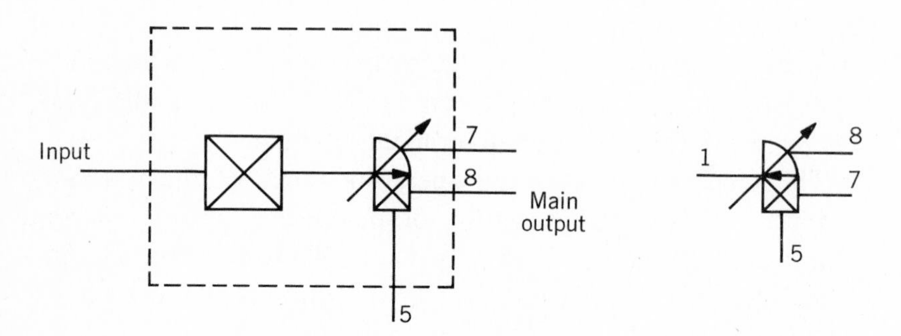

Fig. 11 · 20 Block diagram of the OFF DELAY *unit. (General Electric Company)*

pressure switches, and the other side of the contact is connected to the input terminal *P* on the d-c original input. The logic output of the device is at the terminal marked *L* as shown in Fig. 11·22. A zero-volt connection must also be made to the device at the terminal marked *O*. A built-in monitor light is incorporated in the 125-volt d-c original input, which is on the 125-volt side of the circuit and indicates whether or not voltage is present at the input terminal. Approximately 5 milliamperes (ma) flow through the pilot device when its contacts are closed.

Each input is separate and snaps into a grooved mounting track. The entire unit is encapsulated and uses wire-clamp terminals. The zero-volt bus must be used to make the connection from the LEPS, in order to minimize voltage drop and provide proper spacing of each unit. One bus accommodates up to six inputs.

Figure 11·22 shows the schematic of the 125-volt d-c original input. With the output terminal L connected to the input of a logic element, —4 volts will be present when the unit is not in the ON condition. The NPN transistor cannot conduct base to emitter or collector to base when the pilot device is open, because no current path to —125 volts d-c exists. When the

Fig. 11·21 125-volt d-c original input. (General Electric Company)

pilot device closes, the input terminal P receives —125 volts d-c which turns the neon indicating light on and commences charging capacitor $C1$ to a voltage slightly above —4 volts. The NPN then commences conduction, since its emitter is then more negative than its base. This allows the PNP transistor to conduct emitter to base and also emitter to collector, which causes the base of the NPN to become even more positive than its emitter.

This results in a rapid turn-on of both transistors and a resultant ON signal at terminal L. With the NPN in full conduction, its emitter decreases in voltage to less than 0.5 volt while still being in full saturation. This also causes capacitor $C1$ to discharge its previous charge, so that when the pilot device again opens, it will be essentially discharged and ready again to go through the above sequence. This technique makes this input

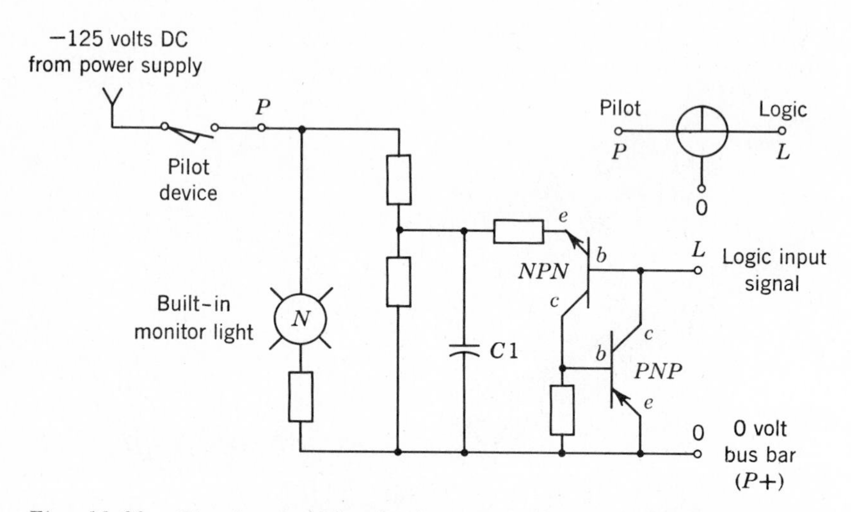

Fig. 11·22 Circuit of 125-volt d-c original input. (General Electric Company)

device very insensitive to transient electric noise and line disturbances, and increases the reliability of the system by providing the logic with an input only when a valid input signal exists. The logic output of the device, terminal L, can supply up to 12 input signals to the logic elements.

Figures 11·23 and 11·24 are examples of static control panels and show the proper grouping of original inputs along the left side of the cabinet.

The 115-volt a-c original-input device is necessary to convert a 115-volt a-c contact closure to a usable input signal to static control. The 115-volt a-c signal is stepped down through a transformer, rectified, and filtered, causing a transistor to become saturated; then the output of the transistor in again filtered.

Figure 11·25 shows the physical appearance of the a-c original input.

Fig. 11 · 23 Solid-state-logic control panel. (General Electric Company)

Figure 11·26 shows the schematic of the a-c original input. Essentially, the circuit is identical to the d-c original input but is preceded by a transformer and full-wave rectifier. When the pilot device closes, 115 volts a-c is supplied to the primary of the transformer. This voltage is reduced, rectified, and filtered,

Fig. 11·24 Large static control panel. (General Electric Company)

Fig. 11·25 A-c original input. (General Electric Company)

causing point *a* to rise to slightly above −4 volts d-c. The NPN transistor then commences conduction (since its emitter is then more negative than its base). This allows the PNP transistor to conduct from emitter to base and also emitter to collector, which causes the base of the NPN to become even more positive than its emitter. This results in a rapid turn on of both transistors and a resultant ON signal at the output terminal.

Approximately 20 ma flows through the pilot device when its contacts are closed. The output terminal can supply up to 12 input signals to the logic circuitry.

The 115 volts a-c for the a-c original input should be supplied from the power-supply side of the incoming a-c line filter described later.

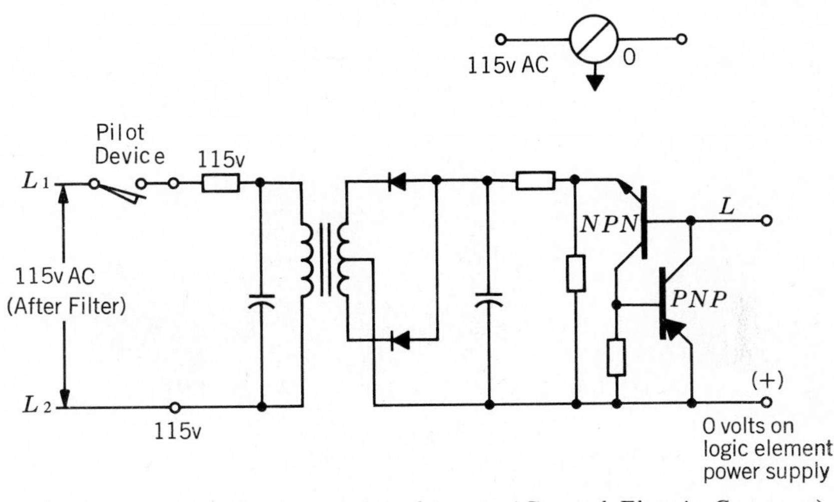

Fig. 11 · 26 Circuit of a-c original input. (General Electric Company)

Two resistance-sensitive input devices (Fig. 11·27) are available to detect a change of resistance, one being a plug-in element for use with photocells and the other a panel-mounted unit for probes or applications requiring greater sensitivity and isolation from system ground potential.

The plug-in module without a built-in monitor light and one with a built-in monitor light have two input terminals 2 and 1 and the resulting output at terminal 4. Terminal 1 is system zero volts, which is ground, and most applications are thereby limited to detecting the change of resistance of a photoconductive cell as light strikes the cell. The device yields an ON signal at terminal 4 upon decreasing resistance and will remain in the ON condition until the resistance raises somewhat above the trip value. The best application is when a considerable resistance change occurs rapidly, as the input is not designed to be an analog-sensitive input. The resistive load must not generate a voltage, as the input device will supply the small potential of a volt or two to the load, and a change of resistance is the parameter being sensed. The photoconductive cell, generally the

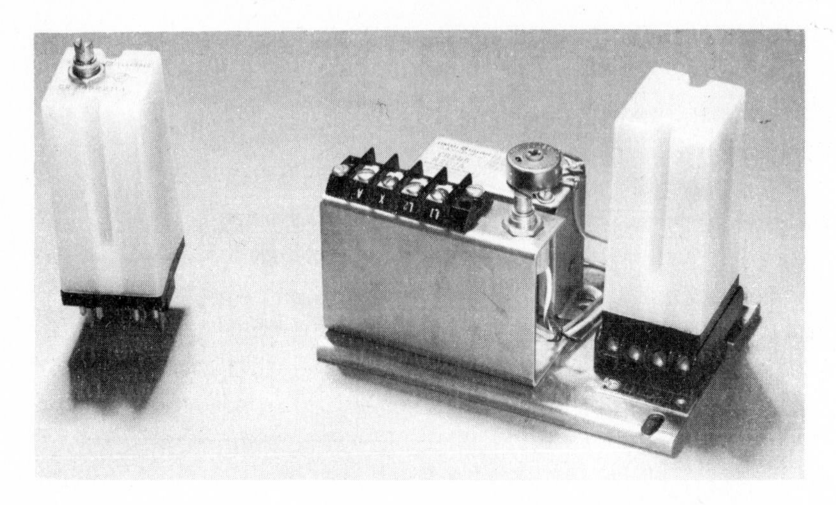

Fig. 11·27 Resistance-sensitive inputs. (General Electric Company)

cadmium sulfide type, is excellent, and the two leads from the cell can be directly connected to the input with shielded wire. This device's trip point is adjustable by a built-in potentiometer from 50 to 4 kilohms.

The panel-mounted input requires 115 volts a-c to be connected. It is more sensitive and in addition has a differential adjustment so that the dropout point may be varied with respect to the pickup point. This self-contained unit has a form C reed switch as an output, which should be connected to the standard 125-volt d-c original input to provide an input to static control. Its operating range is adjustable from 300 kilohms to 500 ohms, and the resistive-load terminals (A and X) are isolated from system ground. The wires from the load to the device should be shielded, and the load can be photocells, probes, or other inputs whose resistance changes greatly upon actuation. Again, the voltage must not be generated in the load, but the resistance-sensitive input supplies about a volt potential across the load.

Each device is very useful to the control designer in providing a means for taking a solid-state photoconductive cell directly into the static control or for using the resistance of the material between two probes to be the switching means, with the safety of a low voltage at the probes.

The proximity switch combined with a CR115D8 power supply can supply signals into static control through the proximity-switch input. This solid-state limit switch can detect ferrous or nonferrous metal without physical contact with the material and, without any moving parts, can supply an ON signal into a static unit. The driving capability of the switch is limited to one input, and normally a proximity-switch input is utilized which can itself provide a 50-unit output-driving capability.

The CR115D proximity switch has four wires colored black, green, red, and white connected to its encapsulated amplifier. These wires connect to the CR115D8 power supply, which supplies the 30 volt d-c for operating the sensing head. All wires should be connected to the power supply, and the red and white wires are to be of the insulated shielded type with only one end of each shield connected to logic common, 0 volts, on the power supply. The white wire is the direct input to the proximity-switch-input module. If time delay is required, it must be accomplished in the static control with the DELAY module.

A single CR115D8 power supply can accommodate up to five proximity switches simultaneously with independent operation of each switch. The proximity-switch input must be used when connected to counting or register-type circuits. Two proximity-switch inputs are packaged in one module. The input terminals are 1 and 5 with respective output terminals 4 and 8. Each output can drive up to 50 inputs of any logic elements.

Occasionally in counting and register circuits a fixed and accurate pulse rate is required. The most convenient reference is the 50- or 60-cycle incoming a-c voltage which is available as ± 12 volts at the *LD* terminal of the LEPS. With this a-c voltage connected to terminal 1 of the sine- to square-wave converter, a square-wave output is available at terminal 4. An ON signal and an OFF signal will each exist for approximately 8 msec.

A NOT output is also available at terminal 3, but the standard output and NOT output cannot be connected to an OR to yield twice the frequency. If the two outputs are each connected to a single shot, and those outputs connected to an OR, twice the line frequency of pulses would be available, with the ON signal existing for approximately 100 μsec for each pulse.

The sine- to square-wave converter output terminals can each drive up to 12 other logic inputs. The element consists of a converter and a unit reset in one module. The unit reset is an independent function.

11·3 GENERAL ELECTRIC OUTPUT AMPLIFIERS

The output of any logic function must be amplified to switch the power required by external loads. A variety of voltage and current ratings are available to economically facilitate the energizing and deenergizing of typical power devices, depending on the power involved. The voltages commonly utilized are 24 volts d-c, 115 volts a-c, and occasionally 90 to 180 volts d-c. Figure 11·28 illustrates the four plug-in module forms and three panel-mounted output amplifiers. Amplifiers cannot be paralleled to increase rating, and must have the proper voltage and polarity applied to their terminals.

The 6-watt d-c output amplifier can switch a d-c load whose current does not exceed 0.25 amp and any d-c voltage up to its nominal rating of 24 volts. The load is usually inductive, such as a pilot-operated solenoid valve, and a counter-electromotive-force (cemf) diode is incorporated in the amplifier to reduce the effect of the cemf voltage generated when an inductive load is deenergized.

Figure 11·29 shows the schematic of the 0.25 amp 24-volt d-c output amplifier, frequently referred to as the 6-watt output. An emitter-follower circuit is employed with the PNP-1 transistor. When an OFF signal, —4 volts, is at terminal 1, PNP-1 transistor conducts from emitter to collector, which in turn causes sufficient voltage to appear at the base of PNP-2 transistor to cause it to conduct from emitter to collector. With PNP-2 transistor in full saturation and conducting, its collector voltage is essentially zero volts, thereby positive-biasing the base of the

Fig. 11 · 28 Output amplifiers. (General Electric Company)

power transistor and assuring it is not in saturation. The diode in series with the emitter of the power transistor causes the emitter to be more negative with respect to the zero-volt bus than the voltage drop through PNP-2, which is the voltage impressed on the base of the power transistor.

When an ON signal, 0 volts, is impressed at terminal 1, PNP-1 and PNP-2 cease conduction, and the power transistor goes into full saturation. This allows current to flow from the zero-volt bus $P+$ through the amplifier to terminal 4, then to the minus-voltage power bus, usually —24 volts d-c, and through the power load. The CEMF diode must be connected across the load to protect the power transistor from the CEMF voltage generated by the load upon deenergization. Connecting terminal 2 across the load also connects the optional built-in monitor light in parallel with the load to provide a visual indication in the plug-in module as to its actual output condition. This is very valuable in panel check-out procedures.

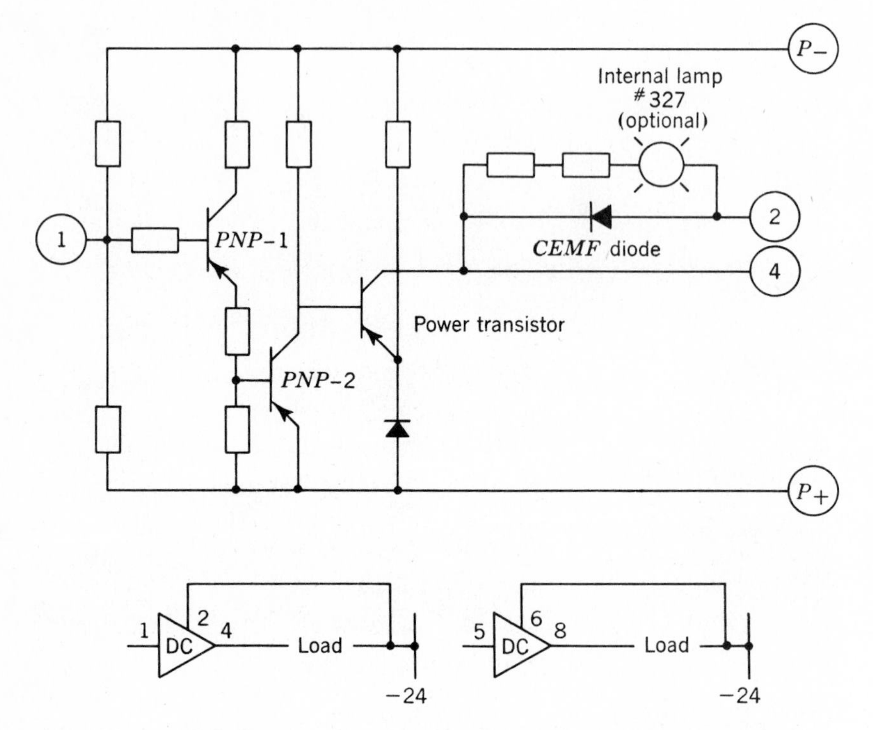

Fig. 11·29 6-watt d-c output amplifier. (General Electric Company)

The maximum rated current of the 6-watt d-c output amplifier is 0.25 amp at a d-c voltage up to 24 volts d-c. D-c power supplies normally have a higher voltage output at no-load, and the maximum voltage limit of the amplifier is 28 volts d-c. The power supply must be filtered.

Two complete and independent 6-watt amplifiers are packaged in a single plug-in module. The optional built-in indicating lamp, if utilized, draws an additional 20 ma from the 24-volt d-c power supply. Approximately a 1-volt drop can be expected in the power circuit of the output amplifier.

The 36-watt d-c output amplifier can switch a d-c load whose current does not exceed 1.50 amp and any d-c voltage up to its nominal rating of 24 volts. The load is usually inductive, such as a pilot-operated solenoid valve, and a CEMF diode

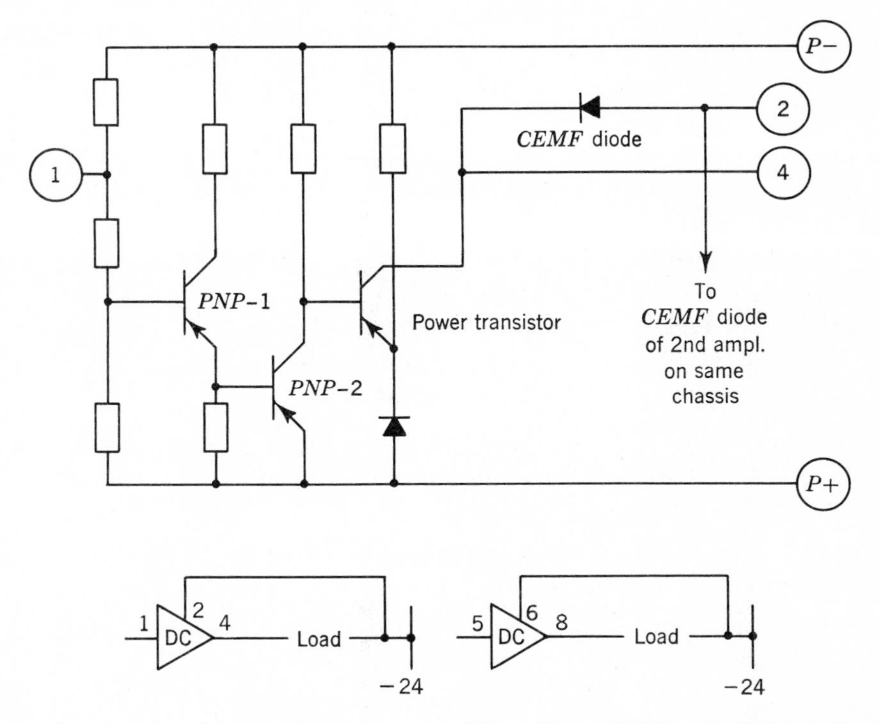

Fig. 11 · 30 36-watt d-c output amplifier. (General Electric Company)

is incorporated in the amplifier to reduce the effect of the CEMF voltage generated when an inductive load is deenergized.

Figure 11·30 shows the schematic of the 1.50-amp 24-volt d-c output amplifier, frequently referred to as the 36-watt output. Similar in operation to the 6-watt output, an OFF signal at terminal 1 causes the PNP-1 and PNP-2 transistors to conduct and the power transistor to have a positive bias to assure its nonconduction. An ON signal at terminal 1 results in the PNP-1 transistor ceasing conduction, and the PNP-2 transistor also ceases conduction. This causes the power transistor to saturate fully and supply power to output terminal 4. The CEMF diode connection at terminal 2 is commoned as a termination point for both amplifiers for the —24-volt bus.

Two amplifiers are mounted on a common baseplate for separate panel mounting. Approximately a 1-volt drop can be expected in the power circuit by the output amplifier. The maximum rated current of the 36-watt output amplifier is 1.50 amp at a d-c voltage up to 24 volts d-c. D-c power supplies normally have a higher voltage output at no-load, and the maximum voltage limit of the amplifier is 28 volts d-c. The power supply must be filtered.

The 1-amp a-c output amplifier can switch 115-volt a-c power to an inductive load whose inrush is 3.7 amp and holding current is 1 amp. This is a solid-state switch utilizing a silicon controlled rectifier (SCR) as the basic switching device.

An SCR is a silicon diode which will not let current flow in either direction through the device unless a small signal is applied to the gate lead, and then conduction occurs in only the forward direction, anode to cathode. After the gate is applied, the SCR continues to conduct even when the gate pulse is removed. The SCR stops conducting when the current through the device essentially decreases to zero and requires another gate pulse to again commence conducting.

Figure 11·31 shows the 115-volt a-c power circuit with one SCR in a diode-circuit configuration. With the SCR in the conducting state, one half-cycle of current would flow to the load from *L*1 to *D*1 to the SCR to *D*2 to the load to *L*2, and the other half-cycle would flow from *L*2 through the load to *D*3 to the SCR to *D*4 to *L*1. This causes full a-c power to flow through the load, and unidirectional current to flow through the SCR static switching device. If the SCR receives a gate pulse, it will conduct for the remaining portion of the half-cycle and require another gate pulse to again conduct for the next half-cycle of a-c power to the load.

The gate-pulse source is a multivibrator circuit which, when an ON signal is applied, generates approximately 2,000 gate pulses per second which cause the SCR to conduct. Each time current through the SCR passes through zero, the SCR ceases conduction, but soon receives another gate pulse to cause conduction very early in the next half-cycle of a-c power.

The RC in parallel with the SCR provides better turnoff characteristics at low currents. The minimum load current is 100

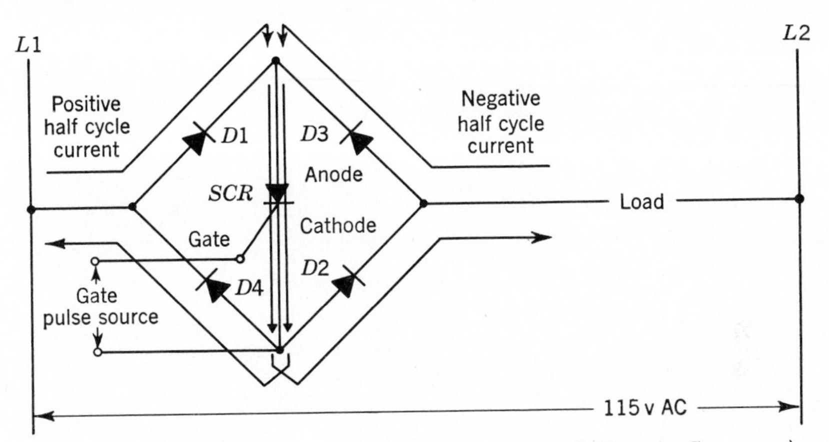

Fig. 11 · 31 SCR control of a-c power. (General Electric Company)

ma. The normal-inrush rating is considered to be 6 times the continuous-current rating, and the 1-amp a-c output amplifier can be suitably protected by a Buss tron KAA-1 fuse or its equivalent.

Figure 11·32 shows the schematic of the 1-amp a-c output amplifier. The unijunction transistor *UJ* in this circuit is the key part of the oscillator which continuously provides 2,000 pulses per second to the two-input AND amplifier portion of this SCR-triggering configuration. With an ON signal at terminal 1, PNP-1 will not conduct when the oscillator circuit also provides an ON signal to its input, which allows the power transistor PNP-2 to momentarily conduct. This conduction of PNP-2 causes current to flow through the air-core pulse transformer *PT*, which gates the SCR into conduction. When the oscillator circuit removes its ON-signal input to PNP-1, this transistor again conducts, which causes PNP-2 not to conduct and no gate pulse to exist. The diode in parallel to the *PT* primary coil is to minimize the CEMF voltage generated upon deenergization of the coil. The diode in the oscillator circuit is to limit the magnitude of the positive-voltage pulse to zero volts, the pulse being sup-

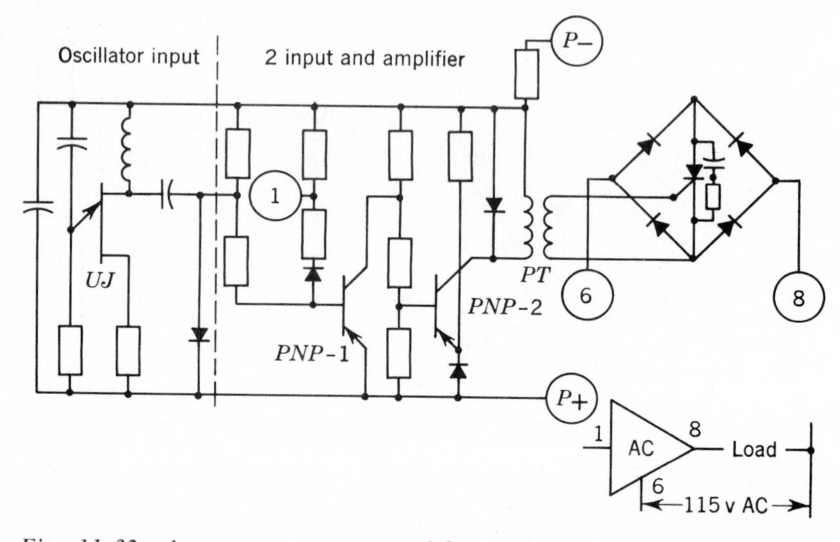

Fig. 11·32 1-amp a-c output amplifier. (General Electric Company)

plied by the combination of inductance and capacitance connected to one of the bases of the unijunction *UJ*. This output amplifier is packaged in a single plug-in element.

The 4-amp a-c output amplifier is a plug-in unit which can switch 115-volt a-c power to an inductive load whose inrush is 16 amp and holding current 4 amp. This is a solid-state switch utilizing a Triac as the basic switching device. The circuit is the same as in the 1-amp device, except that a Triac takes the place of the SCR and the four diodes. Fuse with a KAA-4 tron fuse.

A 10-amp panel-mounted a-c amplifier is also available; see Fig. 11·33.

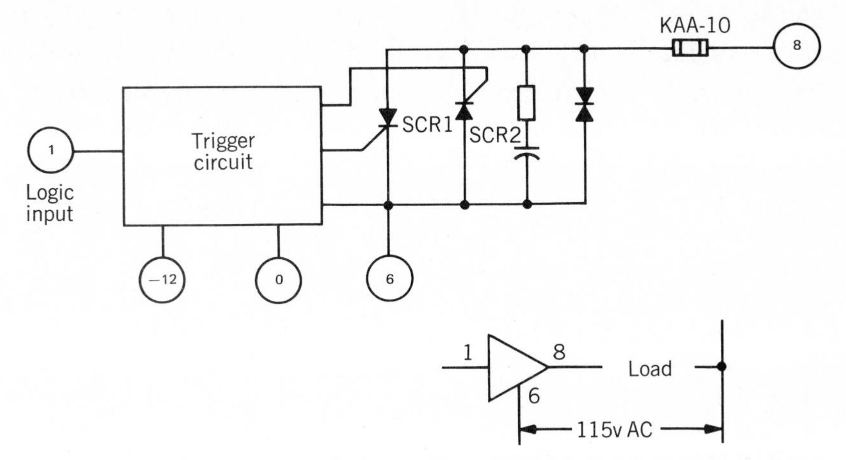

Fig. 11·33 10-amp output amplifier. (General Electric Company)

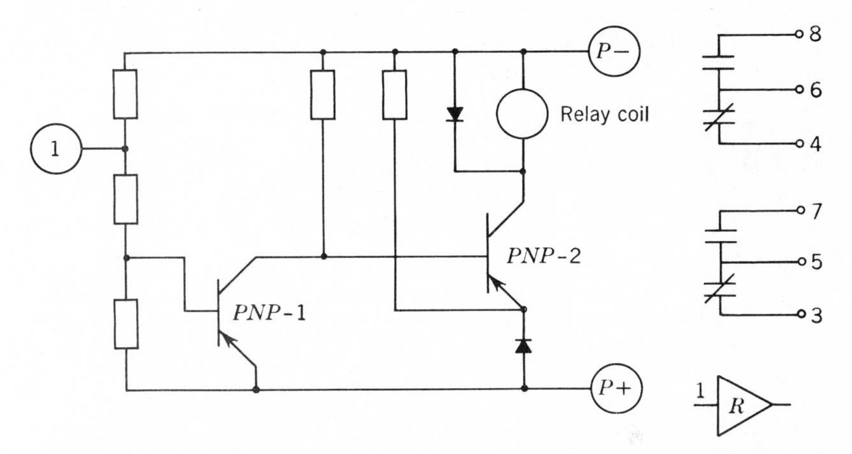

Fig. 11·34 Relay output amplifier. (General Electric Company)

Occasionally an independent-contact output is required which can switch remote 115-volt a-c circuits. The relay output amplifier is a plug-in element containing a transistor-driven relay whose contacts are rated 12-amp inrush, 3-amp carry in 115-volt a-c circuits only, and whose approximate life under rated load is 10 to 20 million operations.

With an ON signal applied to terminal 1 in Fig. 11·34, PNP-1 will not conduct and PNP-2 will go into saturation, energizing the relay coil. With no ON signal applied to terminal 1, PNP-1 does conduct, which makes the base of PNP-2 more positive than its emitter. With PNP-2 not conducting, the CEMF energy in the relay coil is dissipated through the diode in parallel with the coil.

The 180-volt d-c output amplifier is a solid-state amplifier which is utilized in d-c circuits above 24 volts d-c. It uses silicon controlled rectifiers as the switching devices, and is frequently used with small clutch and brake coils. Its maximum current rating is 1 amp. The circuit for this panel-mounted device is shown in Fig. 11·35.

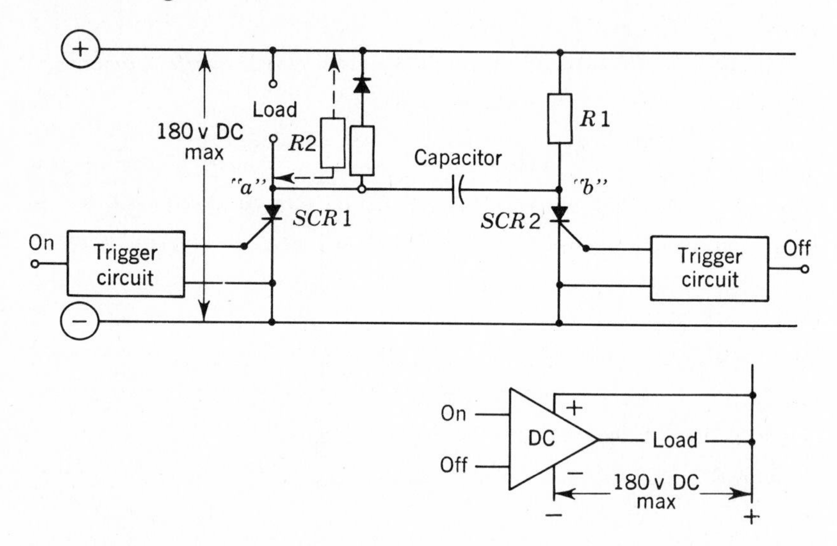

Fig. 11·35 180-volt d-c output amplifier. (General Electric Company)

With this output amplifier, an ON signal must always be supplied to either the ON or OFF input terminal, and the NOT of that ON signal must be applied to the other input terminal. With the built-in NOTs conveniently available in most logic functions, this is simple to obtain.

One SCR will always be conducting, since one triggering circuit always will have an ON signal as an input. Assume SCR-2 is conducting, which will charge the capacitor so that point *a* is positive and point *b* is negative. When the ON signal is removed from the OFF terminal and applied to the ON terminal, SCR-1 commences conduction and connects the positively charged point *a* of the capacitor to the minus bus. This momentarily causes SCR-2 to see no voltage difference across it and cease conduction. The capacitor proceeds to charge to the opposite polarity, and the reverse situation will occur when the ON input is moved to the other input terminal. Power flows through the load when SCR-1 conducts.

With the ON signal removed from the ON input terminal and an ON signal applied to the OFF input terminal, SCR-2 commences conduction, SCR-1 is no longer being triggered, and point *b* of the capacitor, being charged to a positive polarity, is connected in parallel to SCR-1; this momentarily causes no voltage difference across it, and conduction ceases through SCR-1. The cemf energy in the load is dissipated through the resistor diode in parallel with the load. SCR-2 remains in conduction, current flowing from the plus bus through *R*1 to the minus bus. The capacitor simultaneously is being charged, so that point *b* becomes minus and point *a* becomes plus. An inductive load will require a resistor *R*2, sized to cause 50 ma to flow in parallel. The ohmic size will depend upon the d-c voltage used and will be separately mounted. The SCR triggering circuits are identical to those shown in Fig. 11·32.

Remote-mounted pilot lights are common in operator consoles and stations; the economical manner to display a logic-level signal is by means of a pilot-light output amplifier. Four separate amplifiers are packaged in a single plug-in element, with each amplifier capable of switching a 40-ma bulb at a voltage up to 24 volts d-c.

Without an ON signal to the input terminal (Fig. 11·36), PNP-1 is in conduction, and the base of PNP-2 is held more positive than its emitter and does not conduct. A few milliamperes do flow through the load, but this has no effect since *R*1 is much greater than *R*2. With an ON signal applied to the input terminal, PNP-1 no longer conducts, PNP-2 goes to full saturation, and power flows to the lamp's resistive load. Emitter-to-base current flow in PNP-2 goes through the load to its minus power bus, in contrast to the PNP-1 emitter-to-base current flow.

The nominal voltage rating of this amplifier is 24 volts d-c, but d-c power supplies normally have a higher voltage output at no-load, and the maximum voltage limit of the amplifier is 28 volts d-c average, if we assume a 120-cps ripple frequency.

The internal lamp driver for monitor lights built into standard logic elements is an optional feature. It affords visual indication

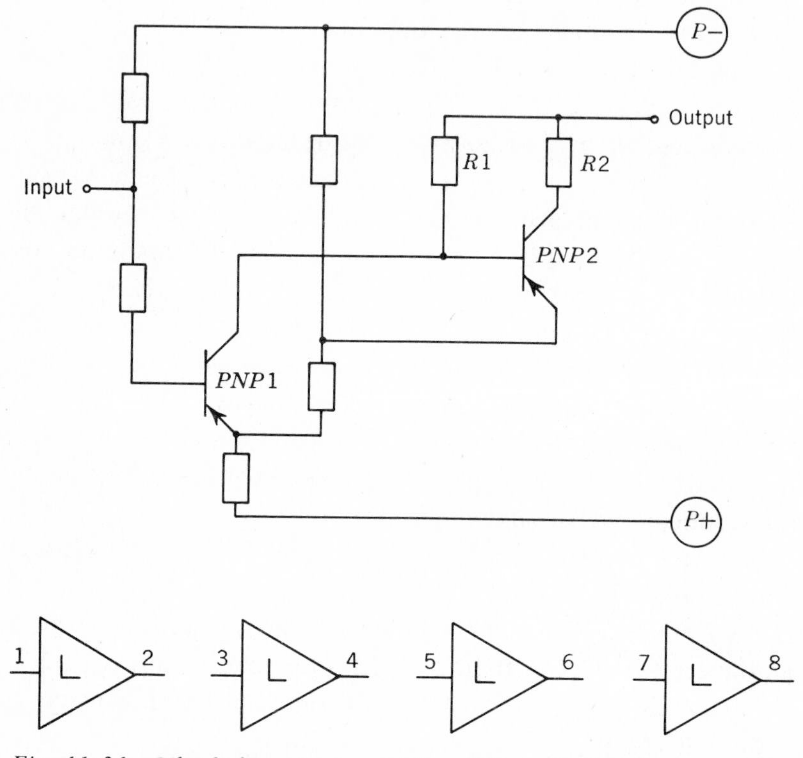

Fig. 11·36 Pilot-light output amplifier. (General Electric Company)

of the output of a logic function and is an excellent aid in checking the sequence of operation and isolating a circuit malfunction. The built-in monitor light indicates the output condition in the module where the logic is performed.

Since the monitor lights are mounted inside the element with a special transistor-amplifier circuit, they cannot be added later to elements not initially equipped with them. A 6-volt incandescent No. 345 miniature lamp with a flange base was used in early designs but has been replaced by a light-emitting diode (LED) to provide long life.

The typical amplifier circuit that switches each monitor light is shown in Fig. 11·37. The input signal for the lamp driver comes from the resistor network, which provides the same signal to the final output transistor of the logic function.

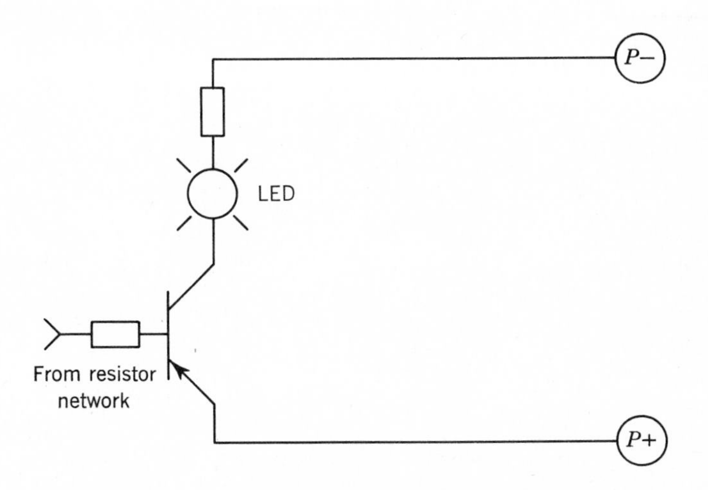

Fig. 11 · 37 Internal light driver. (General Electric Company)

11·4 GENERAL ELECTRIC POWER SUPPLIES

The logic system operates on d-c voltages, and the 115 volts a-c must be converted to those appropriate voltage levels.

The logic-element power supply (LEPS) (Fig. 11·38) converts filtered 115 volts a-c to 12 volts and 125 volts d-c. Zero volts, P+, and −12 volts, P−, are the basic power buses utilized to operate the logic elements, built-in monitor lights, and output-amplifier triggering circuits. The current capacity of each supply is 5 amp at 12 volts d-c.

The zero-volt bus, referred to as logic common, is also common with respect to the —125-volt d-c bus which powers the original input. It has a maximum current rating of 200 ma, which is equivalent to 40 simultaneously conducting original inputs.

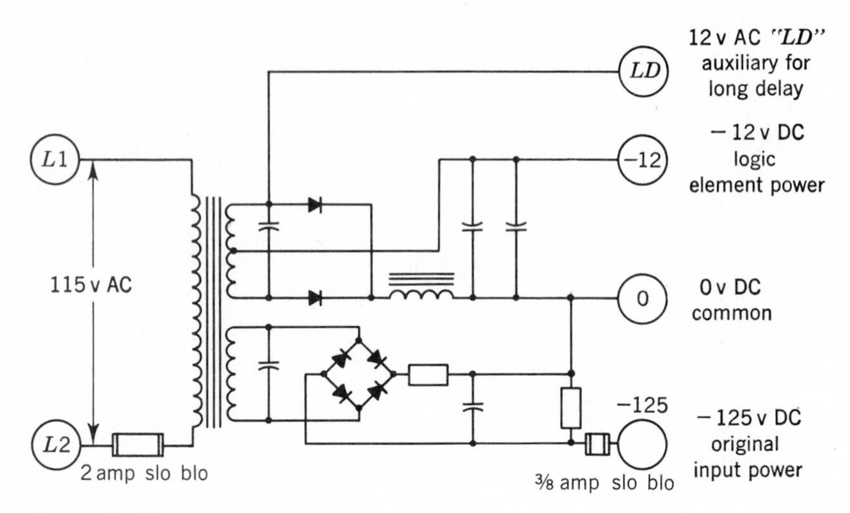

Fig. 11 · 38 Logic-element power-supply circuit. (General Electric Company)

The ± 12-volt a-c source *LD* is an auxiliary connection required by the sine-to–square wave converter.

The capacity of the power supply to drive elements varies with the type of unit and depends on whether monitor lights are used. However, a good approximation can be made by considering the 12-volt capacity equal to a total of 600 load units. Each logic *function* is counted as one unit of load except those listed in Table 11 · 1.

A diversity factor may apply to the relay output amplifier and internal monitor lights as the individual circuit is considered. If more than one power supply is required, the —125-, —12-, and 0-volt buses should be wired in parallel with additional power supplies.

Table 11·1

Function	*No. of units*
Six-input AND	3
Seven-input AND	3
Seven-input AND-NOT	2
OR	3
Off-return MEMORY	4
Retentive MEMORY	7
DELAY	5
Step (off-return) MEMORY	3
Step (retentive) MEMORY	9
Bi-step MEMORY	3
A-c and d-c original inputs	0
Unit reset	2
Resistance-sensitive input (R201A)	3
Proximity-switch input	5
Signal amplifier	5
6-watt d-c output amplifier	6
36-watt d-c output amplifier	6
1-amp a-c output amplifier	2
4-amp a-c output amplifier	2
180-volt d-c output amplifier	4
Relay output amplifier	14 (coil energized)
Relay output amplifier	2 (coil deenergized)
Internal monitor light	5 (lamp energized)

The zero-volt bus, logic common, is to be connected to a good water-pipe earth ground. The 12-volt nominal voltage should not be permitted to go lower than 10.5 volts or above 15 volts. A logic power supply is illustrated in Fig. 11·39.

The incoming 115 volts a-c must be filtered as shown in Fig. 11·40. The choke-capacitance filter is very effective in the 115-volt a-c circuit, and if one side of the line is grounded near the LEPS, only one set is required. A one choke and one capacitor combination is shown in Fig. 11·41 and is available under one number, CR245X103A. One filter set can accommodate up to four power supplies. A 1-amp power supply (120 load units) and a 10-amp power supply (1,200 load units) are also available.

When the logic system is operated by batteries, limit-switch, push-button, or other pilot-device information is also required by the control. It appears simpler to statically convert 12 volts d-c to 125 volts d-c than to utilize a second bank of

Fig. 11 · 39 Logic power supply. (General Electric Company)

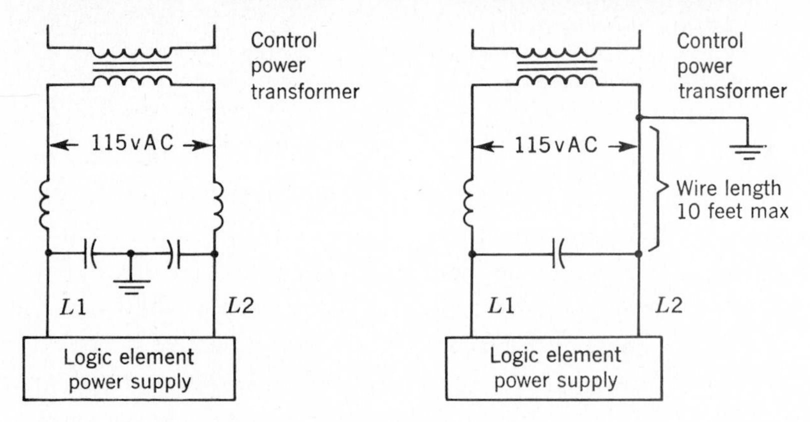

Fig. 11 · 40 Incoming a-c line-filter circuit. (General Electric Company)

batteries. The converter, shown in Fig. 11·42, is a panel-mounted device requiring an input of 9 amp at 12 volts d-c, with a resulting output of 0.8 amp at 125 volts d-c. It can supply up to 160 conducting inputs.

When 24-volt d-c output amplifiers are employed in a static system, a source of 24-volt power must be available. This power supply will normally have 115 volts a-c as an input and the rated 24 volts d-c nominal output. The current required will

Fig. 11·41 Line-filter components. (General Electric Company)

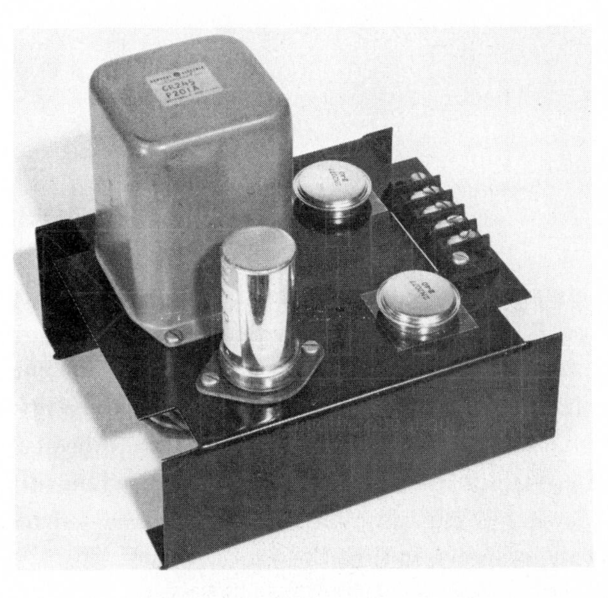

Fig. 11·42 12-volt d-c to 125-volt d-c converter. (General Electric Company)

be determined by the maximum number of loads conducting at one time plus a margin for future loading.

The maximum open-circuit voltage that should be applied to the amplifiers is 28 volts. The power supply must be filtered. When higher voltage is applied, too much current will be drawn by the load, and an overvoltage will also be applied to the unit. The minimum voltage applied depends upon the load, but a good guide can be 22 volts. Each amplifier will have up to a 1-volt drop within the device, which will result in somewhat lower voltage applied to the load.

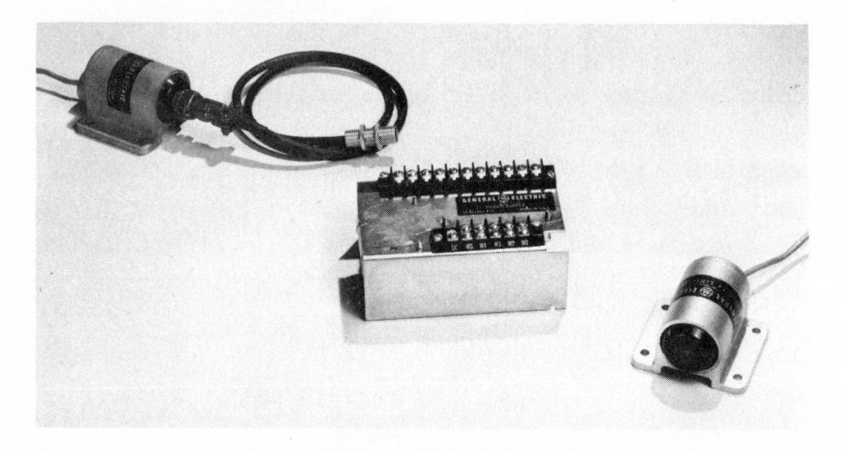

Fig. 11·43 Proximity switches and power supply. (General Electric Company)

The primary characteristics to consider in choosing the power supply are (1) regulation of the 24 volts with loading combined with incoming a-c line variations, and (2) maximum open-circuit voltages as read by a 20,000-ohms-per-volt voltmeter.

When the solid-state proximity switch is connected directly into static control, a special power supply must be utilized. Five proximity switches can simultaneously be connected to a single power supply as shown in Fig. 11·43.

Terminals *L*1 and *L*2 will have 115 volts a-c connected in addition to a terminal for logic common, zero volts from the

LEPS, and terminals for the four wires from the switch. The black, green, red, and white wires of the proximity switch will be connected to the power supply, the white wire being the logic level signal to be connected to the proximity-switch input. The white and red wires from every switch must be individually insulated shielded cables, with each shield connected only to one point, *LC*, at the power supply. All four limit-switch wires must also run in steel conduit back to the power supply if they are exposed to power wires. The black, green, and red wires each have three terminal points on the power supply; therefore no more than two wires need be terminated under one terminal point.

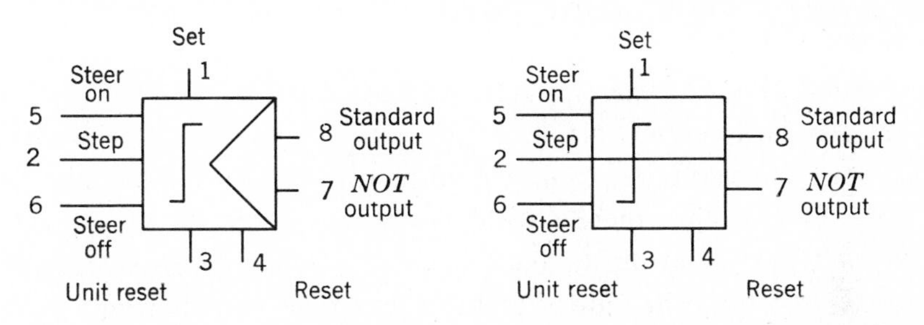

Fig. 11·44 Step MEMORY *symbols and pin identification. (General Electric Company)*

11·5 SPECIAL FUNCTIONS

Any logic control circuit can be built using the five basic logic functions: AND, OR, NOT, MEMORY, and DELAY. But when certain combinations of logic functions are repeated many times in a logic circuit, a special unit can be devised to perform the function as well or better, in some cases, with a reduction in required panel space and at a lower cost.

The *step* MEMORY is a special unit and is most useful in counting circuits and information-handling applications. The *step* MEMORY (Fig. 11·44) is a modification of the *off-return* MEMORY, with a unique turn-on and turn-off input network.

Pin 5 and pin 6 are steer inputs which direct pin 2 in the proper part of the flip-flop when an input signal is applied to the step terminal. The best analogy is that the steering network is like a double-barreled shotgun, with pin 5 and pin 6 being the hammer for each barrel and pin 2 the trigger. The only moment the gun will fire a shot down a barrel is when a particular hammer is actuated and the trigger is pulled. The time relationship in this analogy is accurate in that the particular hammer that is actuated may change up to the moment the trigger is pulled and may again be changed after the trigger is pulled. One pull of the trigger will cause only one shot, and the trigger must be released and pulled again to cause another shot. The firing of the shot occurs when the trigger is pulled, not while being held nor upon release of the trigger. The firing of the shot down a particular barrel will cause pin 8 to turn on and pin 7 to turn off, or pin 7 to turn on and pin 8 to turn off, depending upon which barrel is fired.

With an ON signal at pin 5, steer on, and a step signal applied at pin 2, the internal flip-flop will assume the state in which the standard output, pin 8, is ON. Likewise, with an ON signal at pin 6, steer off, the result would be pin 7 in the ON condition. It is obvious that there should never be an ON signal applied to both pin 5 and pin 6, since the unit would then be arranged to do two opposite things at the same time, which would not be a logical application of the *step* MEMORY.

Pin 3, unit reset, must be connected to the unit-reset module, as previous units required such a connection to their pin 5. Pin 1, set, provides a method to turn on pin 8 by loss of an ON signal to its terminal. This is commonly called setting a flip-flop. Pin 4, reset, can reset the flip-flop by loss of an ON signal to its terminal. The reset state of the *step* MEMORY is when pin 8 is OFF and pin 7 is ON. The set and reset terminals override the steering network in their effect. For example, if pin 1 does not have an ON signal, pin 8 is ON; if pin 2 is stepped, no change in the output of the *step* MEMORY occurs. These terminals provide a method for changing the output condition independent of the steering network; this method is very useful in resetting a counting circuit to zero or placing a predetermined number in a counter circuit and then continuing the counting from that number.

It can be seen in Fig. 11·45 that set and reset override the steering network, and the network is capacitively coupled with the flip-flop. With an ON signal, 0 volts, applied to one steer terminal, there will logically be an OFF signal, —4 volts, at the other; when an ON signal is applied to pin 2, only one capacitor will charge up, causing the appropriate AND-NOT in the flip-flop to commence conducting. This momentary conduction turns the other AND-NOT off, which results in flipping the flip-flop in the new maintained condition. The ON signal to pin 2 must be removed, and then —4 volts applied at the terminal in order to have the steering network properly reorient itself for application

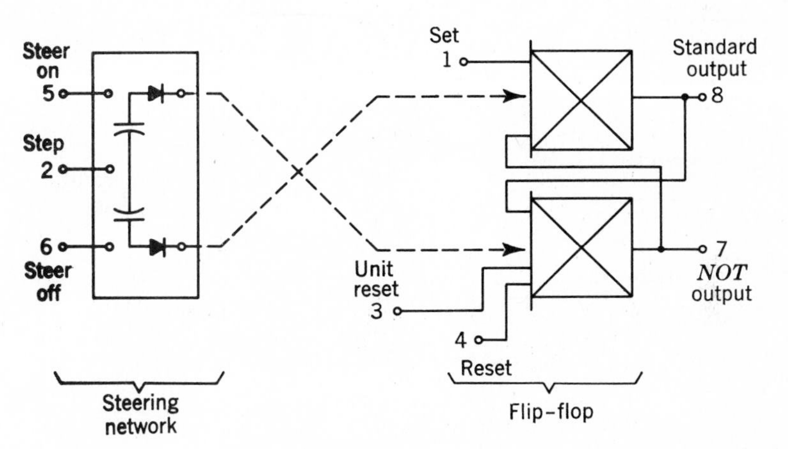

Fig. 11·45 Step MEMORY *block diagram. (General Electric Company)*

of another step pulse. The recommended maximum pulse rate which can be received by the step input, pin 2, is 10 kilocycles per second.

The *step* MEMORY is available in the *off-return* or *retentive* form, with operating speeds and terminal connections the same in both cases. The difference is only what state the flip-flop will assume on regaining system power (12-volt d-c previously having been removed). The OFF state, in which pin 8 is off and pin 7 is on, is given by the *step (off-return)* MEMORY type. The state existing prior to power interruption is given by the *step (retentive)* MEMORY type. The outputs of the *step (off-re-*

turn) MEMORY can each drive up to 12 other inputs. The outputs of the *step* (*retentive*) MEMORY can drive up to seven inputs.

The *bi-step* MEMORY is shown in Fig. 11·46. This off-return type of unit is similar to the *step* MEMORY, except that it has two complete steering networks in place of one. It is useful in add-subtract counters and reversible shift registers, each of which is normally limited to a unidirectional advance.

The terminal for steer off for each steering network has been eliminated in the *bi-step* MEMORY. This function still is required by the element, but a built-in NOT of steer on is included inside each unit, thereby automatically providing a steer-off signal in the absence of an ON signal applied to the steer-on terminal.

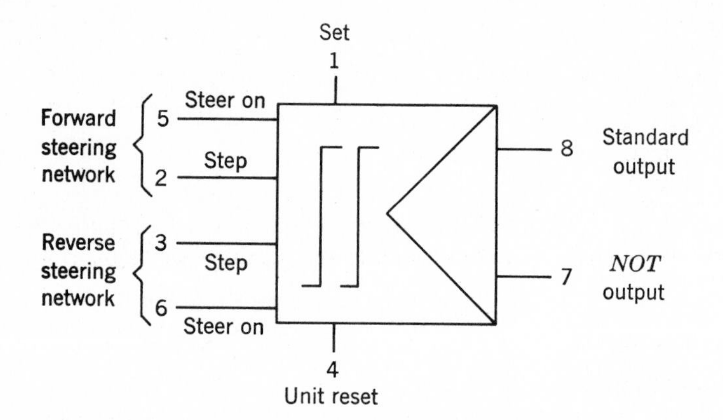

Fig. 11·46 Bi-step MEMORY *symbol and pin identification.* (*General Electric Company*)

The *bi-step* MEMORY has two completely independent steering networks which control a single flip-flop, and one step signal should be applied at one time. Pin 1, set, overrides each steering network by loss of an ON signal, and this results in having pin 8, the standard output, in the ON condition. The unit reset of pin 4 must be connected to the UNIT RESET, and if the reset function of the flip-flop is also required, an external AND-NOT can be used to control this input from two external signals. The *bi-step* MEMORY is available only in the *off-return* form, and each output can drive up to 12 subsequent inputs.

Most contact-making devices have contact bounce as they make a circuit, and it is necessary to remove this bounce when driving counter circuits and information storage registers. The signal amplifier incorporates an antibounce circuit, in addition to being able to simultaneously drive 50 logic functions, compared with the normal 12-unit loading capability.

Upon receiving an ON signal from a pilot device via an original input, an output will lock ON for 10 to 15 milliseconds (msec) regardless of whether the input signal bounces on-off or stays off. If an input signal is still there after 15 msec, an output from the signal amplifier will continue until the input is removed. If the input is removed prior to the 15 msec lock-ON time and remains OFF, the unit will turn OFF after 15 msec.

A dual function is also incorporated in the signal amplifier to take one logic input and provide an output capable of supplying up to 50 input signals into subsequent logic functions.

The single-shot unit is very useful in circuit designs incorporating MEMORYS, since they exhibit a first come, first served input configuration. A maintained signal can be made into a momentary signal by interposing the single shot. It provides an output of approximately 100 μsec for a maintained input. An input must be removed and again applied to yield another 100-μsec output signal.

An adjustable-single-shot unit is also available (Fig. 11·47). This function produces a standard output for a set period of time with a momentary or maintained input signal. It is adjustable from 0.006 to 0.3 sec and can be extended to 12 sec by

Adjustable single shot

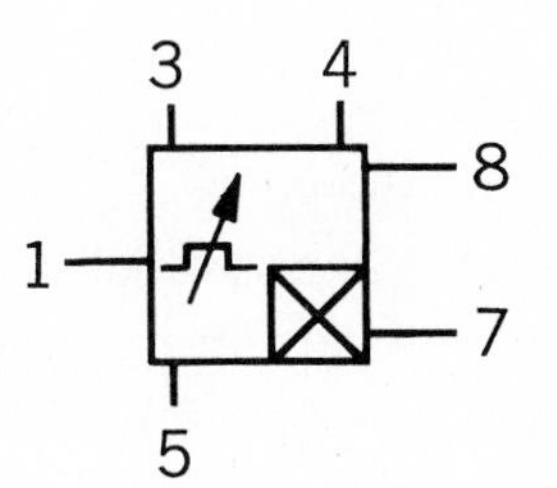

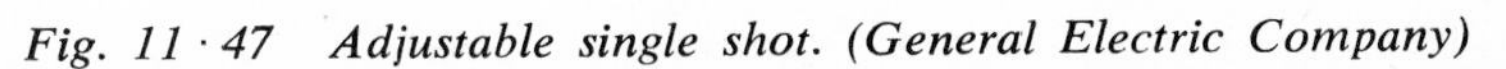
Fig. 11 · 47 Adjustable single shot. (General Electric Company)

connecting an external capacitor to socket terminals 3 and 4. A typical application of this unit would be the energizing of a solenoid valve for a short set period of time regardless of the length of the input signal. This module takes the place of the whole circuit shown in Fig. 11·48.

The unit-reset auxiliary unit is required for all MEMORYS, DELAYS, *sealed* AND logic functions, and *step* MEMORYS to make their output condition dependent upon application of system power. This individual unit can simultaneously be connected to 50 inputs, and supplies a delayed ON signal to these units.

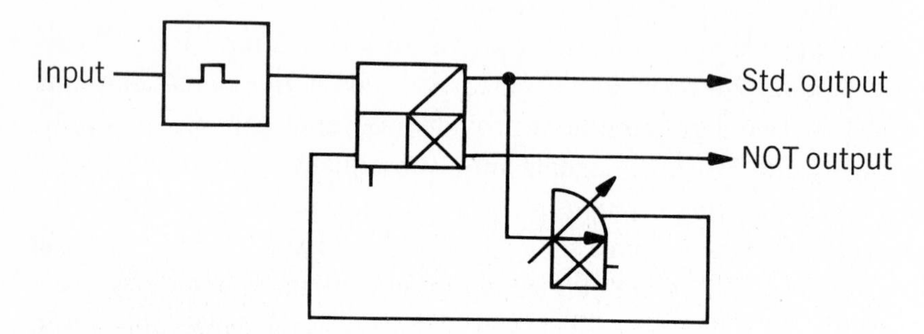

Fig. 11 · 48 Equivalent circuit of adjustable single shot. (General Electric Company)

As system power is applied to the logic panel, the output of this unit is in the OFF condition for approximately 15 msec, and then the unit provides an ON signal continuously at its output, terminal 8. It only provides this delayed ON signal upon application of system power and then is always in the ON state. The unit reset must be used with logic functions containing flip-flop circuits and has a 50-unit driving capability.

The probe test light shown in Fig. 11·49 is the basic test unit for a static control panel which is plugged into a spare socket. Two monitor lights are incorporated within the module, one indicating whether an ON signal or an OFF signal exists at the black test lead.

The black test lead will be used while power is present on the panel and can be employed while the static system is operating without disturbing the operation of the panel.

The red lead is to be used with the terminal marked SIGNAL and is a continuous ON signal to simplify manual insertion of an ON signal into the logic system. It is equivalent to manually

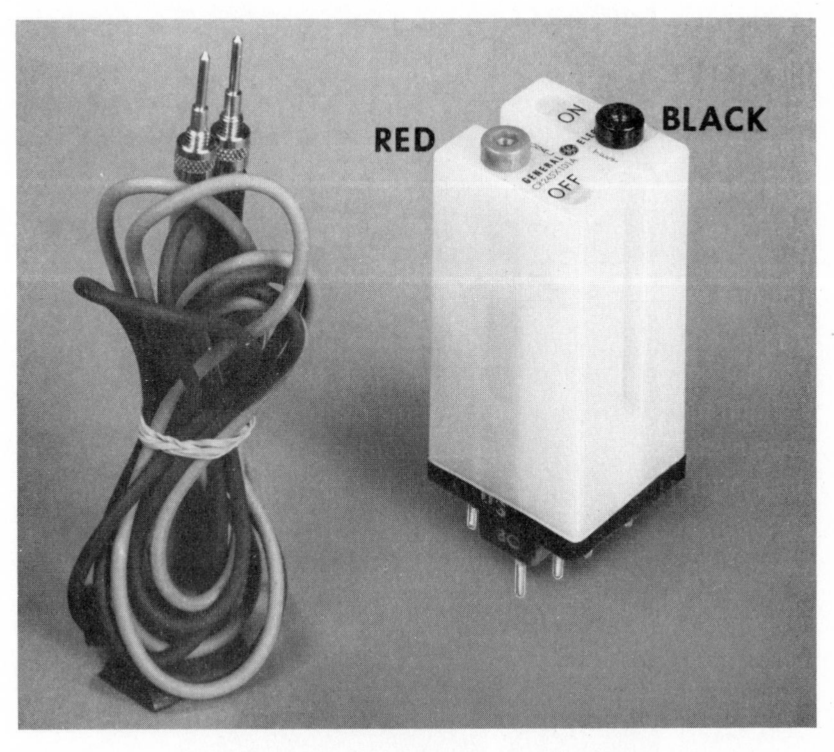

Fig. 11 · 49 Probe test light. (General Electric Company)

operating a relay in a conventional panel and therefore is color coded as red to signify caution in its use. A logic unit can be tested in place by the use of the red lead as an input and the black lead as an output monitor. Each logic input and output signal is accessible in the panel, and the probe test light is a direct, simple, and convenient method to test each unit or the system function. It must be used only on logic-level signals, not on 125 volts d-c, 24 volts d-c, or 115 volts a-c.

By the addition of a temporary jumper wire between terminal 1 and terminal 2, the test lead will act as a memory if an ON signal becomes present and then turns OFF. This is very useful in checking limit-switch continuity into a panel. The test lead is on the logic side of the input, and then the switch

Fig. 11 · 50 Plug-in logic modules. (General Electric Company)

is actuated. The probe light will operate its ON light if an ON signal appears at the test lead and will remain ON. Thus, to check an external switch, the probe test light will monitor the panel while the switch is actuated and then display the results until manually reset. By disconnecting the jumper or touching

(*a*)

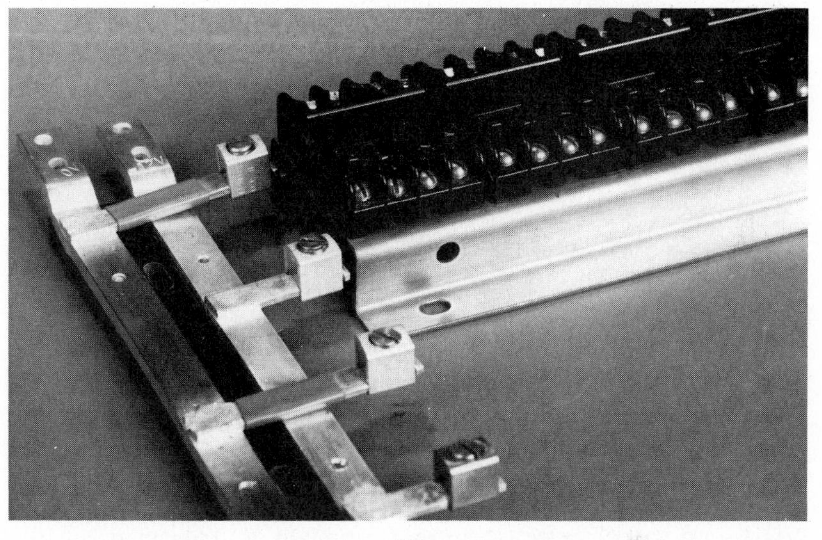

(*b*)

Fig. 11 · 51 (*a*) *rear of base assembly and* (*b*) *power bus bar.* (*General Electric Company*)

the jumper with the red signal lead, one can reset the memory circuit with the ON monitor light.

11·6 ACCESSORY DEVICES

Most logic functions are packaged in a plug-in module. This element is protected by a removable nylon cover, designed to give maximum protection from vibration, shock, and corrosion.

This well-protected and easy to handle element plugs into a base assembly (Fig. 11·50). Each module can be plugged into its socket in only one orientation to assure proper connections and has color-coded nameplates on the top of each cover. The nameplate shows the logic-symbol terminal connection numbers and catalog number.

Sockets for the modules are available on assembled base rails in 5-, 8-, 11-, and 17-unit lengths. Wiring connections to each base are made to screw-type terminals, similar to a terminal

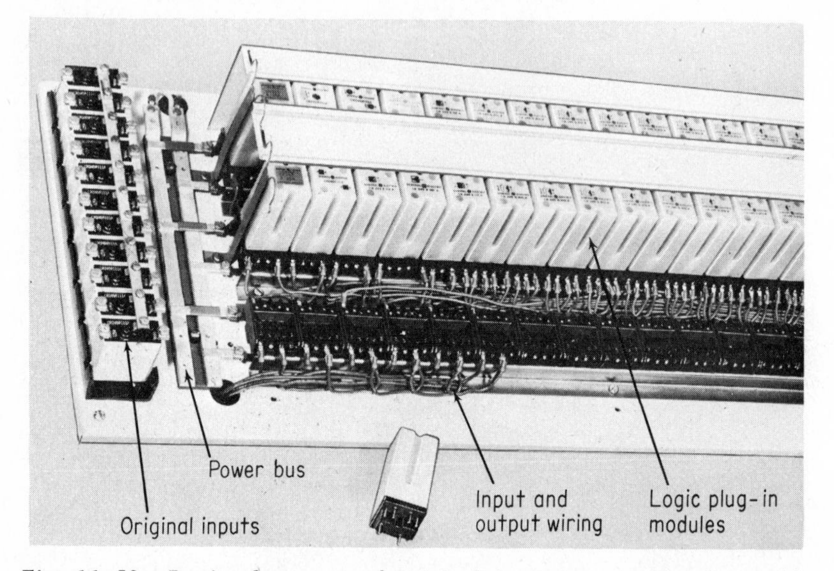

Fig. 11·52 Logic elements and original inputs on panel subassembly. (General Electric Company)

board, and each screw-terminal number is identified in the phenolic molding. Power connections to each socket are made by heavy bus bars as shown in Fig. 11·51, and power terminations are made to the bus bar by sectional terminal blocks.

Top, bottom, and middle retaining covers with associated clips are available to provide an enclosed wiring trough. One set of top and bottom covers would be used in addition to the appropriate number of middle covers to form a neat appearance in a panel (Fig. 11·52).

The mounting-track method of mounting the single 125-volt d-c original input elements allows a few screws to mount 3 ft of track and 34 input points. The elements snap into the track and are retained by two edges on the track that grip the two slots on the element. Spacing is maintained by the bus bar described in the following section.

The standard 3-ft length is slotted to permit easy installation and removal of each element and can be quickly cut to shorter lengths by a hacksaw as desired.

Each zero-volt connection to the 125-volt d-c original input must be made by the recommended bus bar. It has seven holes which can accommodate six inputs plus the connection to a subsequent bus bar. A bus bar is recommended for convenience of wiring and as the simplest way to maintain spacing between inputs and minimize voltage drop in the zero-volt reference.

Summary

The General Electric Company system of solid-state-logic "static control" covered in this chapter is an English logic system providing basic building blocks of AND, OR, NOT, MEMORY, and DELAY. There are several special units available for use in counters and shift registers.

The basic transistor switch is the heart of all logic elements in this system. The patented circuit of the General Electric Company transistor switch is a self-biasing circuit requiring only a single 12-volt supply for operation of the logic element. The General Electric Company system uses a 0-volt d-c signal as an ON signal and a —4-volt d-c signal as an OFF signal.

This system uses the English logic NEMA symbols, with the NOT outputs of MEMORY elements shown by a small NOT symbol as part of the overall basic function symbol. This form of combination symbol indicates that the element has the inverted, or NOT, output available at one of its plug-in pins. This usually reduces the available number of inputs. The basic AND has three inputs, but the AND with NOT output available has only two inputs. This is necessary because the system uses a standard 10-pin plug-in module for packaging.

This system utilizes a plug-in component for all logic elements, and rather conventional wiring techniques are all that are required for field connections. The basic AND, OR, and NOT functions are packaged two in each plug-in unit. Should a component fail in one of these units, all that is required is to replace it by plugging in a new unit without disturbing any wiring. A plug-in test unit is also available and should be included in each panel, as it provides the best possible test equipment for use on this system.

Electrical noise on input signals must be guarded against as in all static systems. General precaution indicates that all input wires should be run together in their own steel conduit and all output wires should be run together in their own steel conduit; logic wires should never be taken out of the control panel; sensor wires (photocells, etc.) should be run with twisted shielded pair; and inductive loads (driven by contact-making devices) mounted near the logic should be suppressed. General Electric Company now has a system with higher noise immunity available called *Series A*. Series A logic elements are so designed that the requirement for running input and output wires in separate steel conduits may be eliminated.

The recommended voltage for use with contact-sensing devices is 125 volts d-c or 125 volts a-c. Power amplifiers are available for static output up to 10 amp at 115 volts a-c.

Review Questions

1. List the special functions other than AND, OR, NOT, MEMORY, and DELAY which are available in the General Electric Company system of components.
2. Draw the proper symbol for each of the special functions.
3. How many voltages would be required for a complete system using direct current on the sensing devices and SCR amplifiers as outputs?
4. What voltage indicates an ON input?

5. What voltage indicates an OFF input?
6. There are two leads on the probe test light, one red and one black. Which lead is used to test for ON and OFF signals?
7. How can the probe test light be used as a MEMORY?
8. What must be used on the 115-volt power lines feeding a static panel?
9. Only two inputs are to be used on a three-input AND element. What must be done with the third input?
10. The logic power-supply voltage is ________ volts d-c.

12
CUTLER-HAMMER, INC. DIRECT STATIC LOGIC

The Cutler-Hammer company system uses the English logic approach which is covered in previous chapters. The purpose of this chapter is to present the details of this system as they apply to and modify material in previous chapters.

The material for this chapter was furnished by the Cutler-Hammer company.[1]

12·1 CUTLER–HAMMER SIGNAL CONVERTERS

We have seen that the function of the signal converter is to convert the higher-voltage pilot signal to a reduced d-c voltage suitable for the input of static logic circuitry. Two basic types of signal converters are commonly used to perform this function; thus a choice of the type of voltage to be applied to the pilot device is available.

[1] SOURCE: Cutler-Hammer, Inc., "DSL Applications Manual," 2d ed., Aug. 1, 1965.

The D-C Resistor Type (Fig. 12·1). This signal converter operates from 48-volts d-c supplied by the power supply. It is suitable for most pilot devices, provided that the contacts have good "wipe" and are subject to frequent use.

The A-C Transformer Type (Fig. 12·2). This type of signal converter utilizes 115 volts a-c across the monitored contacts, and then uses voltage, when present, to produce a 10-volt d-c output signal to the logic. It is recommended, because of its greater energy level, where pilot devices are located in contaminated atmospheres or subject to infrequent use.

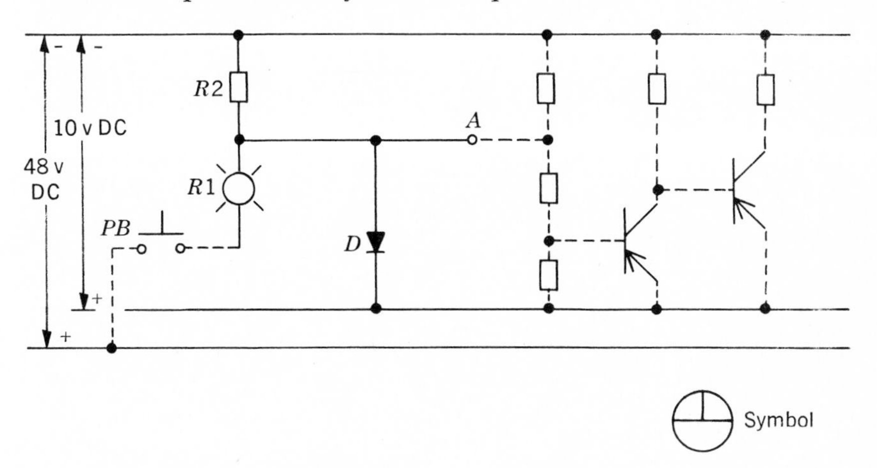

Fig. 12·1 D-c signal converter. (*Cutler-Hammer, Inc.*)

The d-c signal converter with a push-button pilot device and typical static switch dotted in is illustrated in Fig. 12·1. When push button *PB* is closed, current flows through resistors *R*1 and *R*2, which serve as a voltage divider. Note that the lamp functions both as *R*1 and as a state light. The purpose of the resistors is to reduce the 48 volts, in this case to voltage that is compatible with static logic circuitry. This voltage comes from the drop across *R*2 and appears at point *A*. It is, therefore, the input signal to the following logic element. Incidentally, the output of any signal converter can be regarded as a voltage at a given point (*A* in our illustration) which can be sensed by a static switch.

With the push button closed, the voltage that appears across $R2$ must be the same as that impressed across the 10-volt buses. Were this not the case, lost signals could result. Yet the voltage-dividing circuit is designed to present a slightly greater voltage than 10 volts across resistor $R2$. Here is the reason for what may appear an inconsistency. Should the bus voltages drift and cause the signal converter to present a slightly higher than 10-volt signal, the voltage can always be "pulled down" to the desired value. But should the signal voltage go below 10 volts, voltage cannot be added to provide the necessary amount. The clamping diode D performs this function of maintaining a uniform drop across $R2$ by "bleeding off" the excess voltage. The

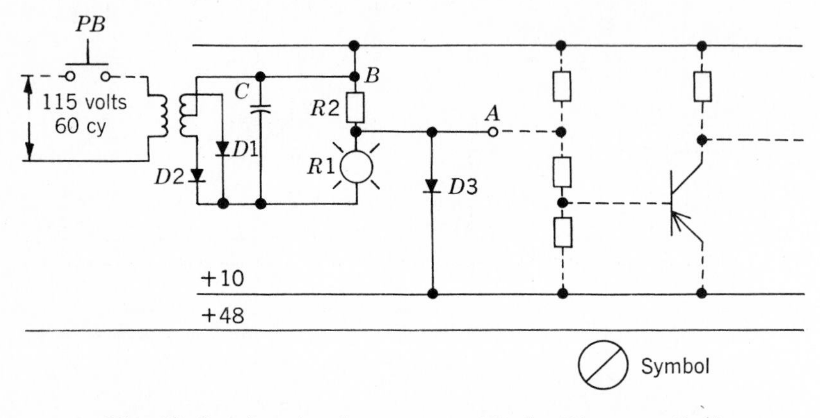

Fig. 12·2 A-c signal converter. (Cutler-Hammer, Inc.)

final result is a signal converter that provides protection from voltage drift.

The a-c transformer-type signal converter (Fig. 12·2) both reduces and rectifies the 115-volt a-c signal voltage to the low d-c voltage required by the logic circuitry. Note that the pilot device, push button *PB*, is in series with the 115-volt primary of the transformer. The latter steps down the voltage. The rectifiers $D1$ and $D2$ plus the transformer produce a full-wave direct current which is filtered by capacitor C. This filtered d-c voltage output is applied to the voltage-dividing resistors $R1$ and $R2$. Diode $D3$, which is connected to the +10-volt bus, passes current whenever the junction of $R1$ and $R2$ attempts to go above the power-supply level. This clamping action keeps the signal level at the same voltage as the power supply fed to the static devices, i.e., 10 volts. It should be noted that when current bleeds off through $D3$, the return path is through the connection at point B. The lamp labeled $R1$ again functions both as a resistor and as a state light.

12·2 STATIC LOGIC ELEMENTS

This section contains the basic building blocks of a control system—namely, the static switches. The elements will be covered

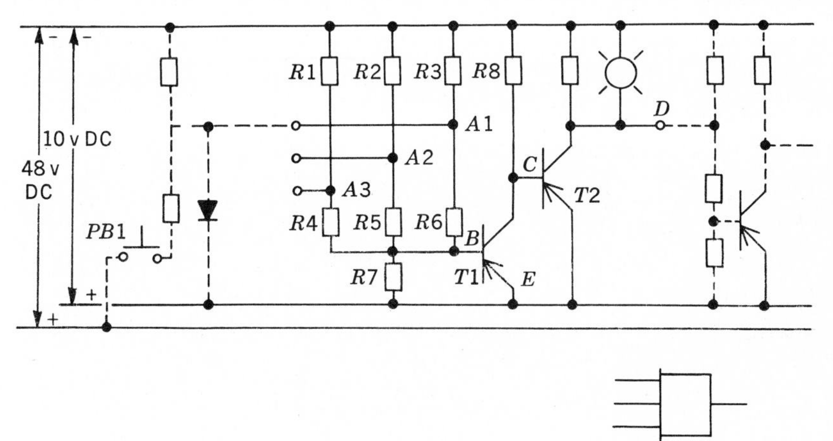

Fig. 12·3 AND *circuit. (Cutler-Hammer, Inc.)*

one by one in detail, and how and why they work the way they do will be explained.

The AND switch is a device which produces an output only when every input is energized. When all three inputs are energized with a +10-volt signal, the output will be 10 volts.

Let us consider the circuit of Fig. 12·3 with all push buttons open ($PB1$ only shown). No input signal can then appear at any one of the input points $A1$, $A2$, and $A3$. This no-input condition permits current to flow through the entire series-parallel voltage-dividing network, consisting of resistors $R1$ to $R7$. Of special significance is resistor $R7$, since the voltage drop across it also appears across the emitter-base circuit of transistor *T*1. Its base, therefore, is held sufficiently negative to cause base current to flow from *E* to *B* in *T*1 and through resistors *R*1 to *R*6. As we have seen earlier, base current causes conduction in the power circuit. Therefore *T*1 conducts, bringing a +10-volt potential to point *C*. With point *C* at + 10 volts, the base path of transistor *T*2 is blocked, causing this transistor to assume its nonconducting or open state. The final result is that point *D*, the output terminal, is connected only to the 0-volt bus and exhibits its no-output condition. When there is no input signal, there is no output signal.

*Push Button PB*1 *Closed, PB*2 and *PB*3 *Open.* With push button *PB*1 closed, one signal input is present; i.e., the signal converter presents +10 volts to point *A*1. This condition may be regarded as blocking the right-hand leg of the voltage-dividing network. But current still flows through resistor *R*7, which still holds point *B* sufficiently negative to permit base current to flow through transistor *T*1 and the remaining two resistor legs. Therefore, transistor *T*1 maintains its conducting or ON state, while transistor *T*2 maintains its nonconducting or OFF state. The "all-or-nothing" concept of the AND switch is demonstrated; that is, one of a given number of inputs to an AND switch cannot, when energized, give an output.

*Push Buttons PB*1 *and PB*2 *Closed, PB*3 *Open.* With the closing of push button *PB*2 (not shown), another input is energized, bringing point *A*2 to +10 volts potential. Now current in the second or middle leg of the voltage-dividing network is blocked. But the voltage drop across *R*7 still holds the base of transistor *T*1 negative. Therefore, base current still flows through *T*1, this time through the remaining leg of the voltage-dividing network. Transistor *T*1 still stays on, and transistor *T*2 still stays off. The all-or-nothing requirement of the AND switch remains intact. Two out of three energized inputs are not enough to turn on the AND switch.

*Push Buttons PB*1, *PB*2, *and PB*3 *Closed.* The closing of push button *PB*3 brings point *A*3 to +10 volts. Now with points

$A1$, $A2$, and $A3$ all at +10 volts, point B too is at 10 volts. Therefore, current can no longer flow from emitter to base of transistor $T1$. The absence of base current, as we have seen, is equivalent to nonconduction in the power circuit. As a result, point C goes negative, permitting $T2$ base current through $T2$ and $R8$. Now transistor $T2$ conducts, turning on the state-indication light and producing a +10-volt output signal at point D. The all-or-nothing concept of the AND switch is realized. With all inputs energized, an output appears.

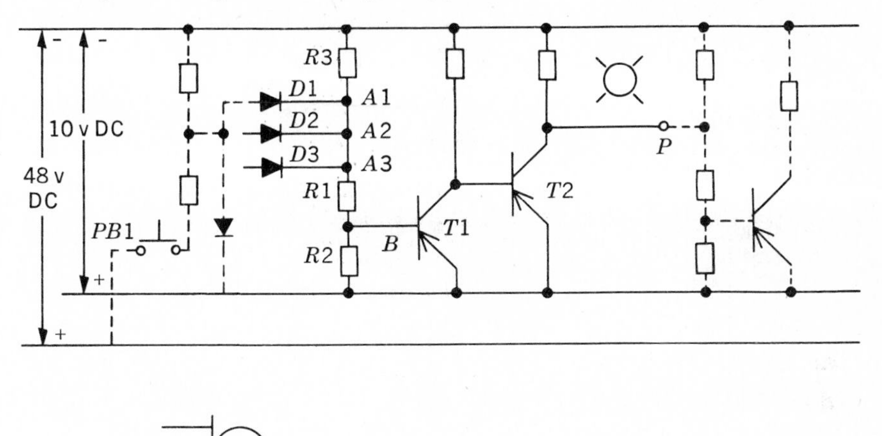

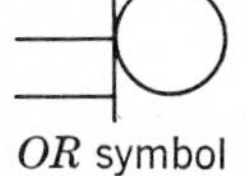

Fig. 12·4 OR *circuit. (Cutler-Hammer, Inc.)*

Should the application not require the full input capabilities of the AND switch, the unused terminals must be connected to the +10-volt bus.

The OR switch (Fig. 12·4) can be likened to relay contacts connected in parallel. Therefore, whenever any one of the three inputs is energized with a +10-volt signal, a +10-volt output will appear. Note that the OR switch diagram is similar to the AND switch, with this exception: the resistors have been replaced with rectifiers.

With push button $PB1$ or either of the other two push buttons closed, the appropriate points $A1$, $A2$, and $A3$ will receive a

+10 volt signal. Therefore, base current cannot flow through transistor *T*1. As a consequence, transistor *T*1 is shut off, and, as we have seen before, transistor *T*2 is turned on. An output appears at point *P*.

The diodes *D*1, *D*2, and *D*3 prevent unwanted feedback. That is, they prevent a feedback circuit from one input to the state-indication light of preceding logic units that may be driving the OR switch. Such feedback could cause false state indication.

The NOT switch (Fig. 12·5) is a device that produces an output only when the input is not energized. But when an input does exist, no output appears. The relay equivalent of a NOT

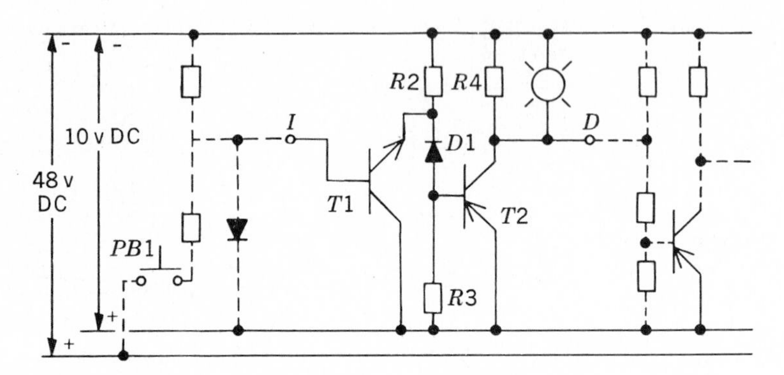

Fig. 12·5 NOT *circuit. (Cutler-Hammer, Inc.)*

switch is a single-pole relay with a normally closed contact and one input signal to the coil. The purpose of the NOT switch is to provide signal inversion where that function is not only desirable but necessary. The NOT switch operates with either an input present or an input absent. The following are descriptions of how the switch works in each of these conditions.

Case 1: Input Present. With push button *PB* closed a +10-volt input signal appears at point *I*. Current will therefore flow from the base to the emitter of transistor *T*1 and through *R*2 to the negative side of the line, causing *T*1 to conduct. Note that transistor *T*1 is an NPN rather than the PNP type. Note also that the arrow points away from the base, indicating that current flows from base to emitter. Now with transistor *T*1 conducting, the junction of diode $D1$ and $R2$ is at $+10$ volts; $T2$ cannot conduct, and there can be no output signal. An input results in no output.

Case 2: Input Absent. With push button *PB* open, no $+10$-volt input signal appears at point *I*. Therefore, the emitter of transistor $T1$ will not be at $+10$ volts. Now current can flow from the emitter of transistor $T2$ to its base, through $D1$ and $R2$. Transistor $T2$ can now conduct and supply a $+10$-volt output. No input results in an output.

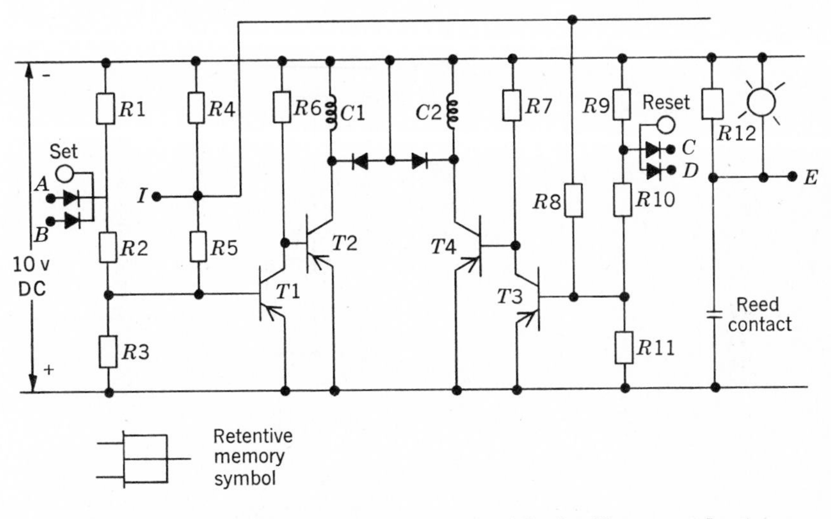

Fig. 12·6 Retentive MEMORY *circuit.* (*Cutler-Hammer, Inc.*)

The primary purpose of the *retentive* MEMORY (Fig. 12·6) is to provide power-loss memory. This is accomplished by a two-coil magnetically latching reed relay which operates similarly to the conventional latched relay. In the *retentive* MEMORY, however, each reed coil is driven by its own logic-performing transistors.

An input signal at either *A* or *B* will block one leg of the AND function associated with transistor $T1$. $+10$ volts applied to the inhibit terminal *I* will block the other leg. $T1$, thus denied a base-current path, shuts off, turning on $T2$. $T2$ energizes the reed-relay coil $C1$, closing the reed contact. Terminal E is the output terminal.

Inputs C and D, in combination with the same inhibit terminal, function similarly through $T3$ and $T4$, except that they cause the reed contact to open. Discharge diodes are provided on both coil circuits.

The set and reset inputs, one associated with A and B, the other with C and D, represent points on the front of the boards. These allow for manual setting of the reed position with a probe

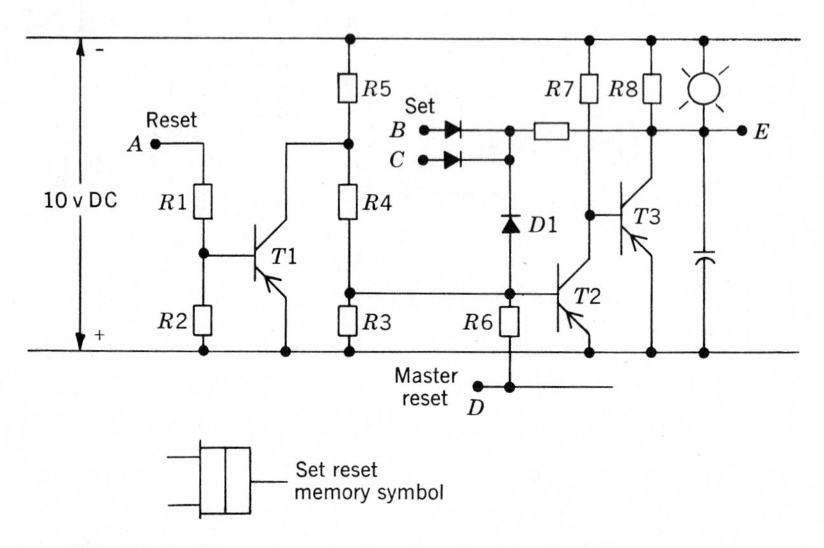

Fig. 12·7 Set-reset MEMORY *circuit. (Cutler-Hammer, Inc.)*

connected to the positive side of the 10-volt supply. Only a momentary signal is necessary to either set or reset the unit.

The inhibit terminal is a master terminal, and as such this one terminal can inhibit the set and reset functions of both switches on the board. If this terminal is not used for logic purposes, it should be tied to either the +10-volt bus or the special reset gate described later.

The *set-reset* MEMORY (Fig. 12·7) performs a simple memory function. A momentary input signal, +10 volts, if applied to the set terminal, produces an output signal; if the input signal is applied to the reset terminal, it shuts the switch off. The master reset terminal, which serves all switches on the board, should be connected to the reset gate circuit described later. This will insure that the *set-reset* MEMORY will assume the no-output condition when power comes on.

If the no-output state is present, transistor *T*2 will be conducting. Its base-current path uses diode *D*1 and resistor *R*8. In the manner we have seen before, the conducting state of *T*2 will block *T*3 base current, holding *T*3 off. *T*1 will also be conducting, with its base-current path through *R*1 and some other logic unit's load resistor. This is essential, as it blocks the *R*4*R*5 base-current path of *T*2 so that terminals *B* or *C* can set the unit.

A setting input calls for either *B* or *C* to be driven to +10 volts by the output signal of some other logic unit. This shuts off *T*2 base current. *T*2 turns off. *T*3 turns on, the output signal appears, and the *T*2 base-current path through *R*8 is blocked. The set signal may now be removed. To reset the unit, terminal *A* is driven to +10 volts. *T*1 shuts off. *T*2 base current appears in *R*4 and *R*5, turning *T*2 on and *T*3 off. *T*2 base current returns to the path *D*1*R*8. The reset signal may now be removed. When the 10-volt power first appears, all the transistors start to turn on. The master reset terminal, held negative by the reset gate, insures that *T*2 has a base-current path and ends up conducting.

The duo-delay timer (Fig. 12·8) is an adjustable, multirange timer. Range selection (0 to 1 minute maximum) is achieved by the addition of external capacitors. A board-mounted potentiometer provides for adjustment. Three modes of operation are possible by varying the external connections to the board. The *E* timer provides time-delay after energization. The *D* timer provides time-delay after deenergization. There is also a combination *E* and *D* timer.

Connecting jumper *A* from the input terminal to the terminal which is associated with diode *D*1 provides an *E* timer. Jumper *B* from the input terminal to the terminal associated with diode

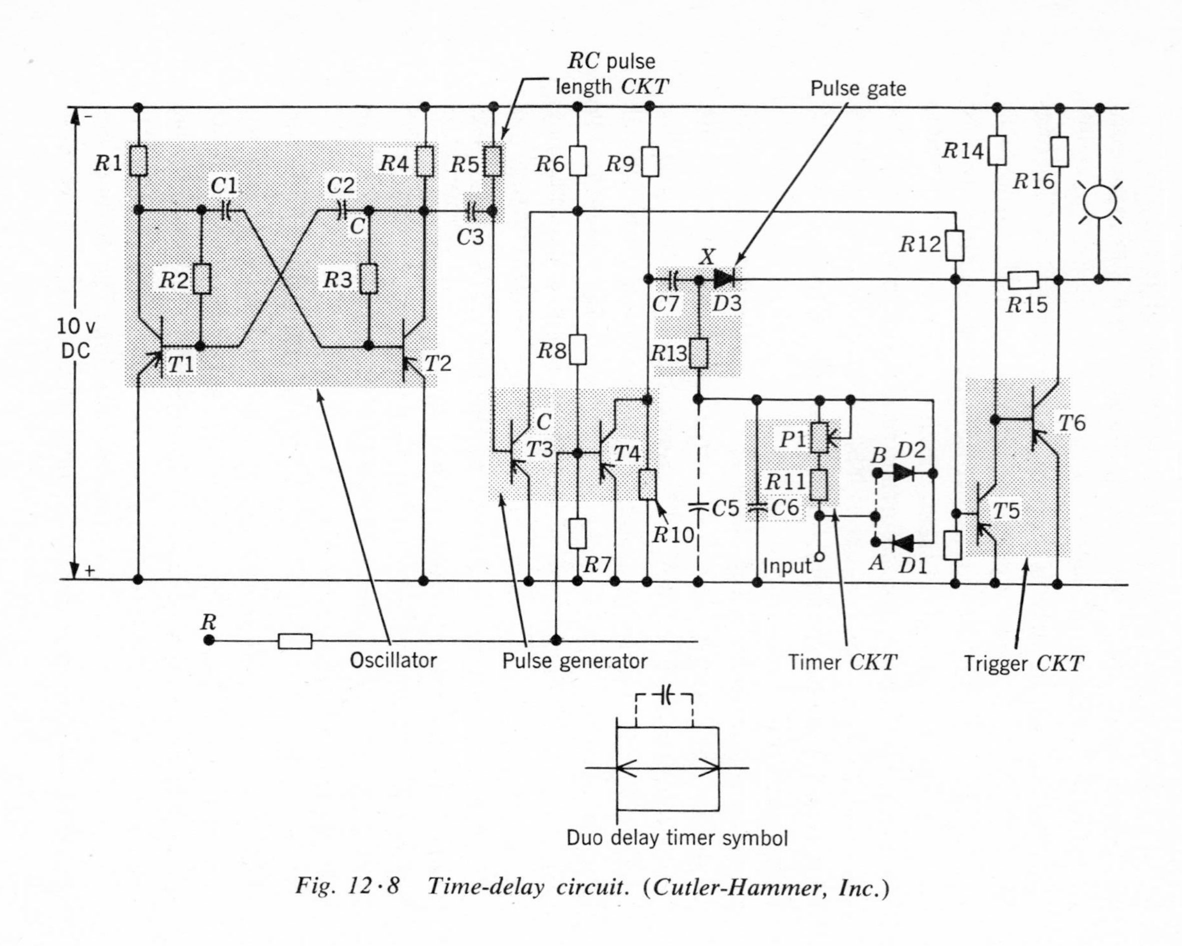

Fig. 12·8 Time-delay circuit. (Cutler-Hammer, Inc.)

$D2$ yields the D timer. Omission of both jumpers produces a combination E and D timer. Only the E timer operation will be discussed.

Transistors $T1$ and $T2$ operate as a multivibrator or oscillator. If one assumes $T2$ is conducting, its base-current path includes capacitor $C1$ and resistor $R1$ to reach the negative side of the line. As we have seen before, the presence of $T2$ base current means that $T2$ assumes the conducting state. This provides a +10-volt output to the RC pulse-length circuit and also holds off $T1$ by blocking its base current at the junction $C2R4$. As $T2$ continues to conduct, its base current causes $C1$ to take on a charge. The charging of $C1$ starts to produce an increasing impedance to $T2$ base current, which starts to shut off $T2$ power circuit. The shutting off of $T2$ causes the voltage at $C2R4$ to drop, which permits $T1$ base current to start turning on the $T1$ power circuit. $T1$ and $T2$ will continue to alternate as long as the 10-volt supply is connected to the board. The output of the oscillator, the +10-volt signal from $T2$, is fed to the RC circuit $C3R5$ where pulse length is established. Transistor $T3$ of the pulse generator is controlled by the pulses from the pulse-length circuit. In a manner we have seen before, $T4$ inverts the signal from $T3$. $T4$ thus produces +10-volt pulses at a frequency set by the oscillator and of a length set by the RC pulse-length circuit. These pulses are sent to both timing circuits on the board. The capacitor $C6$, which may be paralleled externally by $C5$, determines the time range. The variable resistor $P1$ controls the discharge of the capacitor in the E timer.

If one assumes that the input terminal is tied to some other logic circuit, its output load resistor provides a path to the negative bus, and its transistor a path to the positive bus. If we assume no input signal, $C6$ charges. The path is: positive bus, $C6$, $D1$, and through jumper A to the input terminal. The timer is now reset.

When an input signal appears, both sides of $C6$ are connected to the positive bus. $C6$ starts to lose its charge through the input terminal which is externally connected to the positive bus, $R11$, and $P1$. $P1$ controls the rate. As $C6$ discharges, point X is driven more positive.

The next event occurs when $C6$ has sufficiently discharged to permit $C7$ to charge with point X positive. If we assume that $C7$ charges with point X approaching $+10$ volts while $T4$ is not conducting, then when $T4$ starts conducting, the negative side of $C7$ will be driven to $+10$ volts. This causes point X to go to $+20$ volts. $C7$ discharges through diode $D3$, interrupting $T5$ base-current path $R15$ and $R16$. $T5$ shuts off. $T6$ thus starts conducting, providing an output signal and blocking $T5$ base current. The output signal will remain until the input signal is removed. This stops the positive pulses from $C7$ and allows $T5$ base current to use $R12$ and $R6$. $T5$ then turns on, and shuts off $T6$.

The *D* timer operates similarly, except that jumper *B*, hence diode $D2$, controls $C6$ holding discharged when the input is at $+10$ volts. $C6$ is forced to charge through $P1$ after the input signal is removed. This delays removal of the output signal, as $C7$ can only generate pulses when $C6$ is discharged.

Note that terminal *R*, associated with transistor $T4$, when held at zero volts as power is applied, allows the timing capacitor to charge. This prepares the timer for normal operation, and terminal *R* is properly called a master reset terminal. It should be connected to a reset gate as described later.

12·3 CUTLER–HAMMER OUTPUT AMPLIFIERS

The output of static switches is not sufficient to provide energy to power-consuming devices, such as pilot lights, relays, and solenoids. If these devices are to be used, power amplification is required. The various devices used are detailed in this section.

The first output amplifier to be considered is the 10-volt power AND. Referring to the diagram of Fig. 12·9, we note that the portion of the circuit to the left of the broken line is the input side of the three-input AND switch. The change in circuitry which makes this a power AND is found to the right of the broken line. On this logic board, transistor $T2$ is the power type, capable of driving a variety of loads, including two 10-volt pilot lights connected in parallel. Note, too, that the load resistor is absent; the load device itself serves that function. Another change is found in the addition of two diodes. Diode *D*1 provides a discharge path when an induced voltage is created by the deenergization of an inductive load. Diode *D*2 reduces the leakage current through transistor *T*2 to essentially zero during the time that transistor *T*2 is in the nonconducting state.

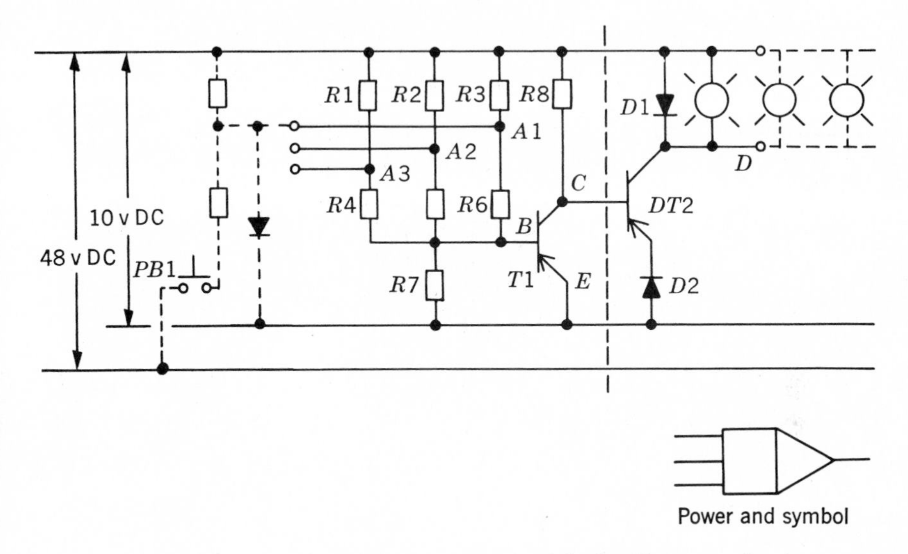

*Fig. 12·9 10-volt power-*AND *circuit. (Cutler-Hammer, Inc.)*

Should the application not require the full input capabilities of the AND switch, the unused terminals must be connected to the +10-volt bus. The maximum capacity of this unit is 5 watts at 10 volts d-c. This switch should not be used to drive other logic switches, as the voltage drop across *D*2 makes operation of the driven switches marginal.

The second output amplifier to be considered is the 24-volt power AND. The circuitry to the left of the broken line (Fig. 12·10) is the same as that of the three-input AND switch. The 24-volt power AND differs, however, in the circuitry to the right of the broken line. Transistor *T*2 is of the power type. Operating at 24 volts, it will drive certain pneumatic solenoid valves. Diode *D*1 provides a discharge path for the inductive load, and *D*2 reduces the leakage current through transistor *T*2 when *T*2 is nonconducting. Resistor *R* drops the 24-volt bus supply to 10 volts to conform with the 10-volt requirement of state-indication lights.

Should the application not require the full input capabilities of the AND switch, the unused terminals must be connected to

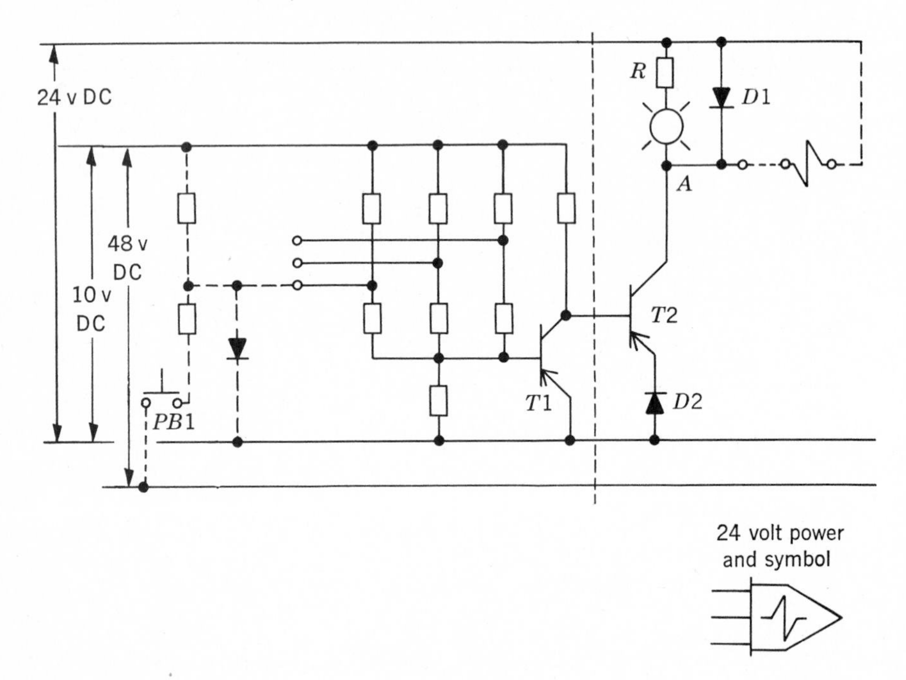

*Fig. 12·10 24-volt power-*AND *circuit.* (*Cutler-Hammer, Inc.*)

the +10-volt bus. The maximum capacity of this unit is 24 watts at 24 volts d-c.

The output relay provides a third type of output amplifier. Where applications require a considerable amplification of both power and voltage, a small panel-mounted relay might be desirable. Two variations are available, differing only in coil voltage. The 10-volt d-c relay utilizes the full capacity of the 10-volt power AND. The 24-volt d-c relay is driven by the 24-volt power

AND. The contact structure and rating is SPDT, 10 amp, and 115 to 230 volts.

The fourth type of output amplifier is the silicon controlled rectifier (SCR) amplifier. Because the SCR requires very little current in its control circuit to cause substantial current conduction in its power circuit, it functions as a very efficient amplifier. It can, therefore, take the output signal of most logic elements and step up the signal sufficiently to drive power contactors. Except for component ratings, the circuit of Fig. 12·11 is the same as those for the 5-amp and 10-amp SCR amplifiers. It

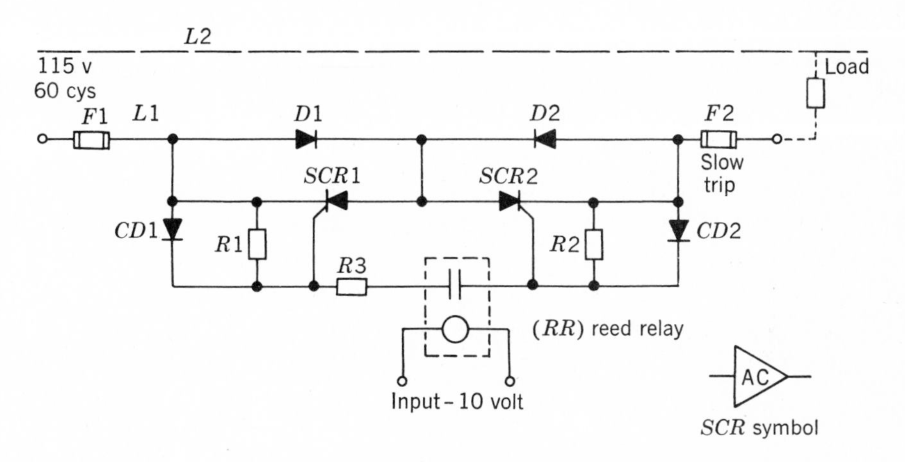

Fig. 12·11 SCR amplifier circuit. (Cutler-Hammer, Inc.)

should be noted that only the 5-amp size has noise-suppression circuits (not shown in the diagram).

Consider the circuit of Fig. 12·11 with the reed-relay contacts closed by a signal from the driving logic element. If we assume that the first half of the 115-volt ac voltage is the positive half, current will flow in the power circuit as follows: through fuse *F*1, diode *D*1, SCR-2, and fuse *F*2. The current through the control circuit for this half-cycle will flow as follows: through *F*1, diode *CD*1, resistor *R*3, and closed reed-relay contacts *RR*, *R*2, and *F*2. The small voltage drop across resistor *R*2 is applied to the gate of the SCR. As a result, a tiny current flows from gate to cathode, causing the SCR to "fire" or conduct in its power circuit.

On the other half-cycle, the current reverses and the power circuit now takes this path: through fuse *F*2, diode *D*2, SCR-1, and fuse *F*1. The control circuit to turn on SCR-1 takes this path: through fuse *F*2, diode *CD*2, reed-relay closed contacts *RR*, *R*3, and *R*1, and fuse *F*1. Now resistor *R*1 provides the gate voltage to trigger SCR-1 to conduct in its power circuit. The purpose of the slow and fast fuses is to make full use of the SCRs while still adequately protecting them.

12·4 MUSTS FOR CUTLER–HAMMER SYSTEMS

This section deals with reset circuits, inhibit circuits, undervoltage protection, and fan-out limitations.

Reset circuits require special attention in transistor circuits. Reset is considered to mean returning a switch to a predetermined state, i.e., resetting a *set-reset* MEMORY to the no-output condition. The case, however, of resetting a switch in the normal course of a functional sequence is considered routine and will not be covered here. A second case, unique to transistors, will be discussed.

When the 10-volt power is first turned on, all transistors in the system start turning on. In elements such as AND, OR, and NOT, the input signals originating with limit switches and push buttons insure that the proper transistor ends up conducting. Memory-type circuits, those that are designed to pick up on momentary signals and then latch up, obviously cannot have the self-correcting type of input configurations. In these elements, with the exception of the *retentive* MEMORY, a resetting action is necessary so that their state is predictable. Where a problem of this type exists, boards are designed with a master reset terminal serving all elements on the board. Such a terminal should be held negative to reset and then positive to permit normal operation.

The reset gate (Fig. 12·12) is a circuit that will, when combined with a NOT switch and upon the application of power, automatically cause the NOT switch to generate the signal required by master reset terminals. It is available as a component for mounting on the rear of the bucket. When the 10 volts d-c first appears, the capacitor starts charging through the base of transistor $T1$. Thus $T1$, as an NPN, conducts, shutting off $T2$.

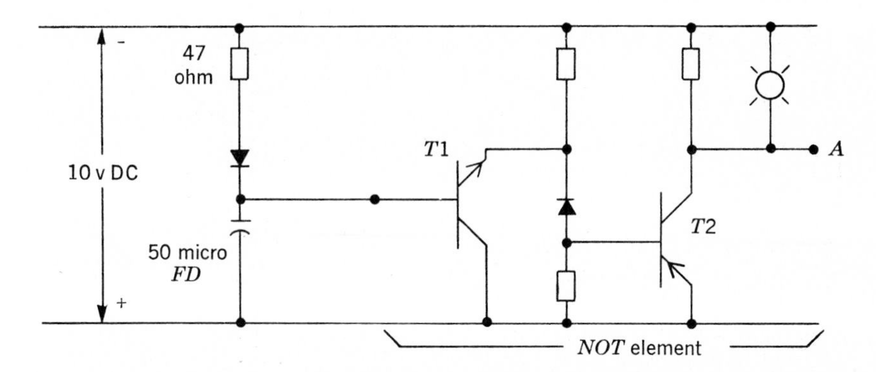

Fig. 12·12 Reset gate circuit. (Cutler-Hammer, Inc.)

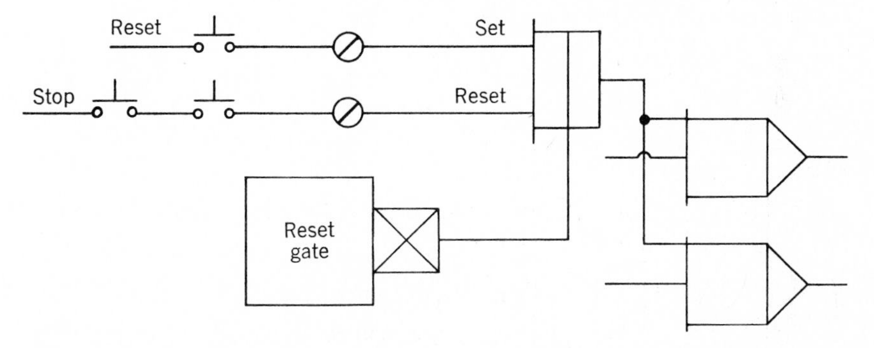

Fig. 12·13 Undervoltage protection using reset gate. (Cutler-Hammer, Inc.)

As the capacitor approaches full charge, it no longer passes $T1$ base current. $T1$ stops conducting, and the delayed +10-volt output signal appears at A.

The 47-ohm resistor and diode provide a discharge path for the capacitor that allows the gate to reset when the 10-volt power is off. The reset gate plus a *set-reset* MEMORY may be used to provide undervoltage protection (Fig. 12·13). When the power comes on, the reset gate holds the master reset terminal of the *set-reset* MEMORY negative. This insures that the set-reset assumes the no-output condition. It further will hold it in the no-output state, regardless of the position of the RESET button, until the the reset-gate output signal appears. After the reset-gate output signal appears, pressing the RESET button will turn on the *set-reset* MEMORY's output signal. Either the STOP button or loss of the 10-volt supply will shut it off.

Inhibit circuits are sometimes required for gating purposes. The inhibit terminal is not a master reset terminal, nor does

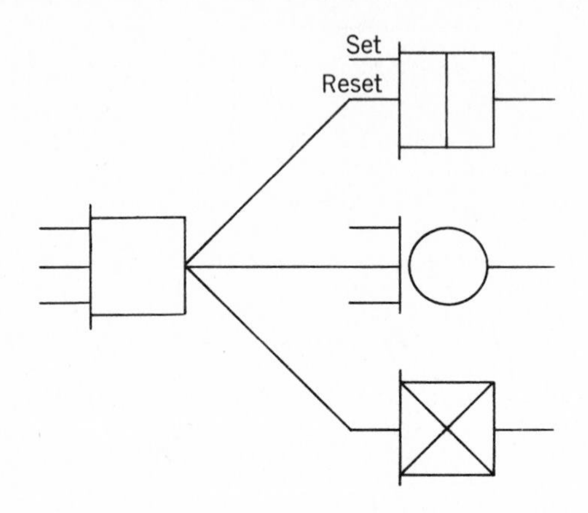

Fig. 12·14 Fan-out from an AND *element.* (*Cutler-Hammer, Inc.*)

it deal with the "power-on" problem. Inhibit functions are strictly for logic purposes and are found only on memory-type switches. The function they perform is to prevent changing state, i.e., from having an output signal to not having one, or vice versa. The general configuration of these circuits is such that the inhibit terminal must be at +10 volts in order to permit change. If not required, these circuits should be connected to the +10-volt bus.

Fan-out limitations must always be considered in the system. The fan-out capability of a given switch is its ability to serve as an input signal to a plurality of other static switches. The term comes from the fact that the signal wires fan out from the output terminal of the given switch to a plurality of input terminals. Since the capacity of a transistor is limited, it is only reasonable to assume that the ability of a transistor switch to drive other switches is limited. In fact, two limits exist. One is the ability of the driving switch to provide current. The second relates to controlling the voltage at the input terminal when no output signal exists.

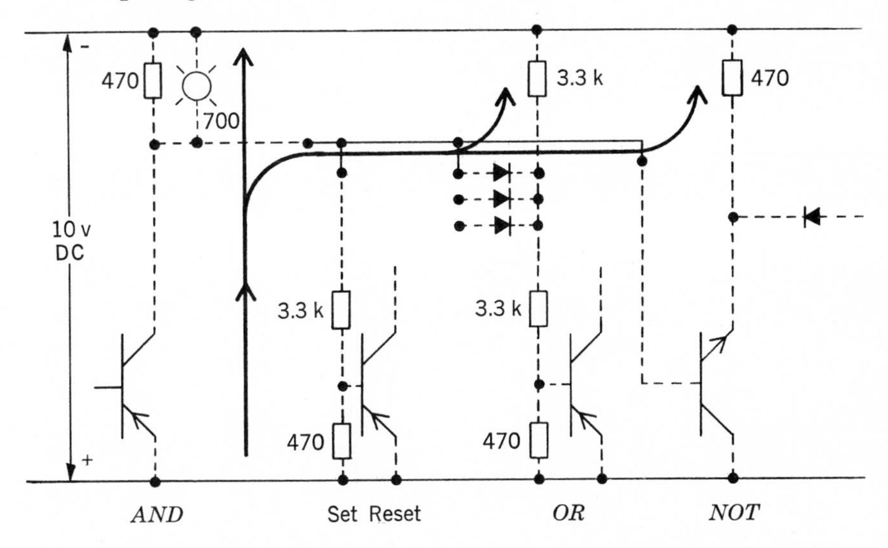

Fig. 12·15 Parallel current paths due to fan-out. (Cutler-Hammer, Inc.)

Let us look at the two cases one at a time, using the configuration of Fig. 12·14. For our purposes the AND is supposed to drive the *set-reset* MEMORY, the OR, and the NOT. As we take each case, this circuit will be redrawn to show the critical portion of each element. For the first case we will assume that the AND switch has an output signal. This means that the transistor shown is conducting. In the circuit of Fig. 12·15, notice how each additional element added increases the parallel paths. Obviously this increases the current that the driving transistor must carry. The maximum collector current recommended is 80 ma. For purposes of calculation the input resistance of the

NOT switch should be considered as 10 kilohms. This is due to the emitter-follower configuration.

The circuit of Fig. 12·16 applies to case 2. The problem is simply that the various input configurations form a voltage-divider network. This network centers about the output terminal of the driving element. When the driving transistor is not conducting, this network sets the actual voltage at the output terminal. It is recommended that the voltage at this point not be allowed to exceed +5 volts. For purposes of calculations the forward

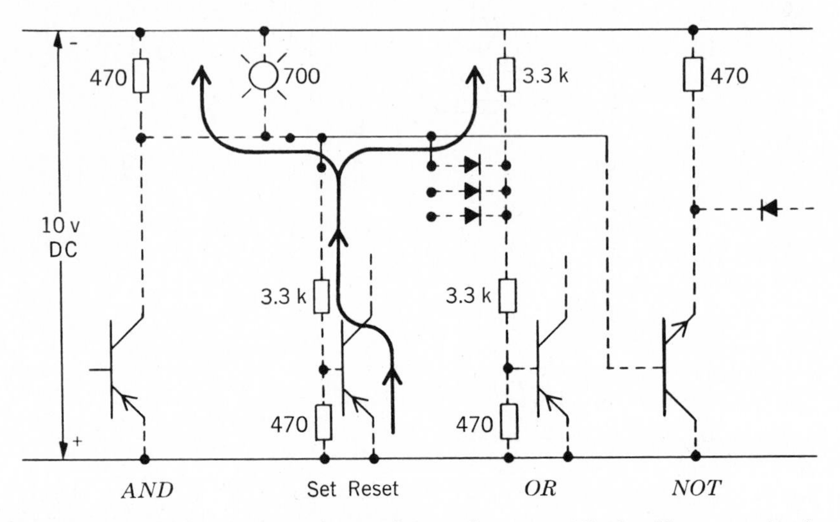

Fig. 12·16 Parallel impedance due to fan-out. (Cutler-Hammer, Inc.)

resistance of transistors and diodes should be neglected. It should be noted that the OR unit does not act as a current source in this configuration, because of its input diodes.

12·5 CUTLER–HAMMER POWER SUPPLIES

Depending upon the components utilized, one, two, or three different voltages may be used in a static-switching control. When three are used they consist of

10 volts d-c for all logic plus certain output amplifiers
24 volts d-c for 24-volt power AND
48 volts d-c for d-c signal converters

Obviously the exclusive use of a-c signal converters in an application eliminates the need for the 48-volt supply, just as proper selection of output amplifiers can avoid the requirement for 24 volts d-c. This discussion will deal with the most complete case—namely, all three voltages.

Regardless of whether one or three power supplies are used to achieve these voltages, they will be related as shown in Fig. 12·17. The reasons for this selection are: first, the logic supply is set at 10 volts d-c, a desirable voltage for transistor circuits. Second, since a greater voltage and power level than the 10 volts can supply is desired across most contacts, a 48-volt signal-

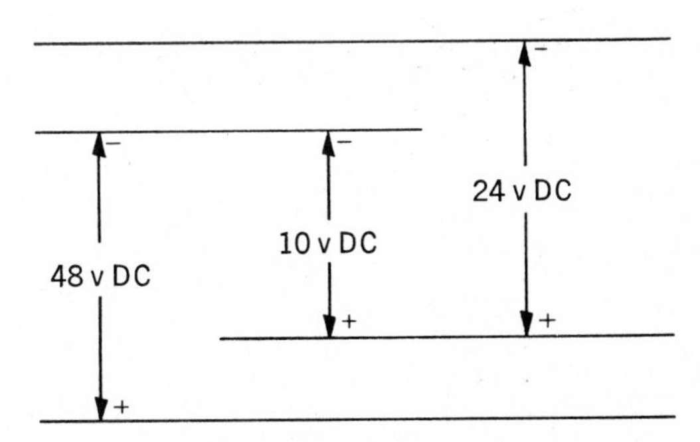

Fig. 12·17 Power-supply relationships. (Cutler-Hammer, Inc.)

converter voltage is used. The 48- and 10-volt supplies have their negative sides made common, as this configuration permits the use of a clamping diode in the signal converter. Third, the 24-volt supply provides an output voltage and power level that is compatible with certain solenoid values, typewriter solenoids, and other devices. Making the positive side of the line common provides the simplest internal circuit for the board which must serve as an interface between two dissimilar voltages.

The proper procedure for selection of power supplies is detailed below. The proper starting place is the 48-volt d-c signal-converter supply. The first step is to determine the maximum number of d-c converters that can be conducting at the same time. The second step is to calculate the minimum required capacity based on supplying 45 ma to each converter and to select an adequate power supply. Guard against excessive overcapacity. Should it be desired to parallel 48-volt d-c supplies, derate them to 85 percent capacity.

Selection of the proper 10-volt power supply requires eight steps. Step 1: Determine both the total number of 10-volt power-AND boards and the maximum number of 10-volt power-AND output signals that can exist at one time. Step 2: Calculate the required capability based on supplying 0.5 amp to each of the maximum number of 10-volt power-AND output signals. Step 3: Determine the number of logic boards, including both the 10- and 24-volt power ANDs. Step 4: Calculate the power requirements for the above logic boards based on 150 ma per board. Step 5: Determine the total 10-volt power requirements by adding the totals from step 2 and step 4. Step 6: Compare the total 10-volt d-c ampere load from step 5 with the ampere capacity of the 48-volt d-c supply at 48 volts d-c. If the 10-volt load does not equal or exceed the amperage available at 48 volts d-c, select a shunting resistor or increase the 10-volt load. Note that since it is possible to get accidental grounds and since the transistors could be destroyed if the positive side of the 48-volt supply gets in contact with the 10-volt bus, the 48-volt supply is designed to collapse under the above loading. Step 7: Based on the 10-volt d-c load requirements determined in step 6, select a power-supply rating. (NOTE: Should it be desired to parallel 10-volt d-c supplies, derate them to 85 percent capacity.) Step 8: Unless the 10-volt d-c load equals approximately 75 percent of the power-supply capacity, reduce the ohmic value of the shunting resistor.

Selection of the proper 24-volt power supply requires four steps. Step 1: Determine the maximum number of 24-volt output signals that are present at one time. Step 2: Calculate the 24-volt load based on 1 amp per signal. Step 3: Select an adequate power supply. (NOTE: Should it be desired to parallel 24-volt power supplies, derate them to 85 percent capacity.) Step 4: The 24-volt power supply should be loaded to approximately

100 percent capacity when the maximum number of inputs are on. Shunting resistors should be used to achieve this state.

Due to the interface between voltages accomplished by the 24-volt power AND and the signal converter, it is important that the 10 volts d-c is first on and last off. To insure this in power-loss situations, two relays are required; these are connected as shown in Fig. 12·18.

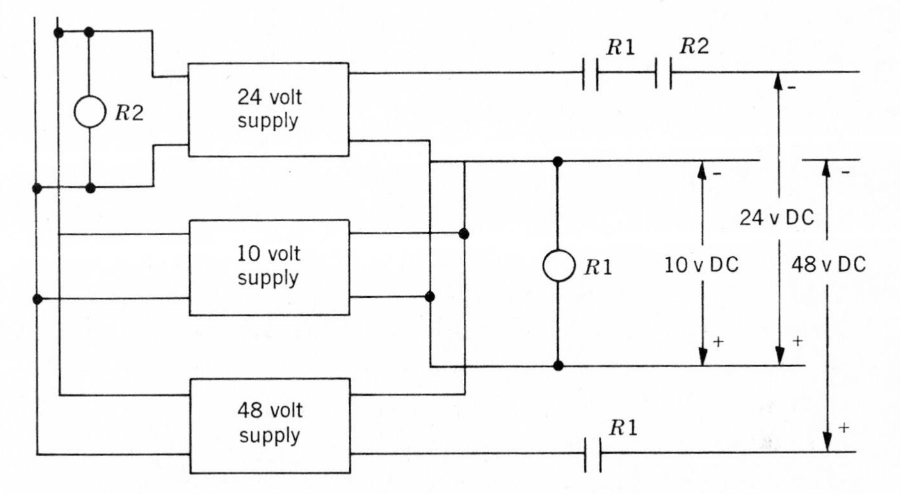

Fig. 12·18 Circuit for power-application relays. (Cutler-Hammer, Inc.)

12·6 MOUNTING AND WIRING CUTLER–HAMMER DEVICES

All static-switching logic elements are mounted on boards such as those shown in Fig. 12·19. Each board contains several switching elements, and each element is an independent switch. All boards are the plug-in type, and since all boards have the same dimensions and conform physically, they can be arranged in buckets to suit circuit convenience. Each circuit point on the receptacle provides for two taper pins connected in parallel to facilitate wiring.

The "bucket" is a sort of "electronic bookcase" for the boards. The ones illustrated in Fig. 12·19 have a capacity of 20 boards. Basically, the bucket consists of a steel framework, 20 receptacles, and the necessary bus work at the rear. This basic configuration is designed to be mounted in a cutout of a steel panel or in a conventional 20-in.-wide relay rack. Because of the separate elements per board, the bucket contains an equivalent of approximately 80 relay circuits.

The standard bucket includes receptacles and two bus bars. In wiring the buckets, it is first necessary to make connection between each plug and its respective *A* or *V* bus bar. A separate

Fig. 12·19 Static panel showing buckets and plug-in logic units. (Cutler-Hammer, Inc.)

wire for each plug is recommended to assure a "solid" voltage supply. The *A* bus bar also serves as the positive side of the line for the 24-volt circuit. The negative side of the 24-volt supply should be brought in by wire to the affected boards. The bus bars on the rear of the bucket should be connected to the voltage source with No. 12 wire. The recommended wire size for logic wiring is No. 22. These wires should be terminated with taper pins for plug insertion and spade terminals for connection to bus bar.

SCR amplifiers are available in two sizes. Components for each size are mounted on a fin (heat sink) which is painted red as a warning that the fin is electrically "hot," i.e., carries 115 volts a-c. The insulating standoffs are first mounted to the steel panel with two screws each. The fin is then attached to the standoffs by means of two keyhole slots. A terminal strip mounted at the base of the fin provides for connections to the +10-volt control voltage and the 115 volts a-c. Again, all connections to the device are provided by screw terminals.

Assembling a DSL system involves two types of wiring, that applied to panel-mounted devices and that used on the rear of the buckets. The panel-mounted devices all have screw terminals and may be used with normal control wiring techniques. The rear of the bucket, however, calls for different wiring techniques. The bus bars are drilled and tapped for 6-32 screws, and spade terminations are suggested here. The 20 receptacles each have provision for taper pins in the form of two rows of 18 sockets, so that two wires may be connected to each point. The pins used are A-MP No. 41663. The recommended insertion tool is A-MP No. 380-310-F-395-005; the crimping tool is A-MP No. 48698.

12·7 THE CUTLER–HAMMER CODING AND DIAGRAM SYSTEM

Symbology and identification coding varies in the electrical control industry. The Cutler-Hammer company has its own system, one that has proved very successful. Its use is suggested since it is built around the marking already present on the boards. This section is devoted to the interpretation of the symbols that appear on Cutler-Hammer company diagrams and the marks that are printed on Cutler-Hammer company logic boards.

Look at the front of a Cutler-Hammer company logic board and you will see several markings (Fig. 12·20). At the top appears a symbol that identifies the function of the board—an AND function in this case. At the bottom appears a function number which identifies a specific circuit. Either the function number or the part number can be used to obtain an exact duplicate board. Boards with the same function number can be used interchangeably.

The familiar convention of "from left to right" is used on Cutler-Hammer company diagrams. That is, inputs are on the left and outputs are on the right. Because more than one element is on a board, identification of terminals is necessary. The letters *A* to *V*, with omissions, identify the specific terminals on the plug of the board. The *A* and *V* terminals of all boards are used for the 10-volt power supply. The symbol for one ele-

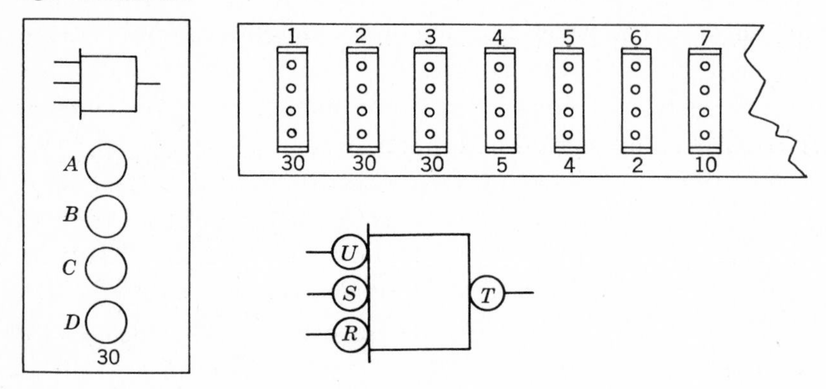

Fig. 12·20 Cutler-Hammer company symbology and identification. (Cutler-Hammer, Inc.)

ment of an AND board is shown in Fig. 12·20 just as it would appear on a Cutler-Hammer company diagram. The letters inside the small circles identify the plug markings. For example, the circled *T* in the sketch indicates that this termination goes to the *T* terminal on the plug. The receptacle on the bucket has markings identical to those that appear on the rear of the plug.

Speedy identification of state-indication lights on logic boards makes for fast troubleshooting. As can be seen on the sketch of Fig. 12·20, lights for each element are identified by the letters *A*, *B*, *C*, and *D*. The fronts of all buckets have slots numbered left to right, at the top. Slots have corresponding numbers on the rear of buckets, at the top. Note, too, the numbers at the bottom of the panel. If the bucket is marked in this manner, when the control is wired, a speedy eye check instantly reveals any board that is not in its proper slot. Though the buckets are not physically identified, they are nevertheless always arranged in an alphabetical sequence from left to right, viz., *A*, *B*, *C*, *D*, etc.

Figure 12·21 is typical of an AND symbol that might appear on a Cutler-Hammer company logic diagram. A simple "reading" of the code is all that is required to systematically trace circuitry and determine the physical location of the portion of control under test. Here, for example, is what the code illustrated above reveals: The AND element can be found in panel 3, bucket

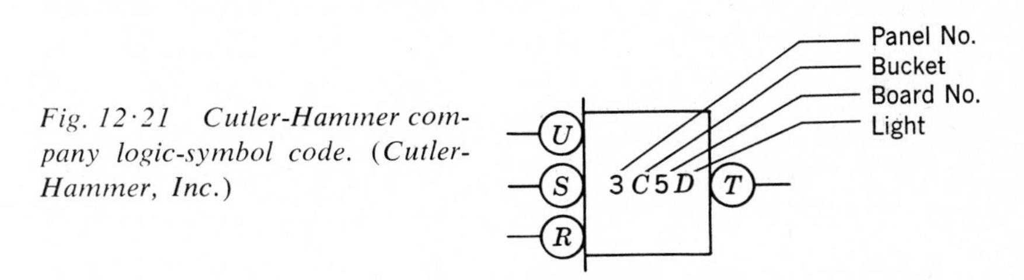

Fig. 12·21 Cutler-Hammer company logic-symbol code. (Cutler-Hammer, Inc.)

C, board 5, and its state-indication light is identified by the letter *D*. With such complete information, the logic diagram is the only "tool" required to effect fast testing for speedy repair or replacement.

Summary

The Cutler-Hammer company static system is an English logic system which is applied by means of plug-in logic elements. The logic elements are packaged four to a board and have individual indication lights mounted on the front of the board. Standard mounting provides for 20 boards in each bucket. Connections are made by means of taper pins which must be installed with the use of a special insertion tool.

The heart of this system is the basic transistor switch, operating from a 10-volt d-c power supply. The ON signal for this system is +10 volts d-c, and the OFF signal is 0 volts d-c. Recommended voltages for contact-type sensing devices are 48 volts d-c or 115 volts a-c.

This system uses the English logic symbols. When the company makes the control-circuit drawing for a factory-wired control system, the symbols are coded with numbers to aid in servicing the equipment.

Fan-out limitations must be calculated on a basis of a maximum of 80 ma load and a minimum output voltage of 5 volts.

Normal installations will require at least two power supplies, one 10-volt supply for logic elements and one 48-volt supply for signal converters. A third supply of 24 volts will be needed if any 24-volt power ANDs are used. Loading for these power supplies should be carefully figured as part of the reliance and safety factors in this system.

State-indication lights are a standard feature of every input and logic element in this system and provide visual checking of any part of the circuit, without other equipment, for normal troubles.

Review Questions

1. What voltage indicates an ON signal?
2. What voltage indicates an OFF signal?
3. The logic power-supply voltage is ________ volts d-c.
4. What must be done with unused inputs to an AND element?
5. What is the function of the diode in the signal converter?
6. This system offers static power amplifiers up to ________ amp, 115 volts.

13

SQUARE D COMPANY ®NORPAK CONTROL SYSTEM

The two previous systems covered in earlier chapters use English logic. The NORPAK system is built around NOR logic and requires somewhat different diagrams. This chapter will take up the details of the NORPAK system. The material for this chapter was furnished by the Square D Company.[1]

13·1 THEORY OF OPERATION

The heart of the NOR logic element is the transistor (Fig. 13·1), which is ideally suited for a logic unit because of its reasonable cost, small size, low power consumption, speed of operation, and excellent performance as a switch. A transistor is a crystalline material that exhibits properties of an insulator in one state and a conductor in another state. Therefore, it behaves as an

[1] SOURCE: Square D Co. bulletins M-292-1, M-295C, M-212-1, M-275A, M-276, M-277, M-278-B, M-286-1, M-285, M288-2.

open contact in the first state and as a closed contact in the second state. Transistor operation is similar to a vacuum tube in that the base controls the current flow between the emitter and collector in much the same manner as the grid controls the current between the anode and cathode. In a PNP-type transistor, because of the inherent properties of the material at the junction of the emitter and the base, a negative voltage on the base allows emitter-base current to flow, while a positive voltage on the base prevents emitter-base current from flowing. Current in the plate circuit of the vacuum tube is controlled by the grid-

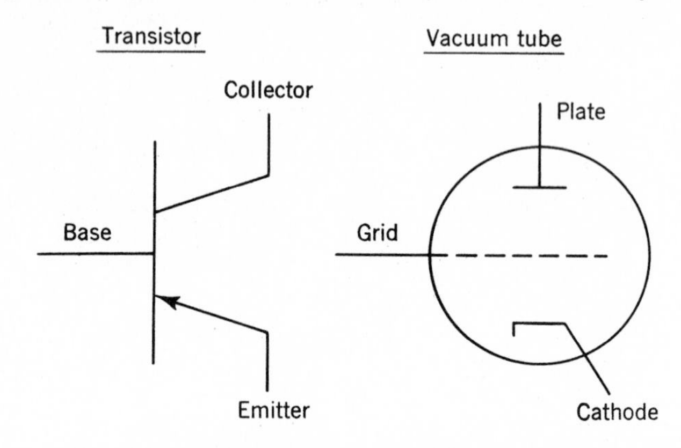

Fig. 13·1 Transistor symbol. (Square D Co.)

cathode voltage; similarly, current in the collector circuit of the transistor is controlled by the presence of current flow in the emitter-base circuit. These conditions result in static switching because no moving parts are required to open or close a circuit. Reliability and exceptionally long life are assured by the use of transistors because they are not subject to wear or deterioration.

Figure 13·2 is the circuit of the basic NOR. With no negative input voltage at *A*, at *B*, or at *C*, the base is held to a positive bias by 20 volts through *R*3. In this condition there is no emitter-base current; therefore, the —20 volts is across the load in series with *R*2, and current *IL* flows. This is called NOR logic since we get an output if neither input A, B, nor C is present. The word NOR actually has its derivation from the contraction of a function OR-NOT. That is, if there is an input at *A or B or C*, the function does *not* have an output. When a negative input exists at one or more of the input terminals A, B, or C, the base becomes negative with respect to the emitter, and current flows from the emitter to the base and out through $R1$. Under these conditions, current flows in the collector circuit; the collector becomes, in effect, grounded (acts as a short circuit across the load), and no load current passes.

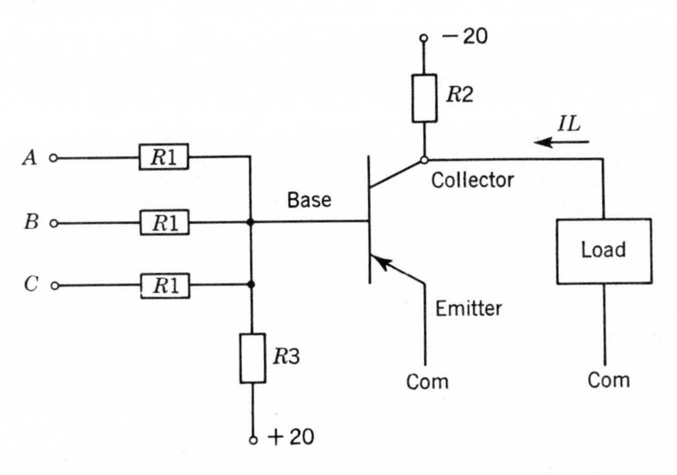

Fig. 13·2 Basic NOR *circuit. (Square D Co.)*

Figure 13·3 shows the logic-symbol equivalent of Fig. 13·2. If we represent the presence of a signal at inputs A, B, or C by 1 and the lack of an input signal by 0, then a 1 at any one or more of the inputs A, B, or C would give a 0 output.

NOR logic elements are quite versatile in that the majority of all logic functions can be accomplished by various combinations of this unit device. In the following illustrations of relay and logic circuitry, a normally open contact is represented by a letter, and a normally closed contact is represented by a letter with a bar above it (A = normally open; $\bar{A}$ = normally closed).

In the AND function (Fig. 13·4) an output is obtained when all of a given number of input signals are applied. In the basic relay circuit both *A* and *B* must be closed to provide an output at *E*. The equivalent NOR logic circuit operates similarly; a signal must be present at both inputs *A* and *B* of their respective NORS

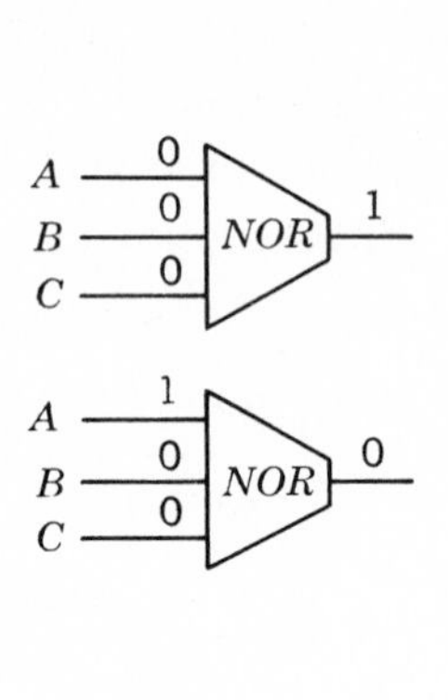

NOR Truth table			
Input			Output
A	*B*	*C*	*E*
0	0	0	1
0	0	1	0
0	1	0	0
1	0	0	0
0	1	1	0
1	1	0	0
1	1	1	0
1	0	1	0

Fig. 13·3 NOR *logic symbol.* (*Square D Co.*)

Basic relay circuit

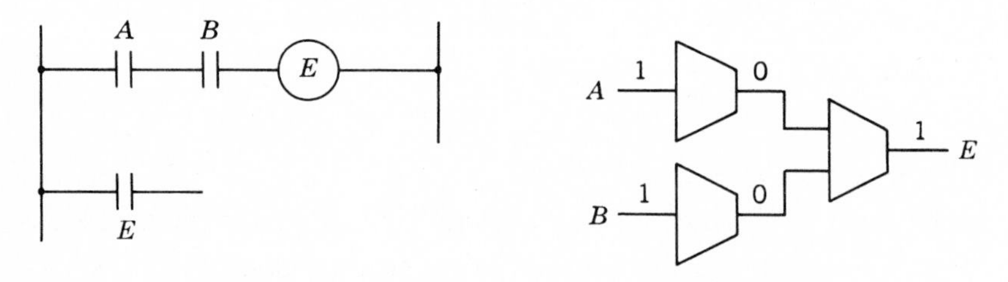

Fig. 13·4 NOR *logic,* AND *function.* (*Square D Co.*)

to provide an output at *E*. With NOR logic, signals at *A* and *B* cause 0 outputs at their respective NORS. These 0 outputs, used as inputs to the third NOR, result in a 1 output at *E*.

The OR function (Fig. 13·5) produces an output when any one of a number of inputs is present. In the basic relay circuit, closure of either *A* or *B* will provide an output at *E*. With NOR logic, a 1 input at *A* or *B* or at *A* and *B* will provide a 0 output which, when used as an input to a second NOR, gives an output of 1 at E. An alternate method of OR logic is provided by diodes. A diode or rectifier is a semiconductor which allows current flow from positive to negative. With reverse polarity the resistance is extremely high, preventing current flow. The above NOR logic circuit can be duplicated by two diodes as shown in Fig. 13·6. This is the Square D company two-input OR unit, usually used in place of two NOR units. The equivalent NEMA logic OR symbol is used to indicate the use of diodes.

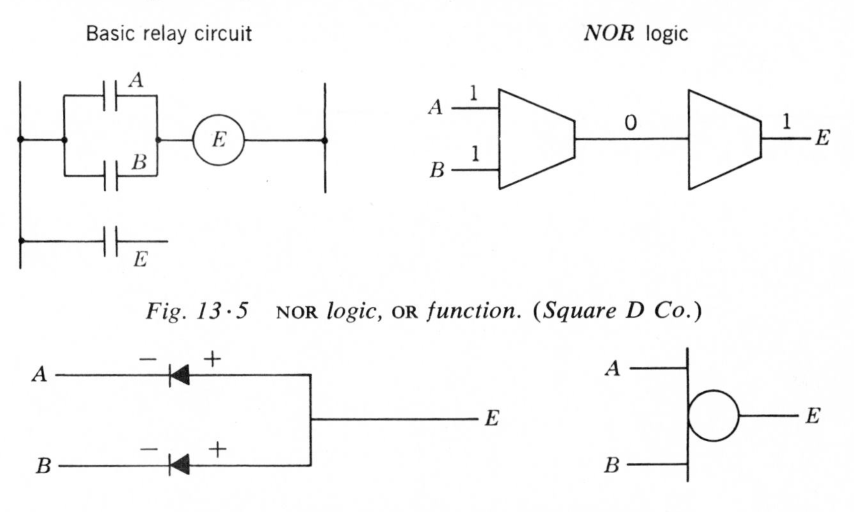

Fig. 13·5 NOR *logic,* OR *function. (Square D Co.)*

Fig. 13·6 Diode, OR *function. (Square D Co.)*

The NOT function (Fig. 13·7) provides an output when no input is present. In the basic relay circuit, an output will be present at E so long as contact A is not closed. With NOR logic, a 0 input at A results in 1 output at $\bar{E}$.

The *off-return* MEMORY function is identical to an electrically held relay. It provides undervoltage (low-voltage) protection. In the relay circuit (Fig. 13·8), momentary closure of push button A will energize coil CR which sets up its own holding circuit by closing the normally open $CR1$ contact, enabling the operator to release push button A. Energy is thus maintained to coil CR. Simultaneously, contact $CR2$ closes and $CR3$ opens. Depressing the normally closed push button $\bar{B}$ (or power failure) deenergizes

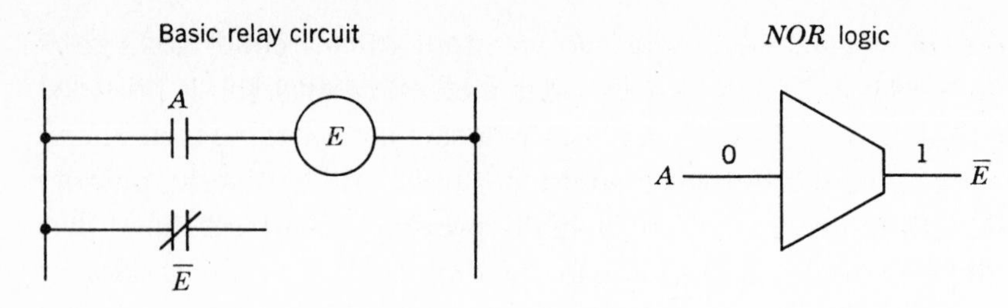

Fig. 13·7 NOR *logic,* NOT *function.* (*Square D Co.*)

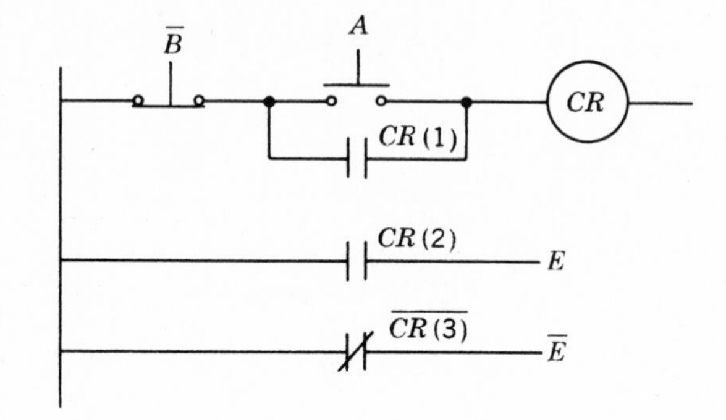

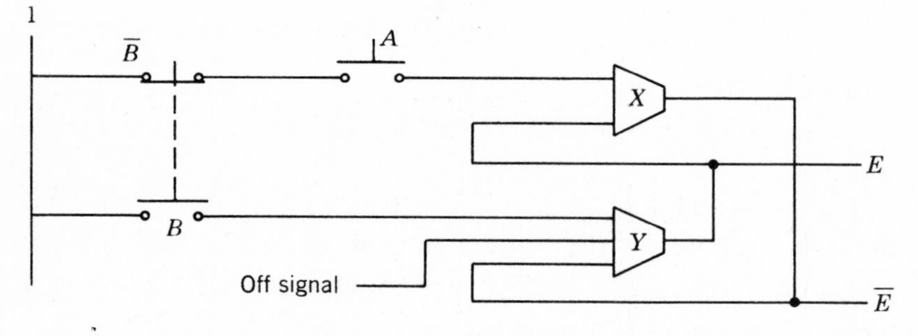

Fig. 13·8 NOR *logic,* off-return MEMORY. (*Square D Co.*)

coil *CR*, allowing *CR*1 and *CR*2 to open and $\overline{CR3}$ to close. If we use NOR logic, momentary closure of push button *A* provides a 1 input to NOR *X*, thus giving a 0 output. The 0 output is used as one of the inputs to NOR *Y*. With the off-return signal absent push button *B* also delivering a 0 signal, a 1 output is assured from NOR *Y*. This 1 output signal provides a continuous output at

E and is also used as an interlocking feedback signal to NOR X so that push button A can be released. Depressing push button $\bar{B}$ changes the MEMORY to the opposite state through contact B. When power fails and is restored, an off-return pulse is provided by a power supply to assure return to the off state (i.e., a short 1 pulse is applied to NOR Y).

The *retentive* MEMORY function is identical to a mechanically latched relay (Fig. 13·9). Momentary energization of magnet coil CRL by closure of contact A closes normally open CR and opens normally closed $\overline{CR}$. This condition is maintained by means of a mechanical latch. Momentarily energizing a second coil CRU by closure of contact B releases the mechanical latch,

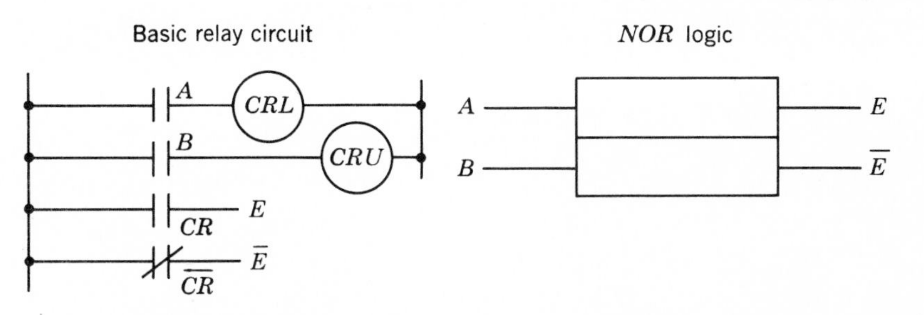

Fig. 13·9 NOR *logic,* retentive MEMORY. (*Square D Co.*)

and gravity or spring action opens normally open CR and closes normally closed $\overline{CR}$. A NOR MEMORY used in conjunction with a magnetic core circuit provides the logic equivalent. The entire function is symbolized by the rectangle. A momentary input of 1 at A assures ON output E. Similarly, a 1 input at B results in OFF output $\bar{E}$. The *retentive* MEMORY retains the condition of the output corresponding to the input last energized. In case of simultaneous inputs to A and B, the MEMORY delivers an overriding OFF signal (i.e., a 1 will appear at $\bar{E}$).

Quite often it is necessary to have an output (or the removal of an output) some time after an input signal is applied. Conventional timers consist of two types—time delay after energization (*on* DELAY) and time delay after deenergization (*off* DELAY). TDE timers may have normally open contacts timing closed (NOTC) or normally closed contacts timing open (NCTO). TDD timers may have normally open contacts timing open (NOTO) or normally closed contacts timing closed (NCTC). The NOR logic equivalent of the TDE version (Fig. 13·10) is such that output E is obtained some time after a 1 input signal is applied at A. To accomplish the TDD timing function, a NOR is merely inserted between the input

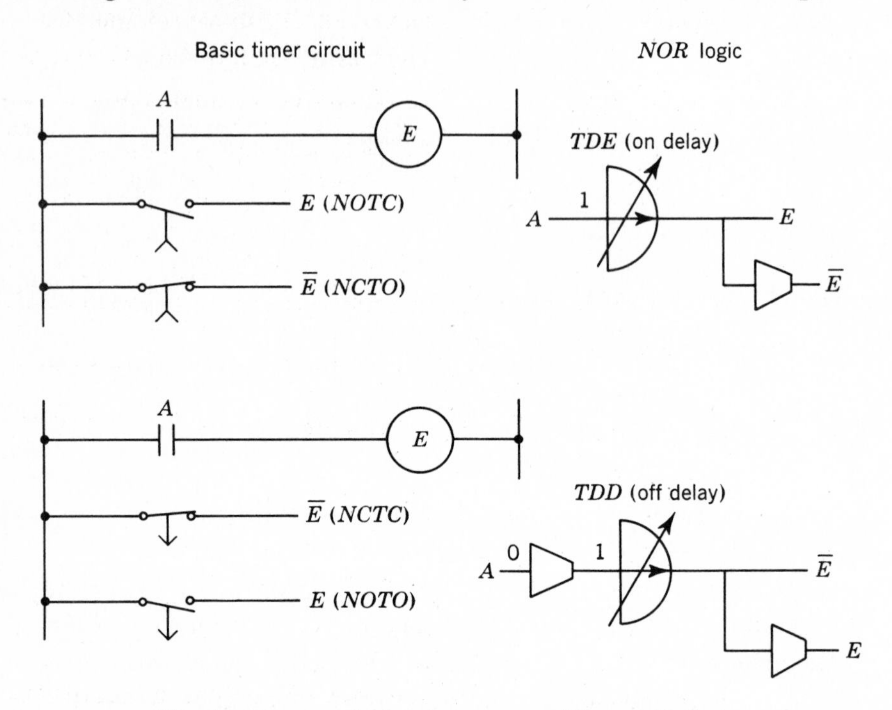

Fig. 13·10 NOR *logic, time* DELAY. (*Square D Co.*)

and the time-delay unit. Complementary signals on both timers are provided by the addition of a single NOR to the output of the basic delay unit.

Figure 13·11 illustrates how basic relay circuits can be duplicated by logic symbols. A relay is designated by a letter, and that letter represents each contact of the relay. A normally open contact is represented by a letter, and a normally closed contact is represented by a letter with a bar above it. (A = normally open;

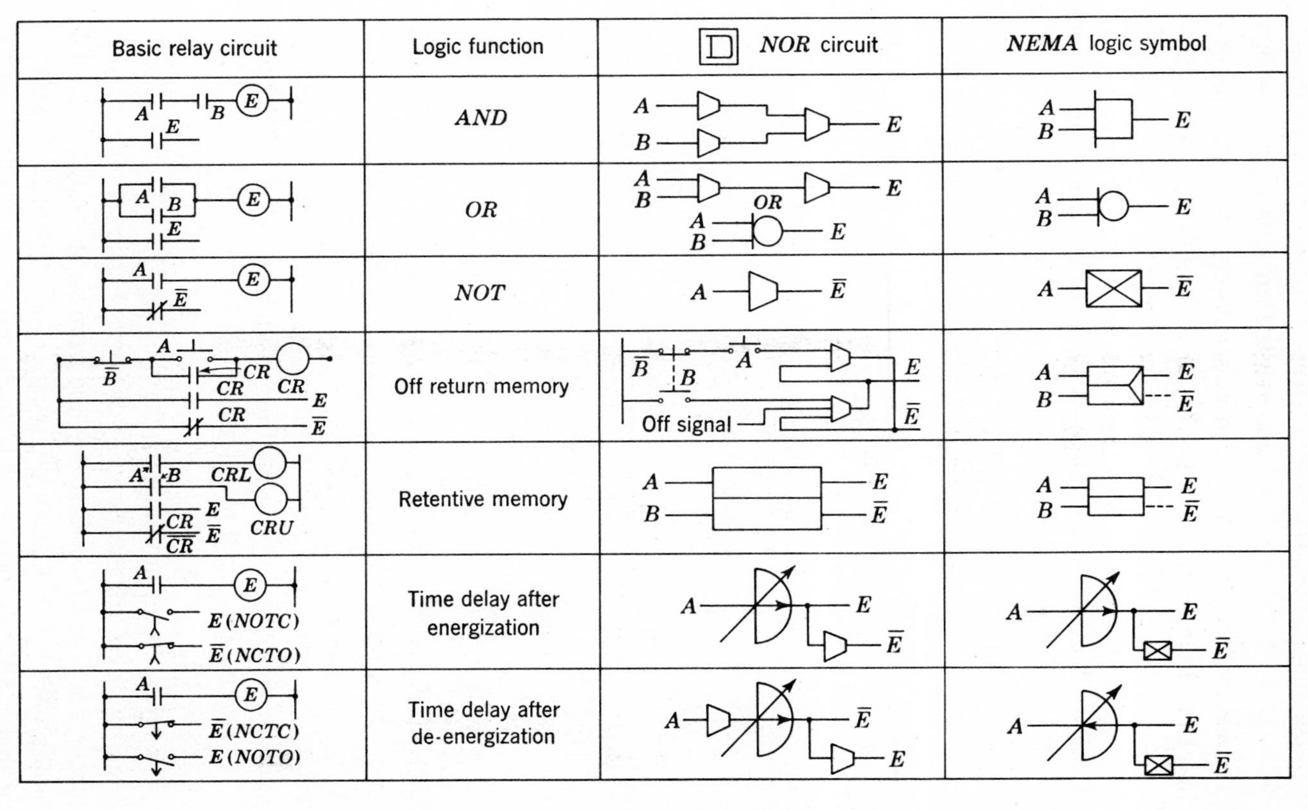

Fig. 13·11 Table of relay equivalents. (Square D Co.)

$\bar{A}$ = normally closed.) E represents the output and $\bar{E}$ is its complement signal.

13·2 DEVELOPING NOR LOGIC CIRCUITS

Several methods can be followed in developing static control circuits. Listed below are some of the most practical in use at this time.

The first method is to convert relay circuits directly to NORS. To convert a relay circuit into NOR logic, substitute for the relay contacts according to the chart shown in Fig. 13·11. After all

Basic relay circuit	*NOR* equivalent	Comment
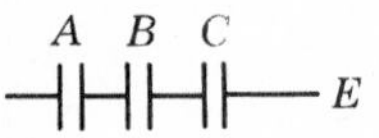 *A B C* *E*	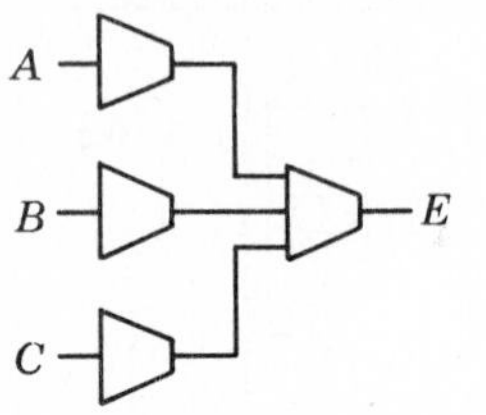 *A B C E*	This is called a 3-input *AND* circuit, using *NORs.*
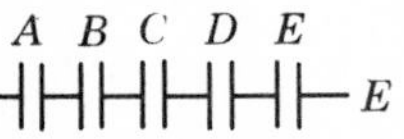 *A B C D E* *E*	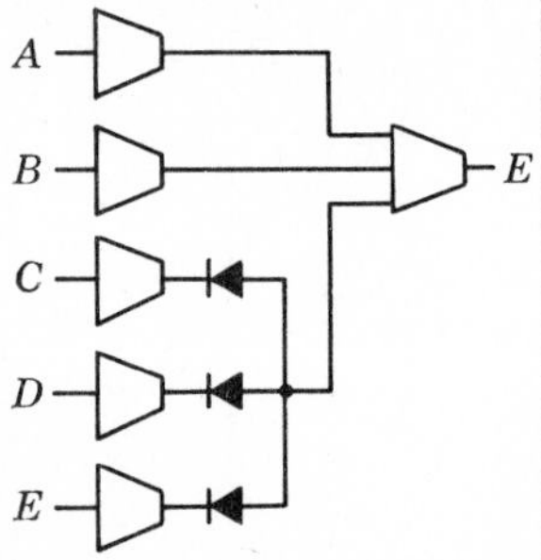 *A B C D E E*	This is called a 5-input *AND* circuit, using *NORs.* Notice that when more than three inputs are needed, additional inputs can be feed into isolating diodes with their output connected to one input of the *NOR.*

Fig. 13·12 NOR *logic, input expansion.* (*Square D Co.*)

the relay functions have been converted, the NOR circuit should be simplified by eliminating redundant functions. Figure 13·11 can be expanded to meet multiple-input circuits as shown in Fig. 13·12.

The relay circuit in schematic form of Fig. 13·13*a* can be converted by direct substitution to NOR logic. During the conversion process it is best to work line by line from right (outputs) to left (inputs). With only a general inspection (Fig. 13·13*b*) we might say that *CR*1 is a two-input AND circuit consisting of inputs *X*

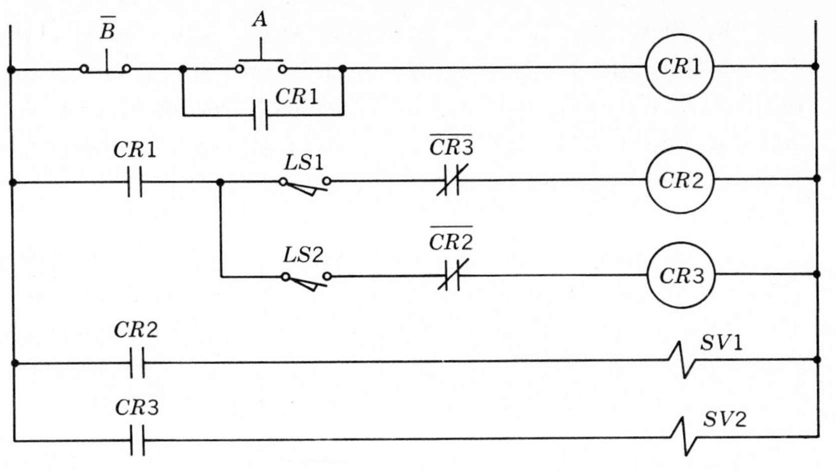

(a)

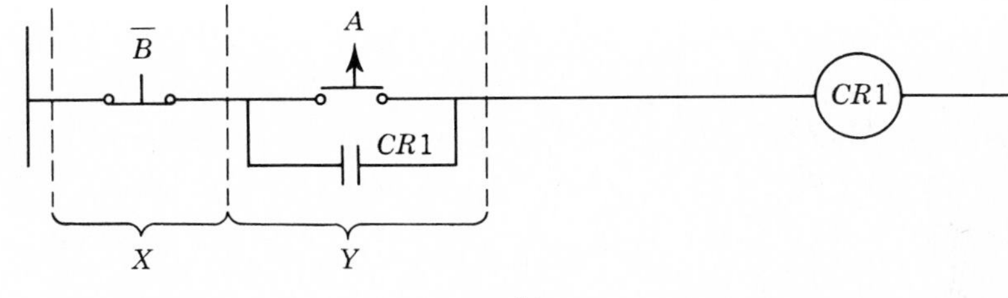

(b)

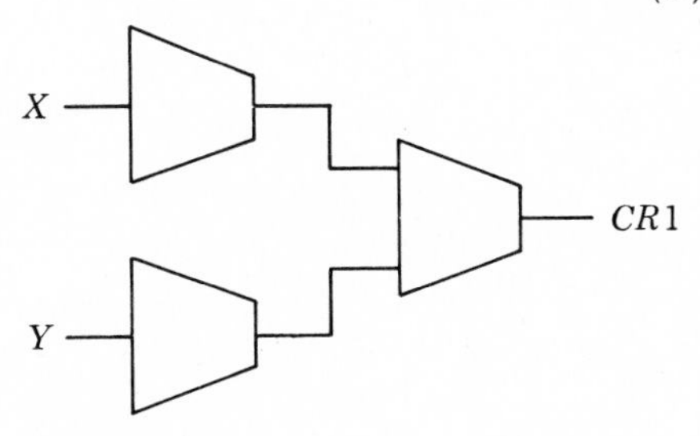

(c)

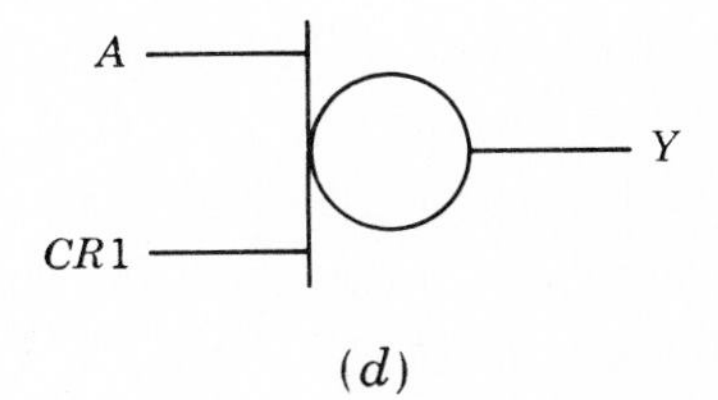

(d)

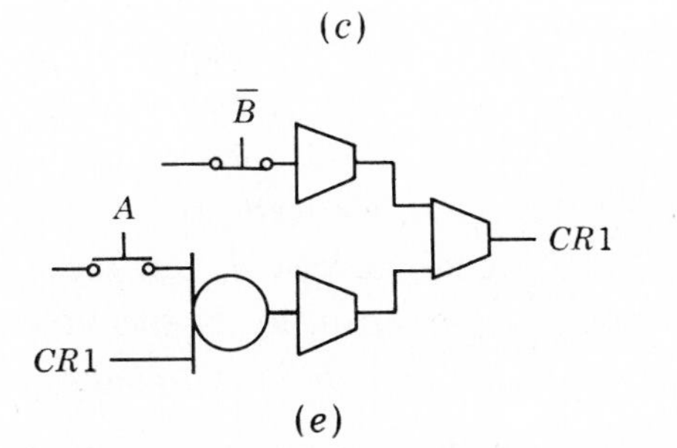

(e)

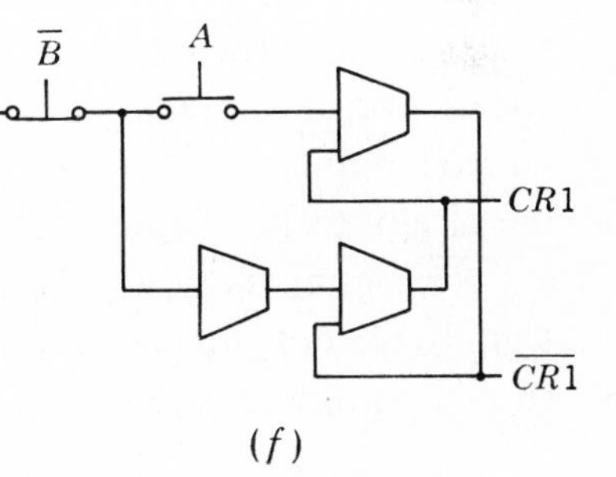

(f)

Fig. 13·13 Direct conversion from relay circuit to NOR *logic, part 1. (Square D Co.)*

and *Y*. First we draw a two-input AND, using NORS (Fig. 13·13*c*). *X* and *Y* would be the inputs, and *CR*1 would be the output. *Y*, however, is a two-input OR with one input being *A* and the other *CR*1 (Fig. 13·13*d*). *X*, of course, is the same as $\overline{B}$; now the pieces can be put together to show the entire circuit (Fig. 13·13*e*).

A closer look at the circuit reveals that one of the inputs, *CR*1, is also the output *CR*1. Some type of feedback or MEMORY circuit is therefore involved. The approach to this circuit could be simplified. MEMORY circuits should be recognized by a holding contact appearing in the same circuit as the relay coil. Returning to Fig. 13·11, we see that *CR*1 is an *off-return* MEMORY. Direct substitution then provides the correct circuit (Fig. 13.13*f*). Note that when signal *B* is absent, the MEMORY should turn off; this is accomplished through the NOR at the OFF input.

We then proceed to the next line (Fig. 13·14*a*). Examination indicates that *CR*2 is a three-input AND consisting of inputs *CR*1, *LS*1, and $\overline{CR3}$. (NOTE: A *CR*2 contact does not appear, and therefore a MEMORY is not involved.) The NOR equivalent is shown in Fig. 13·14*b*.

If we proceed to the next line (Fig. 13·14*c*), we see that *CR*3 is also a three-input AND consisting of inputs *CR*1, *LS*2, and *CR*2. Figure 13·14*d* shows the NOR equivalent.

Consider the next line (Fig. 13·14*e*). *SV*1 is the output device; therefore we simply draw the symbol for an output amplifier and show that the input is *CR*2 (Fig. 13·14*f*). The same is true for *SV*2 (Fig. 13·14*g*).

The next step consists of first putting the circuit pieces together, making sure that all like "labels" are connected together (i.e., connections are made from *CR*1 to all *CR*1 points, from *LS*1 to all *LS*1 points, etc.) as shown in Fig. 13·15, and then reducing the circuit to its simplest form. At this point, we want to connect a $\overline{CR3}$ signal to NOR *b* and a $\overline{CR2}$ signal to NOR *d*, but these signals do not appear to be present. The versatility of the NOR shows to great advantage here, as NORS can be eliminated to reduce the circuit to its simplest form. To eliminate NORS, it is necessary to understand the complementary principle.

Figure 13·16 illustrates the complementary principle with a one-input NOR. A 0 input gives a 1 output. An $\overline{X}$ input gives an X output, and a $\overline{CR3}$ input would give a $CR3$ output. The output of NOR b is $CR3$, and therefore a line can be run directly to the

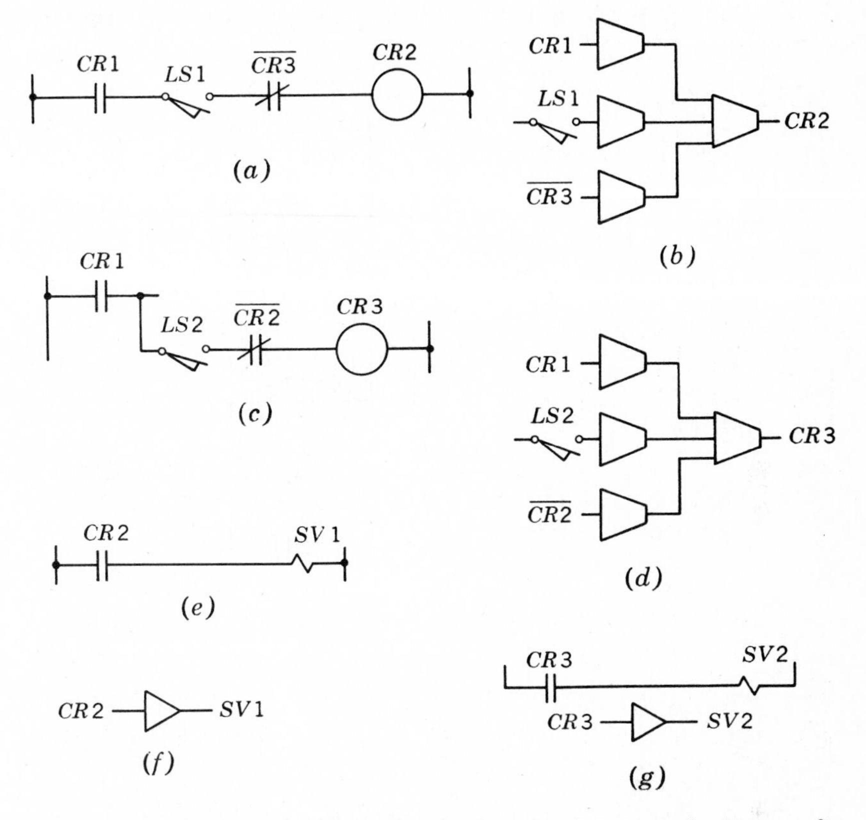

Fig. 13·14 Direct conversion from relay circuit to NOR *logic, part 2. (Square D Co.)*

$CR3$ connection that feeds $SV2$. The same can be done with NOR d for the $CR2$ that feeds $SV1$. Therefore, NORS b and d can be eliminated. Notice too that NORS a and c produce a $\overline{CR1}$ output (i.e., the complement of $CR1$). The NOR MEMORY has a $\overline{CR1}$ output available; hence NORS a and c are not necessary. The final circuit is shown in Fig. 13·17. With a little practice, circuits can be reduced mentally as the conversion is made, but it would be best to work by straightforward substitution until one becomes more experienced.

The second method is to convert conventional logic to NOR logic. If a conventional logic diagram is to be converted to NORS, the following steps should be taken. Step 1: Replace each NOT

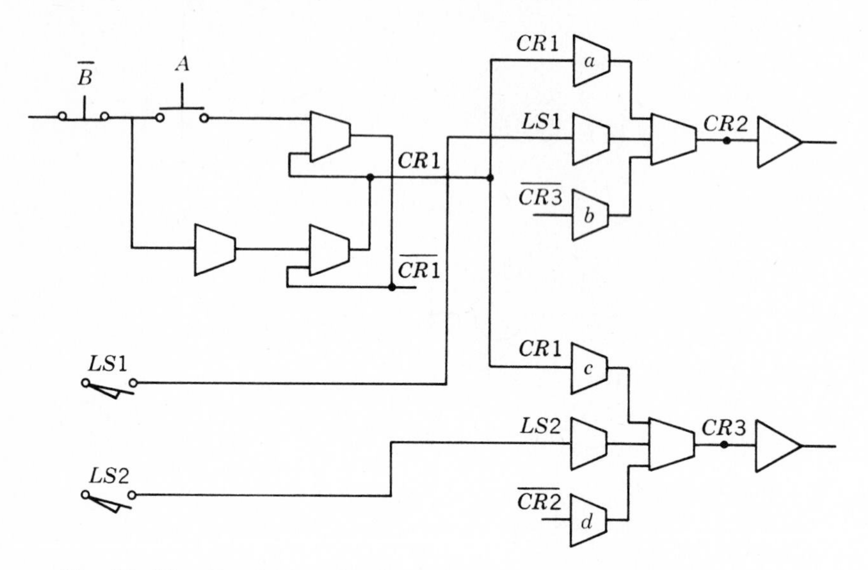

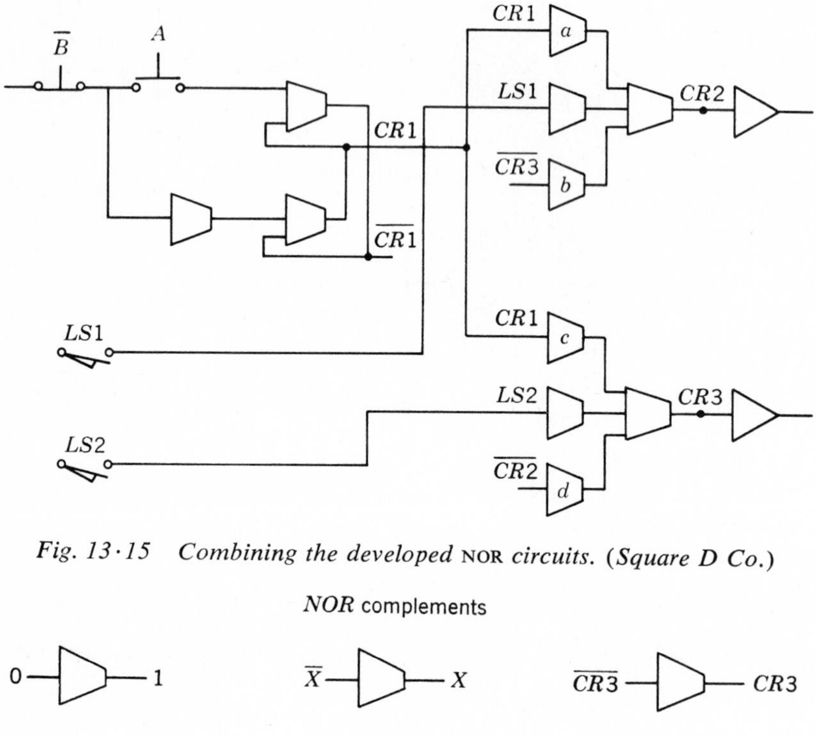

Fig. 13·15 Combining the developed NOR *circuits. (Square D Co.)*

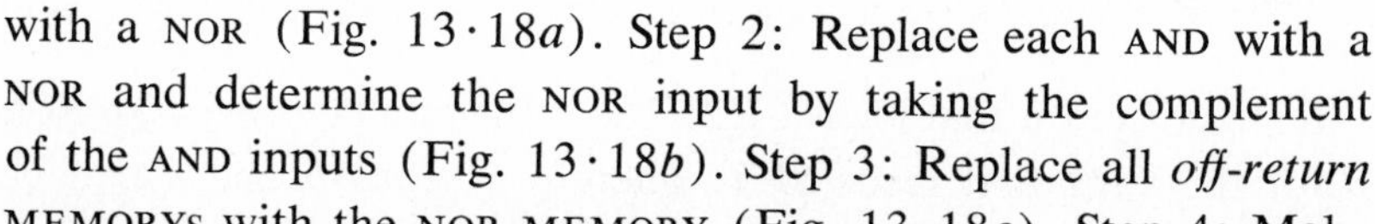

NOR complements

0 — 1 $\overline{X}$ — X $\overline{CR3}$ — $CR3$

Fig. 13·16 NOR *complements. (Square D Co.)*

with a NOR (Fig. 13·18*a*). Step 2: Replace each AND with a NOR and determine the NOR input by taking the complement of the AND inputs (Fig. 13·18*b*). Step 3: Replace all *off-return* MEMORYS with the NOR MEMORY (Fig. 13·18*c*). Step 4: Make all circuit connections, adding NORS only where complementary signals are needed. Step 5: Examine the circuit for redundancies (Fig. 13·18*d*) and remove unnecessary NORS.

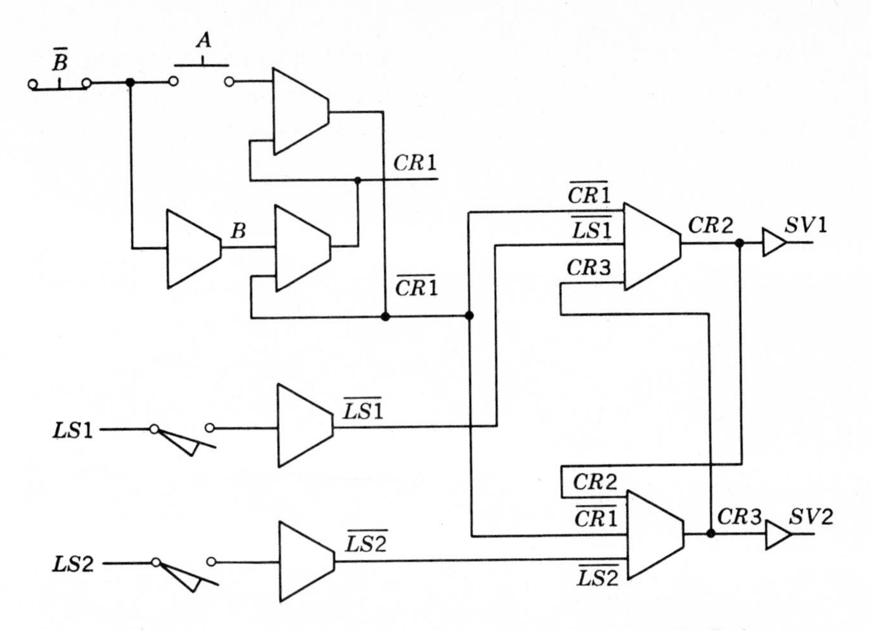

Fig. 13·17 Final NOR *logic development. (Square D Co.)*

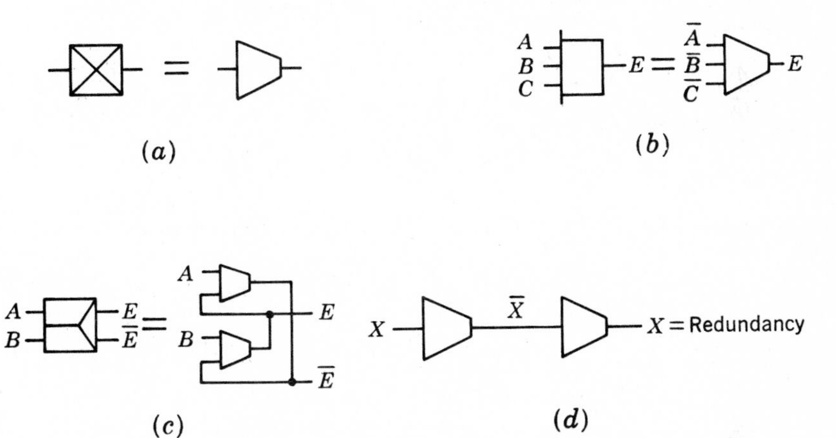

Fig. 13·18 NOR *equivalent of English logic. (Square D Co.)*

Figure 13·19*a* provides an example of an English logic circuit. Figure 13·19*b* covers steps 1 to 3. If we make circuit connections (*CR*1 to *CR*1 points, *CR*2 to *CR*2 points, etc.) according to step 4 and examine the circuit for redundancies (step 5), the final circuit is as shown in Fig. 13·20.

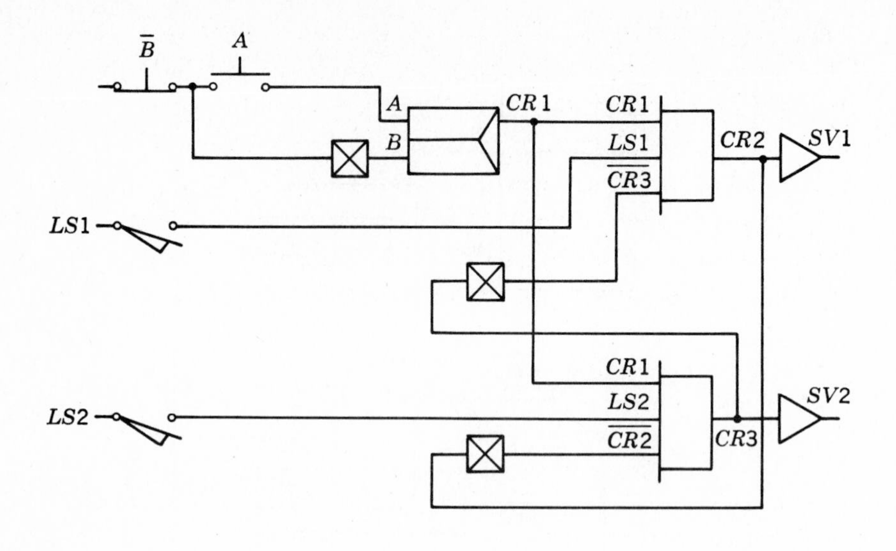

(a)

Conversion to *NORs*

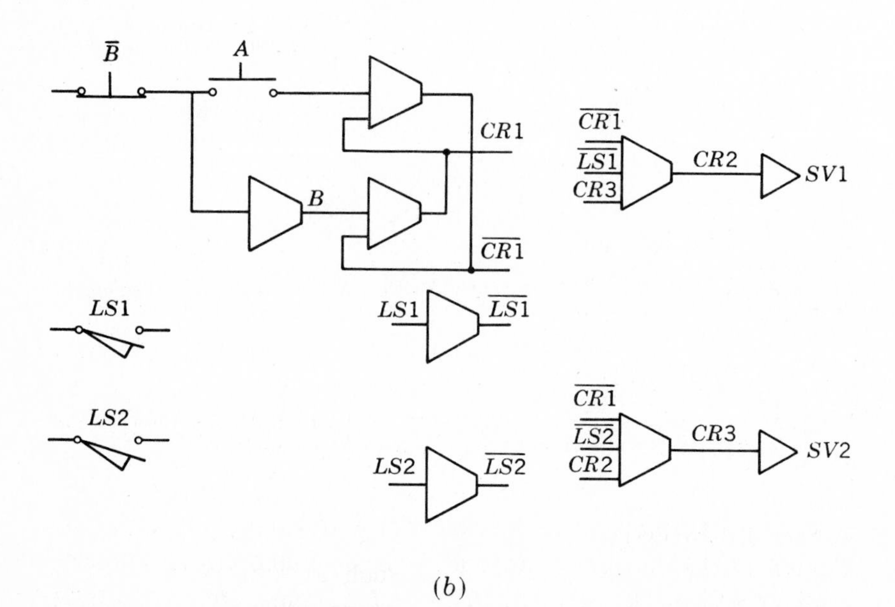

(b)

Fig. 13·19 English logic circuit conversion to NOR *logic. (Square D Co.)*

The third method is to convert word description to NOR logic. Due to the newness of solid-state control and its associated symbols, this is probably the most difficult method, especially for those accustomed to thinking in terms of relay circuits. However, as the use of solid-state control increases, more circuits will be developed by this method, since it is, in reality, the pure approach. Depending upon one's background, two variations are possible. The designer can convert word description to English

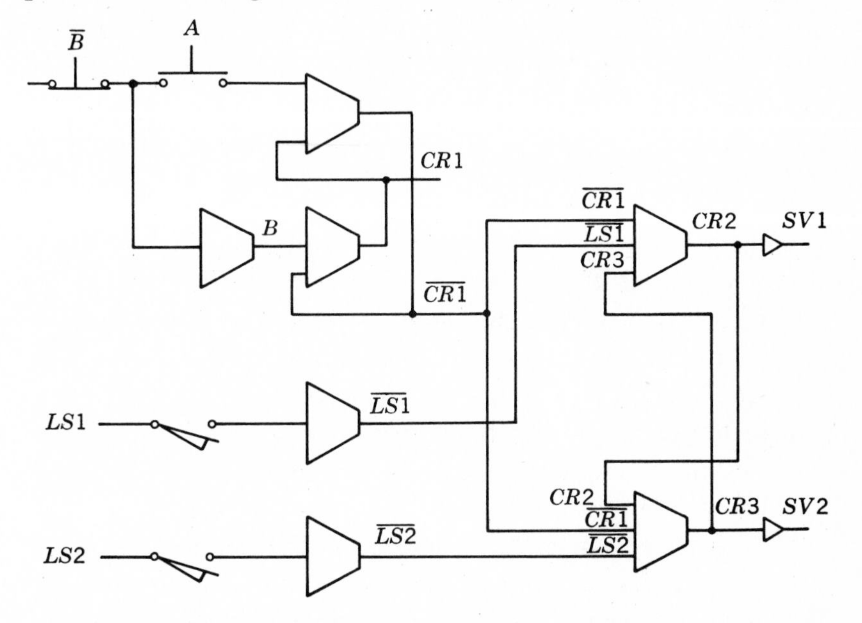

Fig. 13·20 Completed conversion of English logic to NOR *logic. (Square D Co.)*

logic and use the second method for conversion to NOR logic, or he can convert word description directly to NOR logic, as follows. First, try to get an overall picture of the general application. Visualize the machine as it goes through the sequences. If possible, actually view the machine in operation. From the word description, notice key words such as *and, or, not, time delay,* and *momentary* (suggesting MEMORY). Attempt to break up the description into pieces, and then, part by part, work out the complete logic circuit.

For an example, consider the machine that consists of a planer table which moves back and forth by means of a reversing motor. Momentary action of limit switches located at each end of the bed initiates the reversing contactor which causes the table to oscillate. The protective controls desired are: extreme operating limit switches on each end of the bed, motor overload protection, interlocking with the coolant supply, and minimum lubrication pressure.

What is required to provide a ready-to-run position? With minimum lubrication pressure present, the coolant motor operating, the drive-motor overload relays incorporated in the circuit, and the emergency switches normally closed, it is now desired that the machine either run continuously or jog.

We should pause at this point to visualize a table that oscillates back and forth; its movement is controlled by a limit switch at each end of the table, and emergency limit switches are mounted near the operating limit switches to prevent the table from traveling too far in the event the initial limit switch fails. Of equal importance is the fact that some type of logic must be performed to put the table into operation and also to stop the machine by either manual or automatic operation if predetermined conditions are not met.

The ready-to-run circuit could be controlled by a six-input AND. In general, all forms of start-stop circuits consist of the *off-return* MEMORY. In this case, the inputs to the AND circuit are coming from pilot devices, and therefore the AND circuitry can be accomplished by merely connecting these signals in series. The complete ready-to-run circuit would appear as shown in Fig. 13·21*a*.

Consider the work cycle of the machine. When the machine is in the ready-to-run position, the operator presses either a right traverse or left traverse button. For example, if the right traverse button were pressed, the table would move to the right until momentary engagement of the right limit switch. When the right operating limit switch is actuated, one desires the right traverse contactor deenergized and the left traverse contactor energized

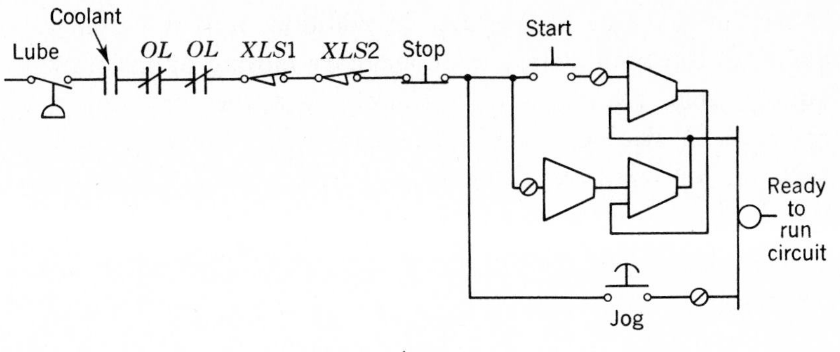

(a)

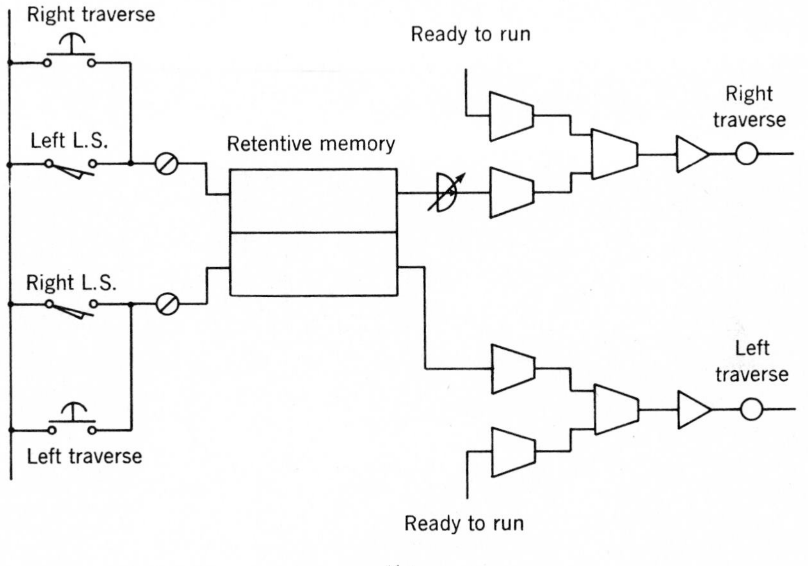

(b)

Fig. 13·21 Direct NOR *logic development, step 1. (Square D Co.)*

so that the table is driven to the left until it hits the left operating limit switch. Actuation of the left limit switch causes the reversing contactor to reverse the motor and drive the table to the right. In some applications it would be desirable to have the table stop at one end of its travel in order to blow off chips (i.e., a time delay is involved).

Examination of the above information reveals that the ready-to-run circuit and either the limit switches or the traverse buttons cause the machine to operate. In addition, it is a momentary action of the limit switch or momentary closing of the traverse buttons that causes action. Therefore, a memory circuit is involved. A decision must be made on whether an *off-return* MEMORY or *retentive* MEMORY should be used. The machine builder in this instance might insist on having the equipment start up where it left off after power failure, in which case a

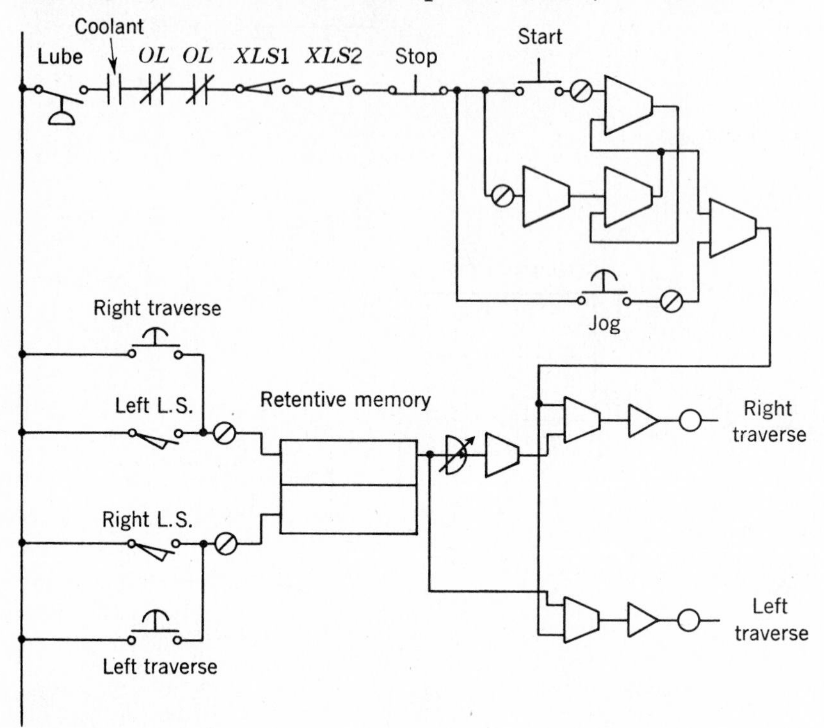

Fig. 13·22 Direct NOR *logic development, complete circuit.* (*Square D Co.*)

retentive MEMORY should be used. The work-cycle part of our circuit is shown in Fig. 13·21*b*. The combined circuit is shown in Fig. 13·22.

13·3 NORPAK LOGIC ELEMENTS

NORPAK circuitry consists of combinations of resistors, transistors, diodes, or capacitors, and hence is inherently static and suitable for encapsulation. Encapsulation is accomplished by packaging the circuitry in molded bakelite cases which are then filled with an epoxy resin. The resulting package is a solid unit which will resist mechanical or thermal shocks and is inert to almost any industrial atmosphere. Mounting holes are provided in each corner. The package may be mounted in any plane at any surface.

The logic elements are packaged in either a size I or a size II bakelite case. A third bakelite case is used to package accessory items such as signal converters and output amplifiers. The

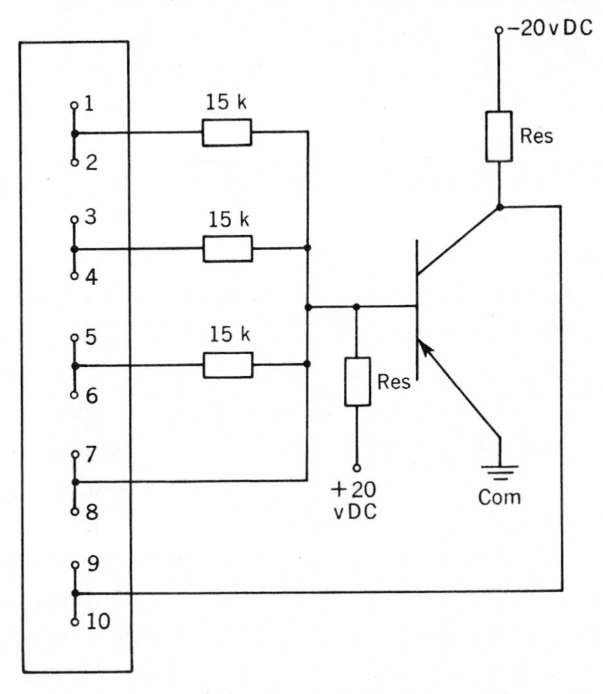

Fig. 13·23 NOR *circuit with terminals shown.* (*Square D Co.*)

size I case has space for seven nylon wafers lettered from *A* to *G*, and the size II case has space for 21 wafers lettered from *A* to *Y*. Each wafer has 10 taper-pin terminals, each numbered from 1 to 10. The terminals are paired together in single electric connections as 1–2, 3–4, 5–6, 7–8 and 9–10, resulting in five separate electric connections. This provision of paired terminals permits each electric connection to become a junction point.

The NOR circuit consists of a transistor and five resistors as shown in Fig. 13·23. Three standard inputs are connected to terminals 1–2, 3–4, and 5–6. A special input direct to the base of the transistor is provided at terminal 7–8. This special input connection is to be used only where instructions so state. The output signal from the collector of the transistor is connected to terminal 9–10. Thus, all the connections for a given NOR are made to the same wafer, and each wafer represents a separate NOR element. The power-supply connections —20 volts d-c, common, and +20 volts d-c are connected internally to a bus bar which is brought out on the *A* wafer. If there are some unused NORs in a Pak, the —20-volt collector voltage and the +20-volt bias voltage will create a collector-to-base voltage of 40 volts, which might possibly damage the transistor. This possibility can be eliminated by connecting any input terminal (1–2, 3–4, or 5–6) to common, thus reducing the junction voltage to within safe limits.

The amount of power required at the input of a NOR to cause the NOR to switch is defined as one unit of logic load. Each standard NOR is capable of powering four other NORs and therefore has an output rating of four units of logic load. All NORPAK elements are rated in these units of logic load.

The type L1 (Fig. 13·24) consists of six standard NORS as described above. The Pak is a size I package. The *A* wafer is red, indicating the power-supply connection of +20 volts at terminal 1–2, common at terminal 5–6, and —20 volts at terminal 9–10. The remaining six wafers each represent a NOR element. Each wafer is lettered, and its color is white to indicate a standard NOR.

The type L2 consists of 20 standard NORS packaged in a size II package. All connections and ratings are identical to those of the L1 types. Since these are standard NORS, the wafers are all white. This is a more economical unit when 10 or more standard NORS are required.

The type L6 power NOR Pak consists of six power NORS in a size I Pak. The power NOR can supply 10 units of logic load as opposed to the four-unit rating of the standard NOR. The power NOR is used in controlling large-unit load circuits or output amplifiers. The power NOR circuit is identical to the standard NOR circuit except that component values are changed and the input and output connections are the same. The power NORS have blue wafers.

The type L9 universal NOR consists of five universal NORS in a size I Pak. It is designed for high logic-load requirements or to be operated as a small output amplifier to control lights

(*a*)

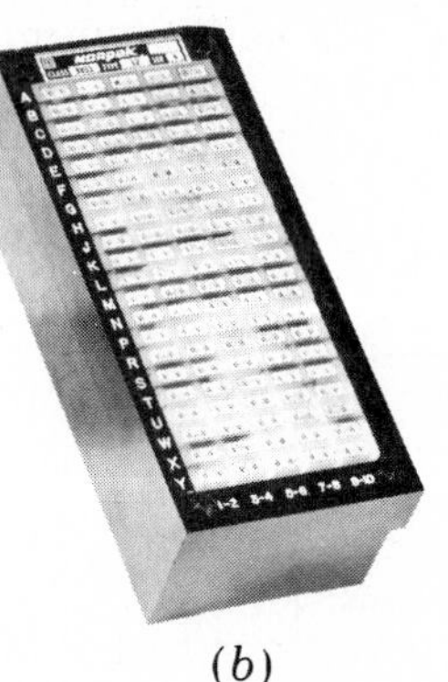

(*b*)

(*c*)

(*d*)

Fig. 13·24 The NOR *Paks. (a) L9. (b) L2. (c) L3. (d) L1. (Square D Co.)*

or relays. The *G* wafer provides the termination of five separate 1.5-kilohm resistors whose opposite ends are connected to the —20-volt d-c bus. This wafer is black, and the universal NOR wafers are blue. The universal NOR circuit is the same as a standard NOR circuit except that component values are changed. In the case of the universal NOR, the collector resistor can be selected to determine the maximum output from 10 to 50 units of load (Fig. 13·25). The required resistance may be selected from the five 1.5-kilohm resistors of wafer *G*, or a separate resistor of the proper size may be connected between the —20-volt d-c bus and the universal NOR output terminal 9–10. The collector current is also a function of the base current in a transistor, and therefore it is also necessary to increase the input power as the output capacity is increased of the universal NOR.

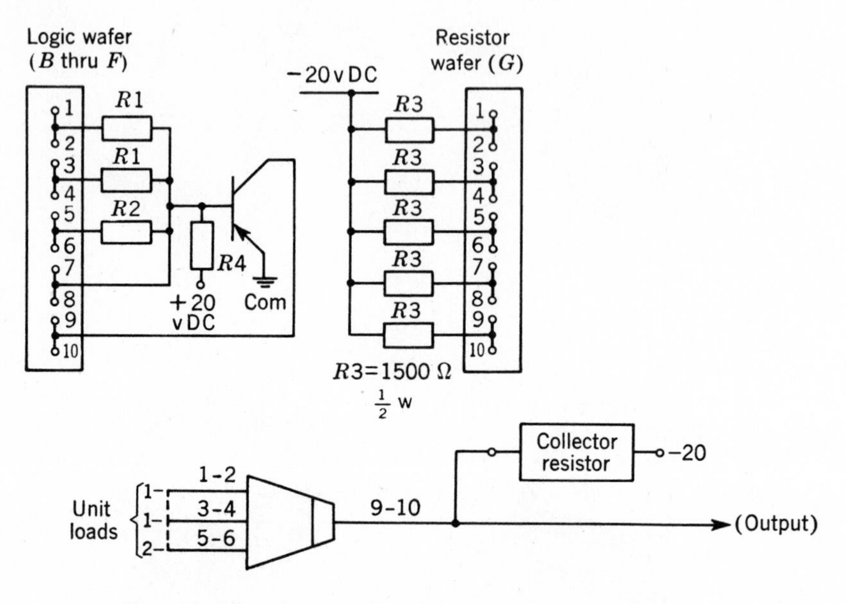

Fig. 13·25. Universal NOR *circuit. (Square D Co.)*

The universal NOR input terminals 1–2 and 3–4 require one unit of load input each, but 5–6 requires two units of load input. By connecting the various terminals together, one can obtain any combination of load input from one to four units. The input terminal 7–8 has a special direct base input to be used only where instructions so state.

The type L3 OR Pak (Fig. 13·26*a*) consists of seven separate yellow wafers in a size I Pak. No red bus module is present, since there is no power supply required for the OR element.

Each wafer contains a one-input OR and a two-input OR. Each OR is capable of handling 50 units of logic load.

The purpose of the OR function is to provide isolation between two or more signals connected to a common point. This is accomplished by using silicon rectifiers in each input leg. Thus a signal can flow in only one direction in an OR circuit. Because

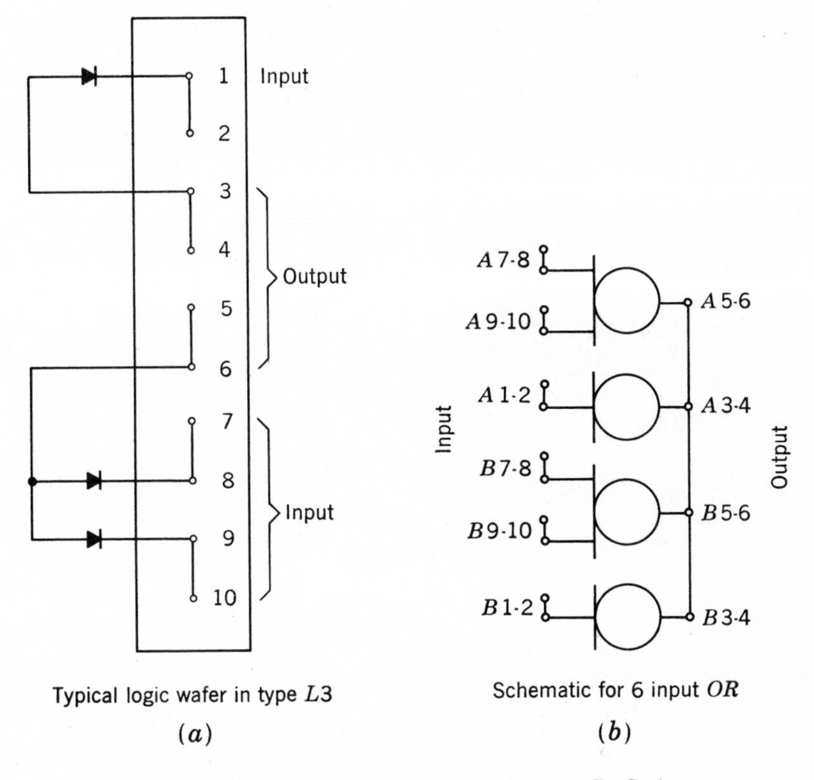

Fig. 13·26 Diode OR *circuit. (Square D Co.)*

of this, the OR cannot be used to control logic elements that require a discharge path such as the types L8, L11, or L12. Multiple-input OR circuits can be connected as shown in Fig. 13·26*b*. One type L3 could supply a 21-input OR if desired. Because relatively few ORs are needed in comparison to the number of NORs, usually it is advantageous to centrally locate the OR Pak among the other logic elements.

If only a few OR functions are required in a system, it may be more economical to consider using two NORS in series to provide the OR functions. Occasionally, it may be desirable to use an OR diode as a discharge path for a d-c relay coil. An example of this is the case of a universal NOR being used as an amplifier to operate a relay coil. In such a discharge circuit, the diode is capable of handling a peak current of 300 ma.

A *retentive* MEMORY may be used whenever it is mandatory, in the event of power failure, that the MEMORY retain its last energized condition upon resumption of power. Thus it performs the same function that a mechanically held relay does in the

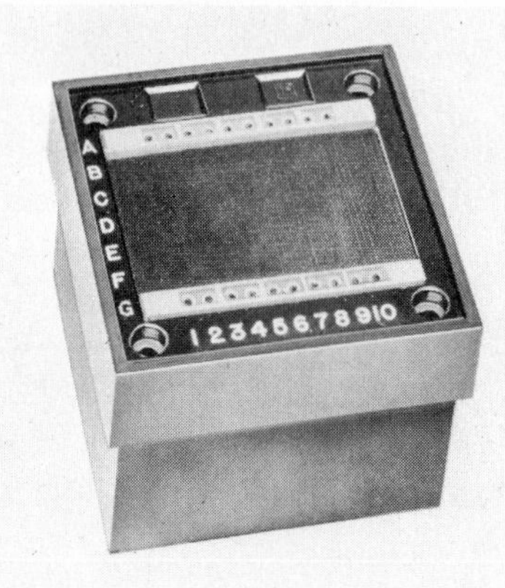

Fig. 13·27 Retentive MEMORY, *package.* (*Square D Co.*)

relay circuit. An alternate method of providing this function is to use standard logic components and provide an emergency battery-operated power supply. The choice is based upon the economics of the application. When only a few functions require a *retentive* MEMORY feature, the type L5 *retentive* MEMORY is best. If a great number of *retentive* MEMORYS are required, it is advisable to consider the emergency power supply. One *retentive* MEMORY is packaged in a size I case (Fig. 13·27). The logic wafer *G* is green in color, and the bus wafer *A* is red. The *retentive* MEMORY requires one unit of logic load on either input and will provide three units of logic-load output.

The *retentive* MEMORY consists of a MEMORY controlled by a saturable core (Fig. 13·28). The saturable core provides the memory feature by gating or directing a pulse signal to the ON or the OFF portion of the standard NOR MEMORY, depending on the state of the core. The state of the core, saturated or unsaturated, is set by the last input signal prior to the power failure. The required pulse signal is obtained from the power supply and applied to the power wafer *A* at terminal 3–4. The pulse signal is 6 volts a-c. A discharge path is needed; therefore the *retentive* MEMORY cannot be driven directly by an OR logic element. The saturable core material will maintain its properties for an unlimited number of years; hence it is an ideal MEMORY element. Thus there is no practical time limit to the retentive feature of the L5 *retentive* MEMORY. If the input signal is

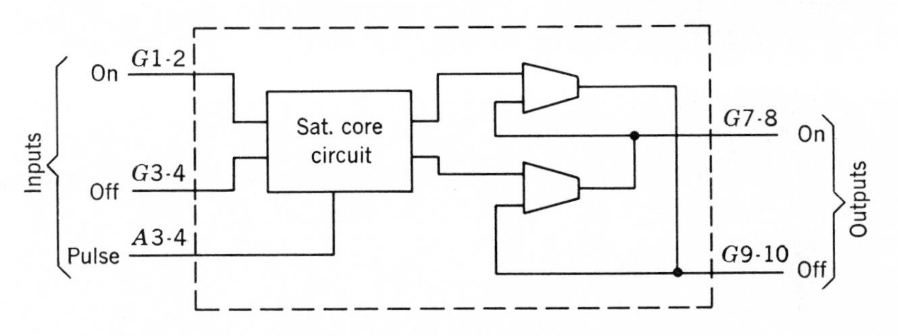

Fig. 13·28 Retentive MEMORY *circuit. (Square D Co.)*

switched, the output signal will switch within 0.005 sec, but the input signal must remain present for at least 0.04 sec to completely switch the saturable core. Thus if a signal is present for less than 0.04 sec, the magnetic core will not retain the MEMORY. If the input device is a contact with bounce, the *retentive* MEMORY output will provide a pulsating output signal after each bounce. If this is undesirable, use bounceless input circuits. If inputs are simultaneously present at both inputs (1–2 and 3–4), the output will become 0 at 7–8 and 1 at 9–10. When the normal circuit configuration is used, this action is known as an OFF override. ON override can be accomplished by designating terminals 1–2 as OFF input and 3–4 as ON input; this results in 7–8 becoming the OFF output and 9–10 becoming the ON output.

The L18 timer (Fig. 13·29) incorporates one complete time-delay-after-energizing function in a single size I Pak. The logic wafer *B* is gray and the bus wafer *A* is red. One unit of logic load is required at the input, and the output will provide three units of logic load. The time delay consists of a MEMORY, a timing network, and two NORS (Fig. 13·30).

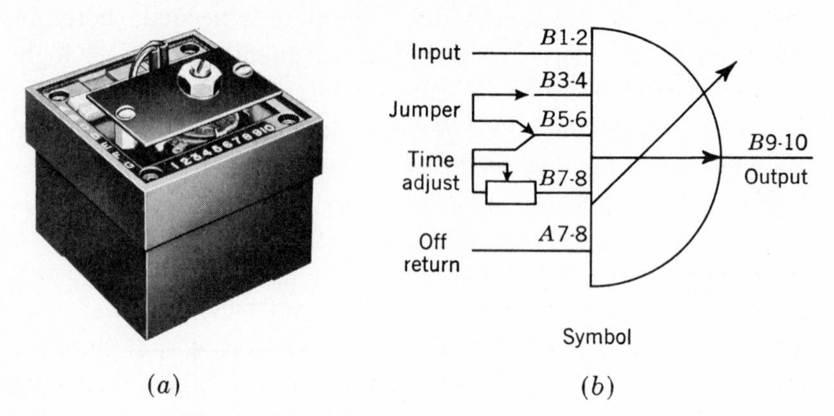

Fig. 13·29 Time DELAY. (*Square D Co.*)

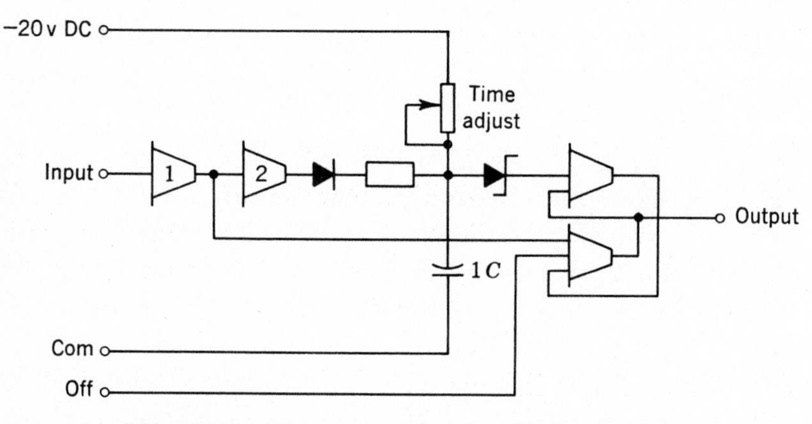

Fig. 13·30 Time DELAY *circuit.* (*Square D Co.*)

The time-delay function is provided by an RC timing network and a zener diode. A logic 1 (−20 volts) signal at the input causes the blocking signal to the MEMORY from NOR 1 to go to 0 and the output of NOR 2 to go to 1 (−20 volts d-c).

This voltage charges capacitor $1C$ through the time-adjust rheostat. The charging voltage on $1C$ continues to rise to the breakdown voltage of the zener diode; at this value the MEMORY turns on, providing a signal at the output. The timing period depends upon the setting of the time-adjust rheostat.

Since the timing network employs an RC time delay, the delay is subject to changes of supply voltage (−20 volts) and the ambient temperature. These parameters will vary from system to system but should be relatively constant for any one system. Thus, the time-delay adjustment pot cannot be calibrated until the timer is being used with a particular system.

Standard pilot devices, such as limit switches, push buttons, selector switches, etc., are usually used as a source of input signals to the static system. To assure the reliability of these contact-making devices, they should be used at voltages higher than the 10- to 20-volt signals that are found within the NOR circuitry. The higher voltages are, of course, easily obtained, but then the problem is to convert them to a usable level for the logic portion of the system.

The most commonly used signal converter is the type N5 (Fig. 13·31*a*), which is a four-input device that uses the 130-volt d-c output from the type P1 logic power supply as its pilot operating source voltage. Each N5 input then reduces the 130 volts to about 20 volts at its output by means of a voltage-divider circuit. As an added feature, the presence of a signal at each input of the converter is indicated by a built-in neon lamp. These are the type NE84 and are easily replaced. Each N5 output will carry five units of logic load while the −130-volt source will drive 20 N5 inputs at the same time. This is equivalent to one unit of −130-volt power per conducting input. The N5 has a built-in capacitor buffering against induced transient signals. Also, it is designed to tolerate a pilot-circuit leakage resistance as low as 40,000 ohms. It is recommended that the type N5 converter be used whenever the pilot input devices are remotely mounted.

Frequently an application requires the pilot devices to be oper-

(a)

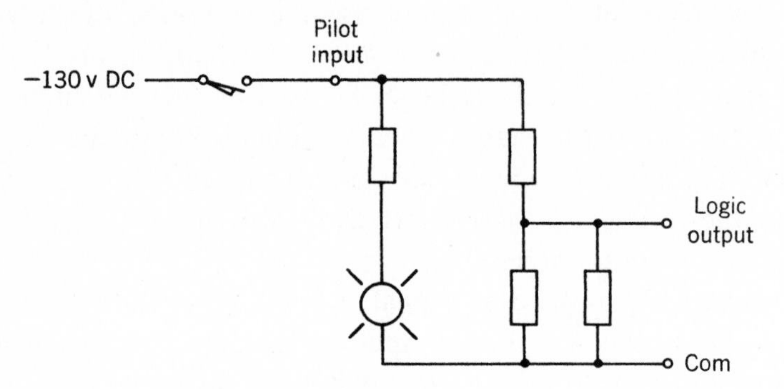

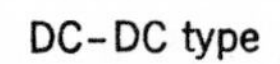

(c)

L1
L2
120 v AC
Pilot
input
Logic
output
Com
L2

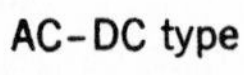

Fig. 13·31 Signal converters. (Square D Co.)

ated on 120 volts a-c. The type N2 signal converter (Fig. 13·31*b*) is a two-input device that converts 110 volts a-c to a —20-volt d-c logic signal. Each input circuit consists of a small 120/24 volt transformer with a rectified and filtered secondary output that can drive five units of logic load. Each of the inputs incorporates a neon light to indicate the presence of an input signal. The filtered secondary output gives the converter an inherent buffer action against induced transient pickup. Its input impedance is 50,000 ohms, however, which would make it sensitive to a severe current-leakage condition. To combat this, a dummy load resistor may be placed across the input terminals. For example, a 3,000-ohm 10-watt resistor across input terminals 1 and *L*2 would allow for a leakage resistance of 10,000 ohms across the pilot device.

A second type of a-c-to-d-c signal converter is the type N3, which is designed to accept a 110-volt a-c half-wave rectified input and convert it into a d-c voltage of about 25 volts. The output is filtered, but its minimum instantaneous voltage drops to 10 volts with one unit of logic load, which is its rated output capacity. The converter is designed to operate across the plate circuit of a thyratron tube, but could be operated from any half-wave rectified source. The filtered output makes this converter rather insensitive to induced interference.

The type N4 signal converter has been specifically designed to operate with the class 9007 type V9, series A or V10, and series B transducers to provide a completely solid-state amplifying and converting circuit with an output suitable to switch a NOR. One N4 is required for each transducer. It has an output capacity of one unit of logic load and requires eight units of —20-volt power, ⅕ unit of +20-volt power, and ⅕ unit of pulse power. (The pulse output of the type P1 power supply is rated at 20 units.) The OFF input requires 3 units. The OFF output capacity of the type P1 power supply is 20 units.

Whenever a contact-making pilot device is used as an input signal to a transfer element that is part of a counting circuit, shift register, information storage system, or the like, there is danger of contact bounce giving false input signals. The speed of response of these circuits makes it quite easy to follow the bounce action and thus give two or more input pulses when only one was intended. Circuits of Fig. 13·32 show how the

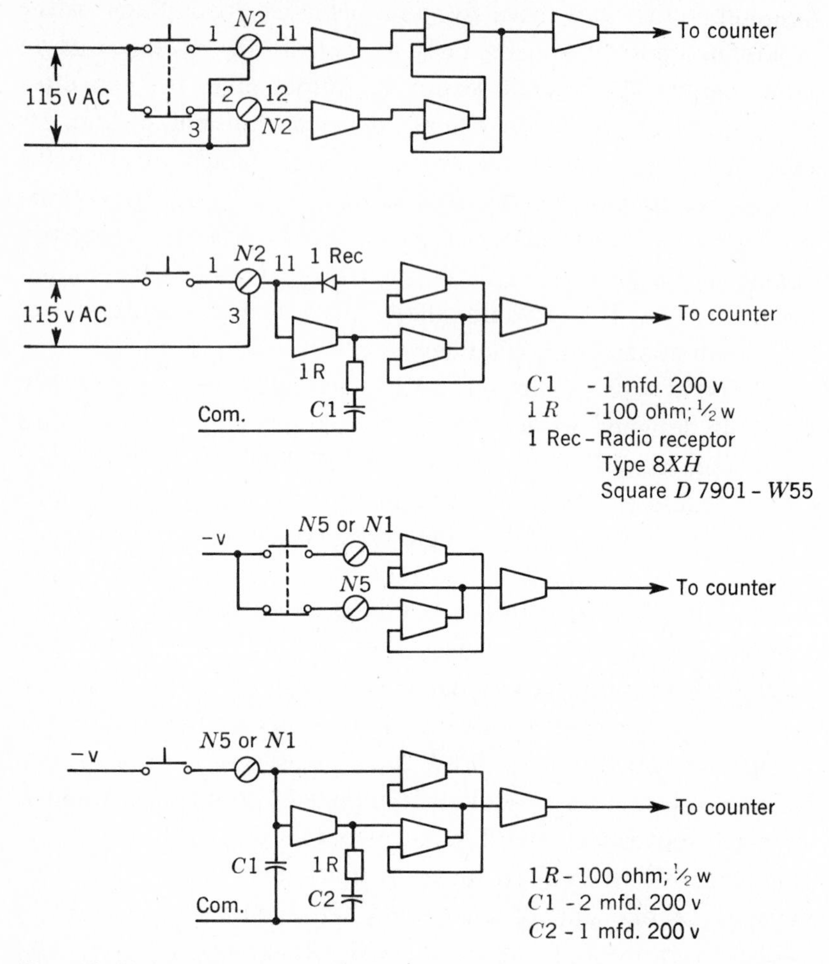

Fig. 13·32 Input circuits used to eliminate effect of contact bounce. (Square D Co.)

bounce effect may be eliminated when using d-c or a-c signal converters for single-throw or double-throw contacts. Slowly opening or closing contacts should not be used. Snap-action switches are preferred.

The type F1 filter Pak consists of 12 identical resistor-capacitor networks encapsulated in a size I Pak. Two filter networks terminate at each of the six black wafers as shown in Fig. 13·33. A red power wafer is used to make a common connection at terminal *A*5–6. No other power connections are necessary.

A filter circuit is designed to provide a buffer against induced or unwanted input signals that could enter the logic circuit through wires coming from points outside the NORPAK panel. Normally, a signal converter serves this function, but if the input signals are already at a NORPAK logic level (−20 volts d-c) the converter is not required; however, protection from stray pickup signals must still be maintained. Logic-level signals could

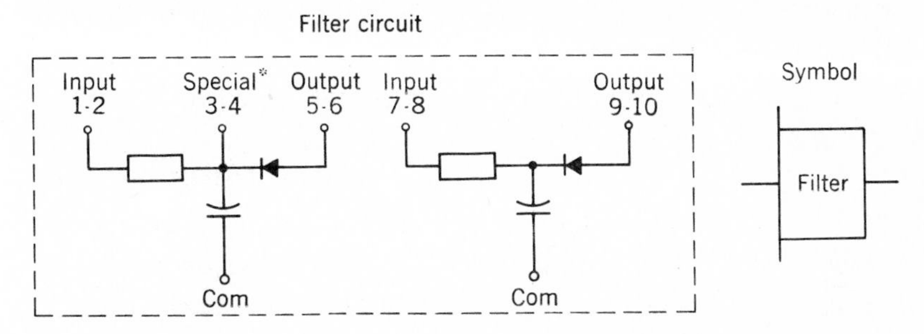

Fig. 13·33 Type F1 filter-Pak circuit. (*Square D Co.*)

originate from another NORPAK panel, a computer, an electrical instrument, or from contact-making devices that are used at low voltages. A separate filter input should be used for each wire entering the logic panel. A logic-level signal leaving a NORPAK control should always be isolated by a separate NOR having no other connections to its output.

The diode included in each of the filter networks has a 5-volt forward drop and a low current capacity. The high forward drop aids in the transient signal protection but increases the delay time of the incoming signal. The 5-volt forward drop thus requires that the NORPAK devices energizing these inputs be derated by 50 percent. The diode may be bypassed, however, in six of the twelve filters by using the special output terminal at 3–4 in place of 5–6. No derating is then necessary. In all cases, the device supplying the input signal must be capable of carrying the output load, since no amplification of the signal is taking place.

The signal delay caused by the filter network is a maximum of 230 μsec when the diode is included. If the diode is bypassed, the delay will be a maximum of 150 μsec. The output load rating is three units when the diode is being used with standard NORPAK inputs. Negative input potentials as high as 80 volts can be used as input signals to the filter circuit as long as the current is held to a maximum of 3.75 ma. For example, a —60-volt d-c signal would require at least 15,000 ohms in the circuit (5 volts is subtracted from 60 to allow for the forward drop of the diode). Since the input impedance to a standard NOR is 15,000 ohms, the current would be held to a safe limit, but only one unit of load could then be connected to the output of the filter.

13·4 NORPAK OUTPUT AMPLIFIERS

The universal NOR can be made to function as either a NOR element or an output amplifier. Fig. 13·34*a* shows both the logic circuit and the output amplifier circuit. When the transistor conducts, it can be thought of as closing a single-pole single-throw switch from collector to emitter. Note that in the logic circuit the transistor (switch) is in parallel with the load, but in the amplifier circuit it is in series with the load. In either case, a 1 input to the transistor causes it to conduct (switch closed). In the logic circuit, the load is shorted out, and the load current is effectively zero. Thus a 1 input to a universal NOR used as a logic element deenergizes the load or produces a zero output. However a 1 input to the universal NOR used as an output amplifier causes current to flow through the load. Therefore, a 1 input to a universal NOR used as an amplifier energizes the load or produces an output. When used as an output amplifier, the universal NOR has a maximum rating of 1.5 watts. Note that as its output rating increases from 0.26 to 1.5 watts, its input load rating increases from 1 to 4 units, since the collector current of a transistor is a function of its base current. Increasing the

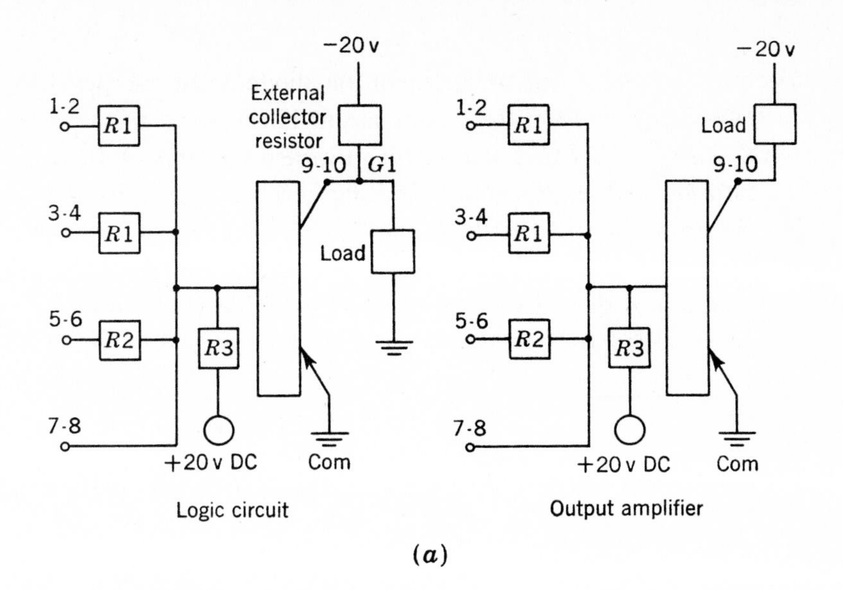

(a)

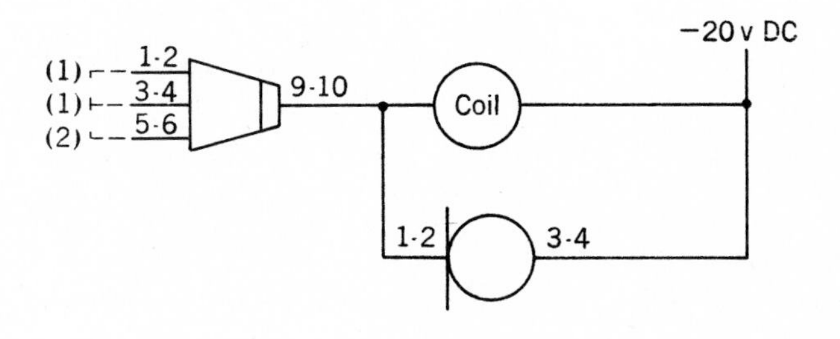

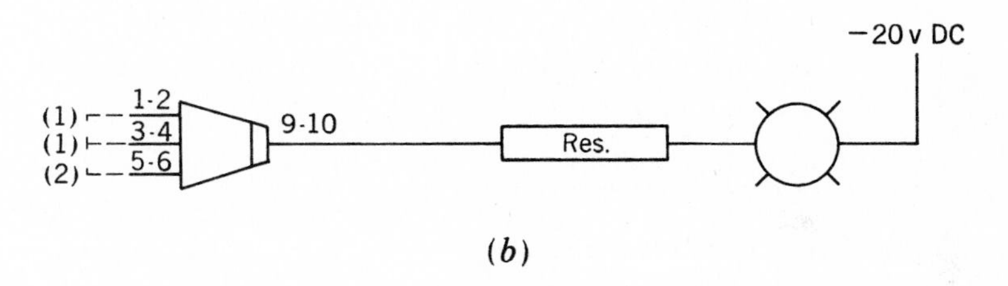

(b)

Fig. 13·34 Universal NOR *circuit. (Square D Co.)*

input load is accomplished by using combinations of the universal NOR's three inputs. The maximum output rating of 1.5 watts is obtained when the load resistance is 270 ohms and all three of the universal NOR's inputs are connected together. When a universal NOR is powering a relay or other inductive load, a discharge path of an OR element or diode must be provided as shown in Fig. 13·34*b*. Incandescent lamps have a low resistance when cold, which can result in inrush currents of 10 times the nominal value. A current-limiting resistor should be placed in series with the lamp as shown in Fig. 13·34*b*.

The type TO3 is a solid-state, epoxy-encapsulated d-c output amplifier. It is designed to operate as a switch in a 24-volt d-c circuit supplied by a class 8851 type A51 or A301 auxiliary

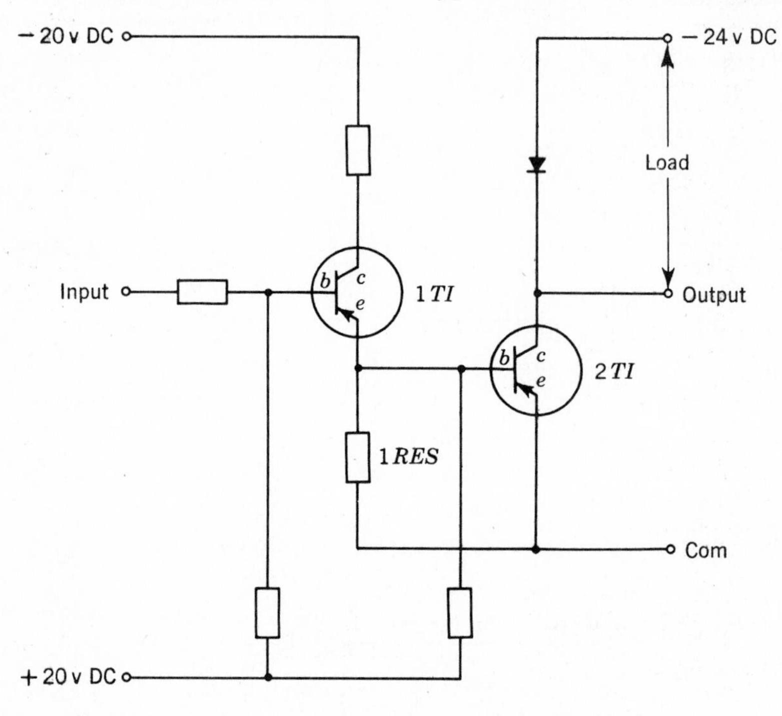

Fig. 13·35 D-c output-amplifier circuit. (Square D Co.)

power supply, and will energize the load when a 1 signal is applied to terminal *B*12. Since the peak voltage of such supplies is 42 volts, the TO3 (actually a 0.5-amp device) is rated at 0.3 amp with a minimum load resistance of 80 ohms. However, if a pure d-c source is used (such as a battery or type P7 power supply), the current rating may be increased to 0.5 amp.

The elementary diagram for a NORPAK d-c output amplifier is shown in Fig. 13·35. This is basically a two-stage transistor amplifier. A logic 1 at the input causes transistor 1*TI* to conduct through resistor 1*RES* to common. The voltage across 1*RES* drives the base of 2*TI* negative, causing it to conduct; this provides a path for load current from the —24-volt d-c source to common, energizing the load.

The type TO4 d-c amplifier is identical to the TO3 except in maximum ratings. The ratings are 28.8 watts at 1.2 amp and 24 volts d-c. Minimum load resistance is 20 ohms at 24 volts d-c. Maximum current with a pure d-c power supply or a type P7 power supply is 2.0 amp.

The elementary diagram for an a-c output amplifier is shown in Fig. 13·36. A logic 1 at the input terminal causes the output

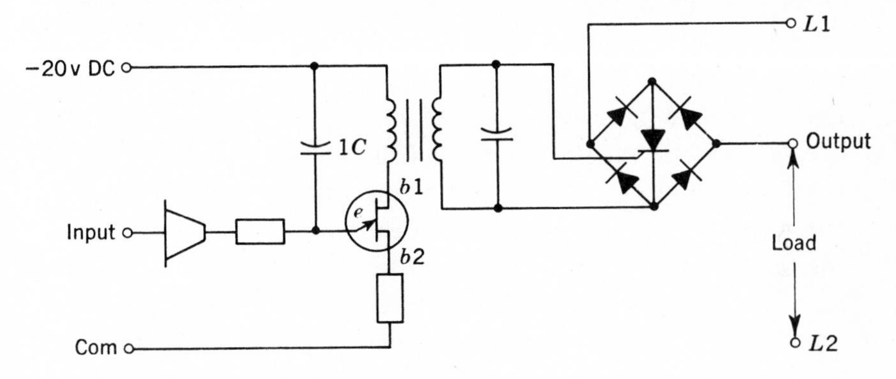

Fig. 13·36 A-c output-amplifier circuit. (Square D Co.)

of the NOR to go to 0, charging capacitor 1*C*. 1*C* charges to the firing voltage of the unijunction transistor. The unijunction then fires, pulsing the pulse transformer, which in turn triggers the gate to the SCR. The SCR fires, providing a path for load current from *L*1 through the bridge rectifier to *L*2 on the positive half-cycle of a-c power and from *L*2 to *L*1 on the negative half-cycle.

As long as the input signal remains, the capacitor will continue to charge and discharge through the unijunction transistor providing continuous a-c voltage to the load.

The type TO10 is an output amplifier designed for use in separate 120-volt (+10 percent, —15 percent) a-c circuits. It has a continuous current rating of 1 amp root mean square, (rms) with a peak inrush current of 10 amp and may be operated from either a 50- or 60-cycle power source. The TO10 will control any Square D Company relay, contactor, or starter up to a NEMA size 2. This amplifier consists of a silicon controlled rectifier circuit encapsulated in a molded bakelite case. Logic and power terminals are located at the top of the case along with a pilot light to indicate the on-off state of the amplifier. The unit is protected internally for line-voltage transients found in supply transformers up to 5KVA. If a larger supply transformer is used, additional surge protection must be added to keep the peak voltage below 400 volts. Leakage current in the output circuit will be 10 ma or less. The type TO10 has an internal firing pulse of approximately 2,000 cps, and therefore it can easily fire into any power-factor load. The type TO10 amplifier energizes the load when a 1 signal is applied to terminal *B*1–2 and requires only one unit of logic load from the controlling logic element. The load should be fused if the device to be operated is a starter, contactor, solenoid, or other electromechanical device wherein there exists the possibility of sticking or jamming. Such sticking may result in sustained peak currents which could be dangerous to the SCR. A Buss Limitron type KAA2 quick-acting fuse, which is rated at 2 amp and 125 volts, is recommended.

The transistorized relay is a device consisting of a transistorized amplifier used to control an output relay. The device includes a power supply which operates from a 120-volt 50- or 60-cycle source. Both the amplifier and the power supply are packaged in an epoxy-encapsulated module. Only one unit of input logic load is required for operation from a NORPAK logic system. The type TO20 is a transistorized relay (Fig. 13·37) offering four different modes of initiation. In all modes of initiation, the 120-volt power is connected to terminals 1 and 2. For operation from a standard NOR, terminal 8 is connected to common, and terminal 7 serves as the input connection. An output of 1 from the controlling NOR will energize the relay, and a 0 output will deenergize it.

There are two general classifications of power supplies used with NORPAK systems—logic power supplies and accessory power supplies. The former is a necessity in all NORPAK systems; the latter is used when particular components become parts of the system. A variety of each type permits selection of only the necessary capacities and functions for a particular system. All NORPAK systems require logic power supplies to provide the power requirements for logic components. Many systems require only one power supply, although in some instances sys-

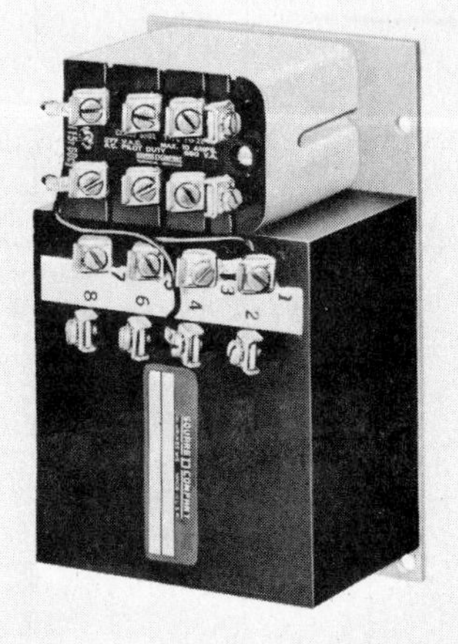

Fig. 13·37 Transistorized relay. (Square D Co.)

tems need more than one power supply because of system size or other factors which present power needs in excess of the capacity of a single basic supply.

The elementary diagram for a NORPAK power supply is shown in Fig. 13·38. This power supply consists of a filtered, full-wave, center-tap bridge circuit for the logic +20-volt and −20-volt d-c voltages. The −130-volt d-c voltage for use with pilot devices is from a filtered, full-wave bridge circuit. All voltages are regulated by means of a regulating-type transformer.

The OFF signal for use with *off-return* MEMORY functions is developed by transistor 1*TI* and its associated resistor-capacitor input circuit. It functions as a single-shot multivibrator providing a short-duration, logic 1 signal to all MEMORYs to insure their return to the OFF state upon loss and reapplication of a-c power.

The pulse signal is a 6-volt 60-cycle a-c voltage for use with *retentive* MEMORYs.

The selection of the type and number of power supplies in a NORPAK system obviously depends upon the power require-

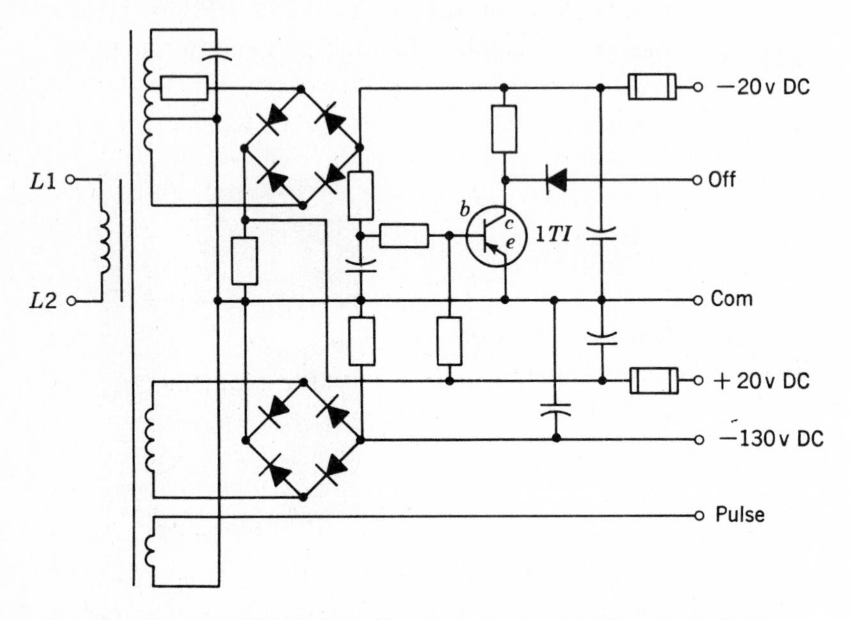

Fig. 13·38 NORPAK power-supply circuit. (Square D Co.)

ments of that system. In general, the −20-volt, and sometimes the +20-volt, system power requirements dictate the particular supply or supplies to be used. Although it is unusual to find a system that requires additional pulse, off, or −130-volt power, a cursory inspection of these requirements should be made. If the results of this investigation reveal requirements which are close to the capacity of one type P1 power supply, an accurate tabulation of the requirements should be completed. More frequently in large systems, it will be found that additional

—20-volt power is required. In this case, the use of a type P7 power supply will provide an additional 1,000 units of —20-volt power. Experience has shown that the addition of this supply is sufficient for all but very large NORPAK systems or those having unusual —20-volt power requirements. Although infrequently, a system may require more than the 1,000 units of +20-volt power to be supplied by the type P1 supply. Such a system could be one which uses a very large number of the type TO3 or type TO4 output amplifiers. The type P7 supply will furnish an additional 20,000 units of +20-volt power. Under these circumstances, the —20-volt terminal of the type P7 would be connected to the system common, and the common terminal would serve as the +20-volt output. When ascertaining the total —20-volt and +20-volt power requirements of the system, one should remember that all NOR elements in a PAK are connected to the supply. Therefore, all NORS, including unused NORS, should be considered in determining the size and number of power supplies. Do not parallel power supplies. If multiple power supplies are required for a system, the common terminals are tied together. The logic load should then be divided proportionately among the applicable outputs of the system power supplies.

13·6 WIRING NORPAK CIRCUITS

Good wiring practice not only means ease of assembly but also is the basis of trouble-free operation. Some circuits are sensitive to external interference, but if the practices discussed here are followed, this problem can be eliminated. The layout of the components cannot be fully planned until the system design has been completed, so that every element has been taken into account. The schematic must then be studied to determine the best layout that will minimize wire runs. The various logic components should therefore be grouped to conform with the circuit functions. One should consider not only interlogic wiring but also wiring from signal-converter outputs to logic inputs, from logic outputs to monitor lights, and from logic outputs to users' terminals and output devices. Grouping the components according to circuit function will frequently result in a few spare logic elements, but this may prove to be an advantage if circuit changes are made or an element must be replaced. If possible, one or two spare NORS should be allowed for in every 20 Pak to facilitate any future changes. It is best to locate such spare NORS at the middle and end of the 20 Pak. Each unused NOR of any type must have one of its inputs connected to common.

Because an OR Pak (L3) contains 21 diodes which are frequently dispersed throughout the system, a central location is most advantageous. *Time* DELAYS (L7, L10) and *retentive* MEMORYS (L5) use few connecting wires and can therefore be placed on the perimeter of the logic area. Transfer Paks (L8) should be mounted adjacent to the MEMORYS with which they are used. Counting or shift-register circuits made up of transfer MEMORY Paks (L11) and BCD counter Paks (L12) should be grouped together, with the input NOR to the counter or shift circuit placed as close to that circuit as possible. Signal converters are usually located along one side of the panel adjacent to the logic elements to which they connect. Power supplies are mounted above the logic components to provide a chimney effect for the heat generated. The output devices are usually mounted to the side or at the bottom of the logic components. If they happen to be magnetic devices (relays, starters, etc.), they are to be kept at least 6 in. from the logic Paks to prevent any possible interference from induced voltages due to magnetic fields.

Logic wiring requires care and thought; because the NOR has a high-impedance input and is switched very rapidly, it is suceptible to induced high-frequency transients. These transients or stray pickup signals are especially effective in circuits using transfer elements in one form or another or where a flip-flop MEMORY happens to be in a position subject to these stray input signals. The effect of stray signals can be eliminated if the layout of the logic components described above is followed. In addition, the following wiring practices should be adhered to. A flip-flop MEMORY circuit should only be made up of NORS of one type (standard NORS, power NORS, or universal NORS) and should be contained in one Pak. Neither of the NORS making up the MEMORY should be used as an amplifier to drive relays or lights. Diodes are not to be included as part of the feedback loop of the MEMORY circuit.

The output of a transfer element is a diode; this allows direct paralleling with other outputs. Each output is to be connected directly to pin 7 or 8 or a NOR, but no more than two transfer elements should be paralleled in such a manner. To accommodate more than two transfer elements, only two outputs are connected to pins 7 and 8 of a NOR with one of its standard inputs connected to —20 volts. When either transfer element pulses, the result will be a —20-volt pulse at the output of the NOR. The outputs of several such NORS are then connected together by means of the usual OR circuit.

When logic-level signals are to interconnect to another remotely located NORPAK panel, these signals should not be taken directly from OR circuits or from MEMORYS but should instead be isolated with NORS (preferably power or universal NORS).

When neon monitor lights are to be used, they should be restricted to use only on or within the NORPAK control cabinet. If the metal case of the neon light is not in firm contact with its mounting surface, the lights may tend to flicker because of slight voltage differences that develop. Incandescent lights should be used whenever a remote indicator is needed.

The +20- and —20-volt power connections to the logic Paks can be made with No. 20 standard wire in series strings "jumpering" from one Pak to the next. The common run of wire, however, should be kept as direct and short as possible. A No. 14 wire should be connected to common on the power supply and then directed down the center of the group of logic components to a series of terminal posts with standoff insulators that are mounted on the corners of the logic Paks. Each of these points then serves as a local common point with a direct return to the power supply. The common connections of no more than three logic Paks should be "jumpered" in series and brought to the local common terminal via the usual No. 20 logic wire. The common terminal at the power supply should then be connected to the chassis with a No. 14 wire under a power-supply mounting screw. The common terminal of a d-c to d-c signal converter (*N*1) should also be connected to the chassis at the mounting screw. Wherever common is brought out of the logic circuit as a users' terminal, it should be connected to the chassis.

Stray interference can be fairly well controlled by following the practices discussed above, but to help in eliminating the paths of this interference, all logic input wires should be separated from the same channel duct conduit or harnesses that carry any power or load wiring. Stray pickup signals are usually the result of transients that are generated by the dropout of a magnet coil. When the coil is being driven by an output amplifier of the type TO3 or TO4, there is a diode built into the device that shunts out any circulating currents resulting from the transient and in so doing provides protection for the transistor and damps out the transient. This protection is lost, however, if any contacts are interposed between the load and the —24-volt terminal; if the contact is opened, the shunt path around the load is broken, and a very high transient could develop. To maintain this protection, terminal *A*3–4 on the static switch should be connected in directly behind the load instead of at the —24-volt terminal.

Whenever the application includes any type of counting circuit, alternator, or shift register that is driven by contact-making devices, the problem of contact bounce must be considered. Circuits designed to eliminate bounce are shown in Fig. 13·32.

Summary

The NORPAK system uses NOR logic and is built around a basic NOR element which is connected in the control panel to make up ANDs and MEMORYs, even ORs.

The logic diagrams used with this system are different from those used in English logic systems but can be developed from English logic diagrams.

An ON signal for NORPAK is −10 to −20 volts d-c, while an OFF signal is 0 volts. The basic power supply provides +20 volts d-c from the base-to-collector circuit of the transistor. In the event that there are some unused NORS, the —20-volt collector voltage and the +20-volt bias voltage would create a collector-to-base voltage of 40 volts which could damage the transistor. Unused NORS should have one input connected to common.

The NORS in NORPAK are encapsulated in standard modules which are then connected into the circuit by jumpers.

The system also includes a test unit which should always be used in servicing the logic section of the panel.

Review Questions

1. List the special functions, other than NORS, available in this system.
2. Draw the NOR logic circuits for AND, OR, NOT and *off-return* MEMORY elements.
3. How many voltages would be required for a complete NORPAK system using direct current on sensing devices?
4. What voltage is used for an ON signal?
5. What voltage is used for an OFF signal?
6. What must be done with unused NORS in a Pak?
7. The logic power-supply voltages are ________.

14 WIRING AND TROUBLESHOOTING STATIC CONTROL

In many ways the problems of installing and troubleshooting static systems are easier and less complicated than those of relay systems. The logic part of the circuit is generally packaged in a panel or panels where it offers ready access to most of the circuit. The modular construction makes unit replacement relatively simple, and trouble location is greatly simplified over a comparable relay system.

14·1 INSTALLATION OF STATIC SYSTEMS

There is one idea which cannot be overstressed in this area of static control: Follow the manufacturer's instructions. Each of the currently available component lines offers reliable, efficient, long-life control, but only when installed as recommended. Each company has engineered its components to work together as a system. They have run exhaustive tests on all phases of the system, including installation. The man in the field can hardly improve upon the proven methods developed by the company.

Some general statements can be made which will apply regardless of which system is being installed. First select good sensing devices. Limit switches and other devices which employ contacts should be selected to provide good contact pressure and good wiping action. Wire each sensing device to an individual signal converter. Never run signal wiring in the same enclosure with power leads. Generally speaking, iron conduit should be used as a raceway for signal wires, but the manufacturer's instructions should be consulted on this. Some systems employ shielded cable under specific grounding requirements.

Should the power supplies be grounded? Which wires should be or can be grounded? Consult the manufacturer's instructions, as there is no general approach to these questions.

If you are building up a panel from components, there are some general principles which should be followed. Group the signal converters so that the wiring from the sensing device does not mix with logic wiring. Place the signal converters so that the wire from the logic side of the logic element is as short as is practical. It is bad practice to feed a logic element in one panel directly from a logic output in another panel.

Group the logic elements within the panel so that a lead between elements can be as short as is practical. Use the size of wire recommended and terminate it in the specified manner.

Amplifiers, relays, and other a-c devices should be mounted to give good isolation from the logic elements and logic wiring. A great deal of electric noise can be generated in these devices, particularly in the relays. Shielding helps prevent interference from electric noise, but isolation is much more effective in most cases.

Good electrical connections are a must in static control. When relays are used for control, the voltage and current in the circuit will allow a great deal of poor workmanship in the joints and connections. Static circuits operate at low voltage levels and at practically no current; therefore they cannot tolerate loose or dirty connections. Bad connections are very hard to locate after installation, and therefore it behooves the installer to be sure each and every connection is made properly when he makes it.

The logic element, whether it be an AND, OR, or NOT, is a very high-speed switch which will react to signals of such short duration that they are merely a flash on the screen of an oscilloscope. Any loose contact or electric noise which provides even a very short pulse to the logic input can cause a malfunction of the system. This type of trouble is very hard to locate unless the signal is applied to a MEMORY element, which will indicate that it has had a momentary input applied.

Any journeyman electrician who has studied the previous chapters of this book and who will read and follow the manufacturers' recommendations should have no trouble installing a static control system properly.

14·2 TROUBLESHOOTING STATIC SYSTEMS

Trouble in any control circuit is first indicated by some part of the machine or process not functioning properly. The cause of the malfunction could be in any part of the system from the sensing devices which initiate the control through the logic elements and amplifier which drive the final action device.

The function of the person who is servicing the equipment is to locate the cause of the trouble and repair it in an efficient manner so as to limit the down time of the system. The serviceman must follow an efficient system if he is to locate trouble and eliminate it with a minimum of wasted effort.

Failure to operate on initial start-up might be caused by errors in wiring. Any system which has been in operation must be wired properly. Therefore, there is no need to check it out wire by wire. Many servicemen with years of experience seem never to learn this fact and waste many hours in circuit tracing.

Consider the circuit of Fig. 14·1. The solenoid does not operate when it should. We will assume that all logic elements are equipped with state-indication lights, as are the signal converters.

The first step would be to make a voltage check at the input of the solenoid or the output of the amplifier, whichever location is accessible without opening connections or disturbing the wiring. The presence of proper voltage at the output of the amplifier would indicate a bad solenoid or broken connections between the solenoid and amplifier. A further check must then be made at the solenoid.

Assume that there was no output from the amplifier and that all logic elements are equipped with state-indication lights. In this case it would not have been necessary to make any voltage

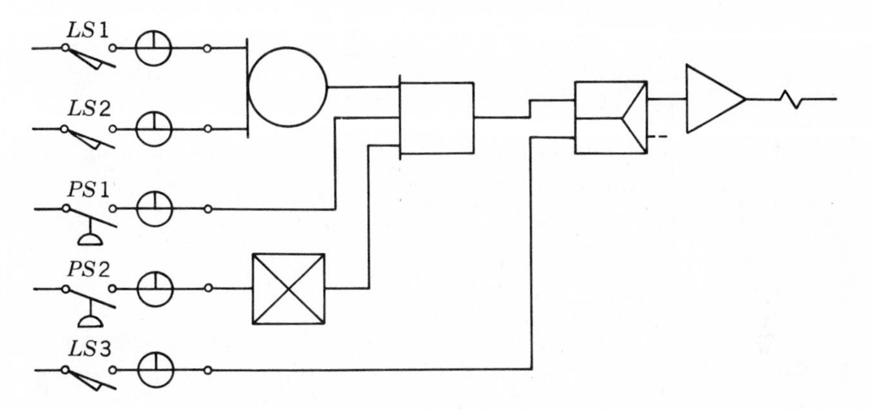

Fig. 14·1 Static control of a solenoid valve.

checks unless the MEMORY element indicated an output. With an output from the MEMORY, the trouble must be in the amplifier if it has no output. A check of the amplifier's input voltage will eliminate any possibility of trouble in the logic wiring between the output of the MEMORY and the input of the amplifier.

The above process can be quickly followed through the logic section by merely checking the state lights on each element. In this case the AND should be on, the OR should be on, and the NOT should be on.

Most cases of trouble will be found in the final action device or the sensing devices. When the signal converters are equipped with state lights, a visual check of these will indicate which sensing device is responsible for the trouble. In the circuit of

Fig. 14·1, the lights associated with *LS*1 or *LS*2 or both should be on. The light for *PS*1 should be on, and the one for *PS*2 should be off. Any light which is not giving the proper state indication will lead you to the actual cause of the trouble.

Suppose the system is not equipped with state lights. This in itself will make servicing the system a slower process but does not materially alter the steps to be used. The panel should be equipped with a tester designed by the manufacturer to be used to test for ON and OFF signals at the input and output of a logic element. It becomes necessary to use the test lead and check the input and output of each logic element from the amplifier to the signal converter until the faulty one is located.

There is no need to check the logic elements until all original inputs are determined to be correct and present at the output of the signal converters. A check of the input to the amplifier will then tell you whether there is any trouble in the logic section. Whenever the input to the amplifier responds properly to changes in inputs to the logic section, the logic elements are operating properly.

Many times it is necessary to check part of the system under input conditions not represented by the condition of the sensing-device contacts. For instance, the NOT in Fig. 14·1 can only be checked when *PS*2 is closed. This condition can be simulated by applying the proper ON voltage to the input of the NOT. This is best done by using the testing device available from the manufacturer as part of the component line for the system.

One of the real advantages of a static control system is that in most cases the source of trouble can be determined while you stand in front of the logic control panel.

The approach up to this point has been to use circuit testers designed especially for the system to be serviced. Signal tracing and service can be performed by taking voltage readings at the input and output of a logic element if good voltmeters are used. The minimum acceptable voltmeter would have a 20,000 ohms-per-volt sensitivity. Do not use neon testers or 1,000 ohms-per-volt voltmeters for this purpose.

If voltage testing is to be used, the serviceman must know what voltages represent ON and OFF in the particular system in question. Of those we have studied, the General Electric Company system uses 0 volts for ON and −4 volts for OFF. The Cutler-Hammer company system uses +10 volts for an ON signal and 0 volts for OFF. The Square D company system uses −10 to −20 volts as an ON input and 0 volts for OFF.

When a zero-volt signal is indicated for the OFF or ON condition, a good voltmeter may well indicate up to about one-half volt due to leakage current through the transistors. This is normal and should cause no concern.

When more than one part of the system is malfunctioning or when the output amplifier seems to be bad, check the power supply. Power-supply troubles account for a large percentage of trouble in most electronic devices and systems and therefore should always be suspected until checked out.

Intermittent troubles are always hard to find and can be caused by bad contacts or poor connections. One factor which must always be considered when one seeks the cause of intermittent trouble is the presence of electric noise. When a system has been in operation for some time without noise problems and then develops trouble which is suspected to be noise, certain things should be checked. If any new wiring has been installed, check it out. New installations of other equipment or wiring near the panel or field wiring for the system should be checked as a possible source of noise. When no changes have been made, check out grounding connections and filters which may have gone bad.

The static elements are probably the most reliable part of the entire system. Be sure to check fuses, switches, and other common electric parts of the system. They probably are the source of the trouble.

14·3 ELECTRIC NOISE

The term "electric noise" has been used many times in previous discussions and deserves some clarification and explanation. This section will be devoted to the subject of noise and its elimination.

Electric noise signals can be generated by almost any power equipment. The opening or closing of contacts under load almost always produces electric noise. The transmission system which feeds power to the building carries electric noise, and every piece of operating equipment connected to the wiring system of the building is a potential source of noise. Electric noise, then, cannot be prevented, but must be kept out of the signals fed into the logic control components.

Electric noise exists in many forms, but for the purpose of this discussion consider it to be essentially a sudden change in current flow or voltage in a conductor. When changes in value of current or voltage are uniform or regular, we refer to this current as alternating current or a-c. We then might describe electric noise as nonuniform alternating current.

An acceptance of the above oversimplified description of electric noise provides us with some rules for keeping it out of the control circuit. Signals of alternating current, and therefore electric noise, are transferred or coupled from one circuit to another by one or more of three ways.

The first way, and probably the most common, is through mutual electromagnetic coupling. The second is through the capacitance between conductors, or electrostatic coupling. The third means of coupling is by means of common impedances between the circuits.

Most cases of electric noise will involve a combination of the three forms of coupling, since every circuit has inductance, capacitance, and impedance in varying proportions. We will consider each form of coupling separately for simplification.

Electromagnetic coupling occurs whenever the circuit containing the noise is run parallel and close enough to the signal circuit to provide coupling through their mutual electromagnetic fields. The most effective means of reducing this form of coupling is by separation. The degree of coupling decreases with the square of the distance between the conductors. Electromagnetic fields tend to cancel each other in circuits with twisted-pair wires;

therefore if either or both the noise circuit and the signal circuit employ twisted-pair wiring, the coupling will be reduced. The third method of reducing electromagnetic coupling is by means of shielding. The noise which is the source of trouble is generally low frequency and therefore requires a ferrous material as an effective shield. Rigid conduit of the common electrical variety is ideal for this purpose, provided only that signal wires are run in the pipe. The amount of electromagnetic coupling between two circuits is proportional to the distance for which they are parallel; therefore reducing the distance will reduce the coupling and the noise problem. Each case of noise coupling is different and may require analysis if it is severe, but most can be eliminated by using good installation practice.

Electrostatic coupling is caused by two conductors run parallel and at different voltages. The conductors form the plates of a condenser. The air or other material between the conductors forms a dielectric. The most effective means of reducing this coupling is by shielding the signal wires. Separation of the conductors and reduction of the distance where the two circuits are parallel will also help.

Common impedance coupling occurs only in circuits which share some impedance. The most common source of this type of circuit is the temptation to use a common wire to feed one side of many sensing devices. The most effective way to eliminate this form of coupling is to use two wires from the panel to each sensing device.

The use of common arc-suppressing circuits on relay contacts and inductive loads can prevent or reduce the generation of noise signals at their source.

A summary of noise suppression would indicate that static systems should be wired with care. Separate circuits should be used for each sensing device. The use of twisted-pair leads for sensing or signal circuits might be indicated in some cases. Ridged conduit for field wiring seems to provide the best likelihood for trouble-free operation. Remember, consult the manufacturer of the system in question for specific details.

Summary

The installation of static control systems requires only normal electrician's skills plus a little care. Follow these rules: Make every joint carefully. Never run signal wires in conduit with power wiring. Keep signal wires and logic wiring separated in the panel. Follow the manufacturers' recommendations. Avoid noise sources.

Whenever possible, use the testing devices available from the manufacturer of the system in question. If it is necessary to service with a voltmeter, use one with at least a 20,000 ohms-per-volt rating.

15 SHIFT REGISTERS AND COUNTERS

Previous chapters of this book have been concerned with the application of static devices to logic control in which the decision section of the control system was completely static in nature. There is another ever-growing field of application—the use of static MEMORY-type elements for counting and sorting of material or operations.

A complete analysis of this area of control and the components used is subject matter enough for a complete book. This chapter will give you an introduction to some of the basic concepts.

15 · 1 COUNTERS

In any industrial control system, it is often necessary to count up to a certain number and then have an action occur. Such a case might be a palletizer in which it is desired to load x products into the pallet, move out the pallet, reset the counter, and repeat the cycle with a new pallet in place.

Various counting methods have been developed to handle almost any application, and it becomes necessary to choose the type of counter best suited for the job to be performed. There are basically three types of counters used today in such applications, the decade counter, the binary counter, and the ring counter. Pure "counting" is usually done with a decade or binary counter; the ring counter is most often applied for sequencing and is the solid-state equivalent of a stepping switch. For pure counting, decade counters are usually chosen because they are easy to understand and apply, easy to decode, easy to read out (visually), and by definition easy to adapt to the decimal system. They also make clean subtract or add-subtract counters. Binary counters, in comparison, use fewer stepping (transfer) elements but take more logic circuitry to decode. They also are harder to read out and do not lend themselves easily to downcounting.

Any kind of counter may be easily constructed by making use of one basic element, the *step* MEMORY (refer to Sec. 11·5, "Special Functions," for information on operation of this element). (Add-subtract counters are made with the *bi-step* MEMORY.) The *off-return step* MEMORY is shown in Fig. 15·1.

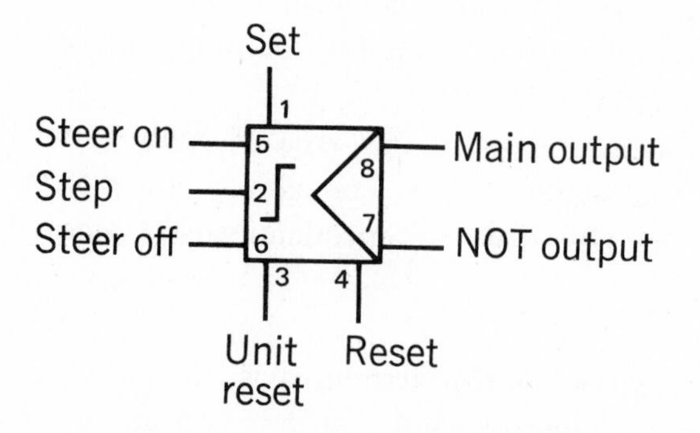

Fig. 15·1 Off-return step MEMORY. *(General Electric Company)*

Step MEMORYS when connected in chains can store as well as transfer information when the chain is properly inputed. They may also be manually set and reset.

Following are descriptions of the decade, binary, and ring counters. In all such applications, it will be noted that the stepping signal (if originating from a contact-making device) should first be brought through a signal amplifier to filter out contact bounce. The system of numbers or counting used in this country is called the decimal system. A counter used to count in the decimal system is called a *decade counter.* Since there are nine numbers plus 0 in this system, one would think that an indicator or counter would require at least ten MEMORYS to count and indicate numbers from 0 to 10. This is not necessarily so, as an examination of Fig. 15·2 will show.

Figure 15·2*a* shows five MEMORY elements, each equipped with an indicating light to show when it is on. Define all lights off as zero. An ON pulse to the first MEMORY will turn on its light and indicate the count of one. An ON pulse to the second MEMORY will turn on its light and indicate a count of two. This process will continue through the count of five, which will be indicated by all lights on.

At this point we can apply an OFF pulse to the first MEMORY, which will turn its light off and leave the other four on. This

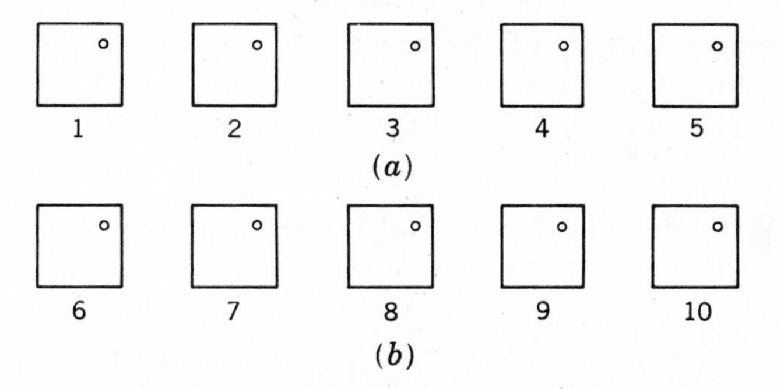

Fig. 15 · 2 Decade counter, block diagram.

indicates a count of six. Successive OFF pulses applied to MEMORYS 2, 3, and 4 will turn their lights out in sequence to indicate counts of seven, eight, and nine. An OFF pulse applied to MEMORY 5 would turn its light off and provide a 10 count, but it would look the same as zero. The decimal system of counting requires that when we count to nine and add one more count, we register a zero and carry a one. The same thing can be accomplished by applying an OFF pulse to MEMORY 5 and at the same time applying an ON pulse to MEMORY 6 (Fig. 15·2*b*). The indication of count 10, then, is MEMORYs 1 to 5 OFF and MEMORY 6 ON. MEMORYS 1 to 5 have the ability to count to 10 and therefore are referred to as a *decade,* even though in themselves they cannot indicate a 10 count as differentiated from a zero count. This follows the decimal system of numbers in that the highest digit which can be indicated in any single column is nine.

Ten MEMORYs, as shown in Fig. 15·2*a* and *b,* will allow a count to 100 in the following manner. MEMORY 6 is left with its light on, and ON pulses are applied in sequence to MEMORYS 1 to 5 for the indication of counts of 11 to 15. OFF pulses are then applied in sequence to MEMORYS 1 to 4 to indicate counts of 16 to 19. The OFF pulse applied to MEMORY 5 must also pulse MEMORY 7 on to indicate the number 20, that is, MEMORYs 6 and 7 both on. The next pulse turns on MEMORY 1 again, and the process repeats itself until the OFF pulse applied to MEMORY 5 has turned on MEMORYs 8, 9, and 10 in sequence, indicating counts of 30, 40, and 50. The next nine pulses will turn on MEMORYS 1 to 5 and then turn off MEMORYS 1 to 4 to indicate counts of 51 through 59. The next pulse will be number 60 and will turn off MEMORYs 5 and 6 to indicate a 60 count. This process is continued until MEMORY 10 is turned off by the 100 pulse. When higher counts are desired, additional decades are added; the fifth MEMORY in each decade transfers its OFF pulse to the next decade, just as you carry numbers in adding written figures.

Practical decade counters might have the indicators separately mounted. They might be arranged horizontally or vertically, or a readout tube or device might be used to indicate the count.

The MEMORY elements used in this type of counter are somewhat special and carry different names when made by different companies. The *step* MEMORY discussed in Chap. 11 is one such component. The essential requirement is that the MEMORY element have the ability to receive a signal and store that information until it receives another signal, and then transfer that information to another MEMORY in series.

15 · 2 THE GENERAL ELECTRIC COMPANY DECADE COUNTER

In the decade counter, five *step* MEMORYS are used to count from 0 to 9 and are coded as shown in Fig. 15·3.

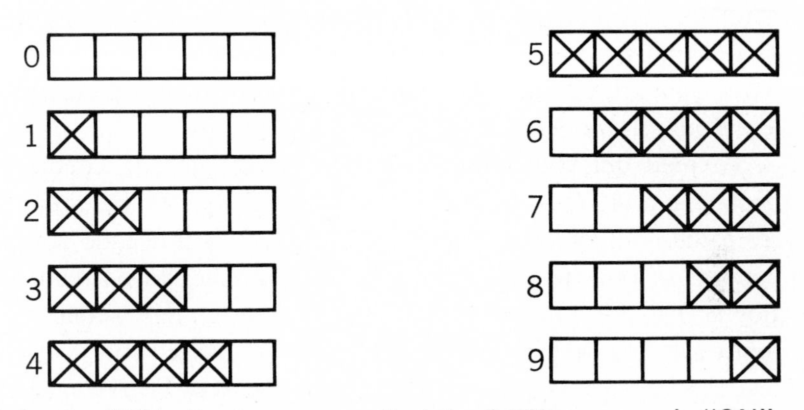

Note: An "X" in the box means that the STEP memory is "ON".

Fig. 15 · 3 Decade counter, coding. (General Electric Company)

To count to 99, two banks of five *step* MEMORYS are used (the second bank is coded to represent 10, 20, 30, up to 90); to count to 999, three banks are used; and so on.

The circuit for the decade counter is shown in Fig. 15·4. Five *step* MEMORYS are used to count to 9. The units are connected in conventional chain fashion (each unit steering on or off the next unit) except that the outputs of the last unit are reversed when they are fed into the steer inputs of the first module. Thus the standard output is fed into the steer-off terminal, and the NOT output is fed into the steer-on terminal. The set terminals (pin 1) are all tied to 0 volts. With this connection, the circuit counts by the *step* MEMORYs, turning on in sequence and then turning off in the same sequence. First count *A* turns on; second count *B* turns on; third count *C* turns on; fourth count *D* turns on; fifth count *E* turns on; sixth count *A* turns off; seventh count *B* turns off; eighth count *C* turns off; ninth count *D* turns off; tenth count *E* turns off. The counter is now back in the start condition.

With the tenth count, the NOT output of *E* unit may be used to step the second decade row (if it is desired to count above 9). Each time the top row counts to 10 the second row is then stepped, and therefore the second row counts 10s. More decade rows may be added for hundreds, thousands, etc., as required. (Note that the stepping of subsequent rows does not take place when panel power is first applied because of the unit reset, which locks out all *step* MEMORYs for 15 msec.)

To read out coded information from the decade counter, a decode circuit must be used for each number to be read out from the counter. The decode circuit for 0 to 9 decade counter consists of one two-input AND for each number to be read out. For a 0 to 99 decade counter, the decode circuit is one four-input AND for each number (two inputs from each row); and so on. Figure 15·5 shows which outputs need to be fed into the AND for each number 0 to 9. For example, to decode the number 6 a two-input AND would be required with the connections as shown in Fig. 15·4. If a momentary readout only is desired and if the step signal is momentary, the step signal may be ANDed with the inputs to the decode ANDs.

Adjustable preset readout. If the readout of the counter is to be adjustable (by means of a selector or thumbwheel switch), each switch contact signal may be brought through an original input and ANDed with the inputs to the proper decode AND. If many selectable points are desired, a more economical method utilizing the 1200A interface amplifier is shown in Fig. 15·6. A two-deck thumbwheel switch and two-input AND are used for a 0 to 9 counter; a four-deck switch and four-input AND would be

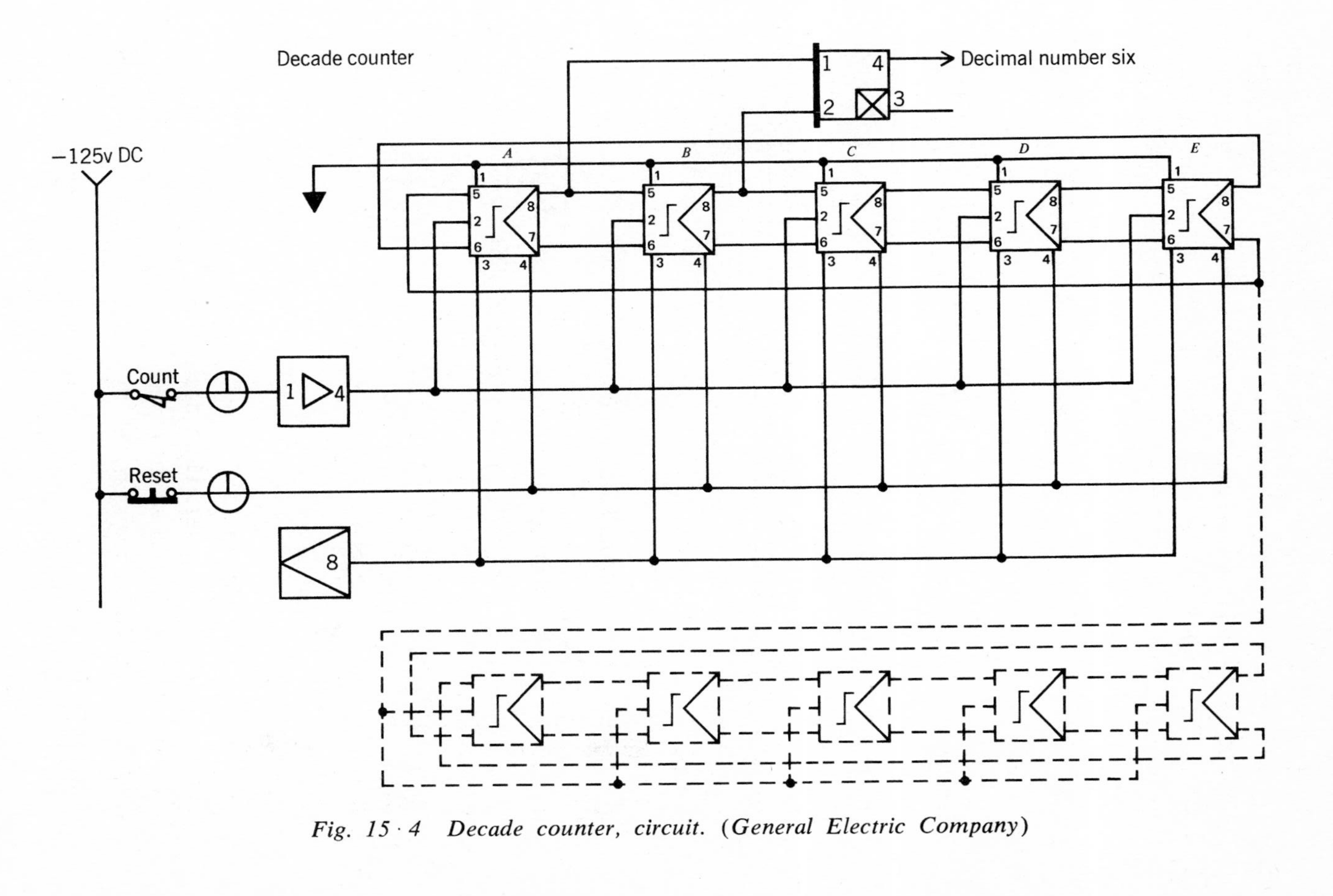

Fig. 15 · 4 Decade counter, circuit. (General Electric Company)

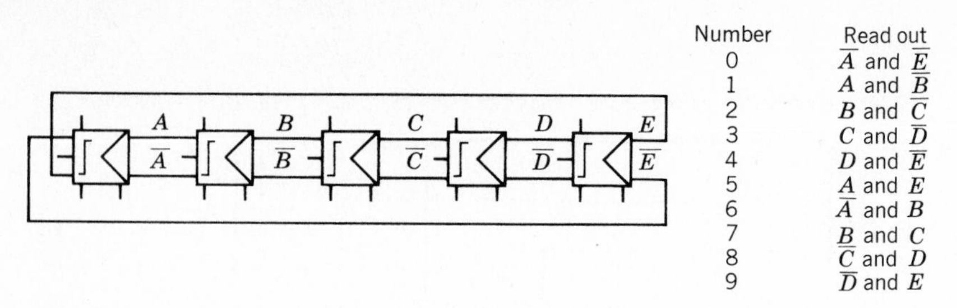

Number	Read out
0	$\overline{A}$ and $\overline{E}$
1	A and $\overline{B}$
2	B and $\overline{C}$
3	C and $\overline{D}$
4	D and $\overline{E}$
5	$\underline{A}$ and E
6	$\overline{A}$ and B
7	$\underline{B}$ and C
8	$\overline{C}$ and D
9	$\overline{D}$ and E

Fig. 15 · 5 Decade counter, decode circuit. (General Electric Company)

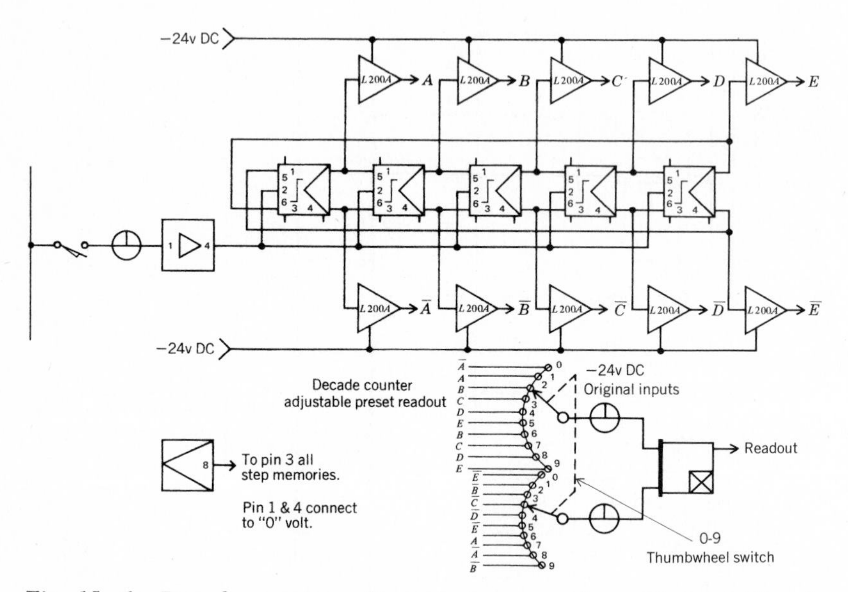

Fig. 15 · 6 Decade counter, adjustable preset readout. (General Electric Company)

used for 0 to 99; and so on. Of course, a source of 24 volts d-c is required.

Down counter. Figure 15·7 shows the connection for a down (subtract) decade counter. Here each *step* MEMORY steers the one just before it, and so when stepped the counter counts backward. The first count registers 99, the second count 98, and so on continuously to 0.

Down counter with adjustable preset input. Sometimes when

it is required to count a preset number of events and then initiate an action, it may also be required that the preset point be adjustable. In such cases, rather than have a large number of selectable decode circuits, it is more economical to set in the required count, count down, and initiate the action when the counter reaches 0. Such a circuit is shown in Fig. 15·8. Here, setting in the proper number is made easy by use of decade-coded thumbwheel switches, and only one decode AND is needed (to read out the count of 0). The beauty of this approach is

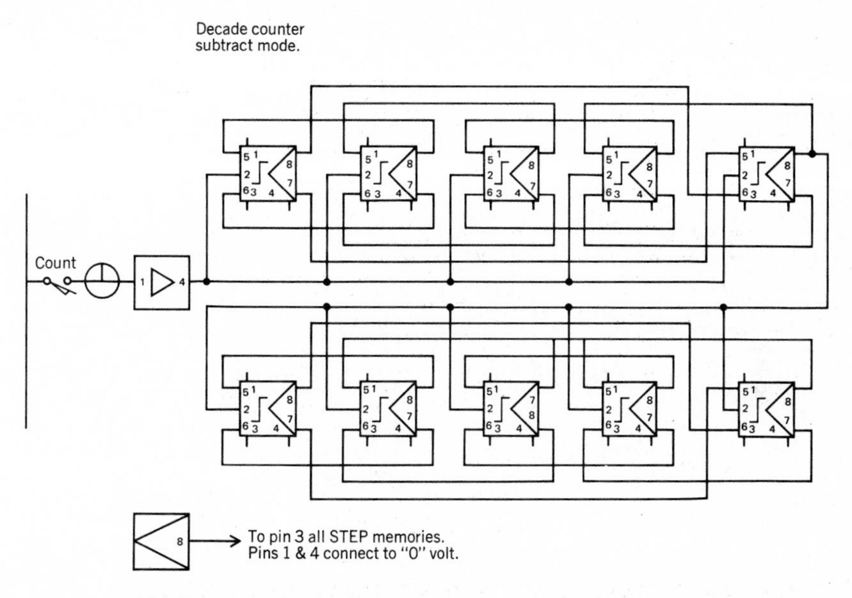

Fig. 15 · 7 Decade down counter. (General Electric Company)

that the decade counter is based on the decimal system (it counts in units, tens, hundreds, etc.) and therefore large numbers may be easily set in with the appropriate number of thumbwheel switches. In Fig. 15·8, for example, the number 82 may be set in by turning the tens switch to 8 (sets the lower decade row to 80) and the units switch to 2 (sets the upper decade row to 2). When the counter reaches 0 and the action is initiated, the thumbwheel switches are permitted (for 100 μsec) to set the counter again.

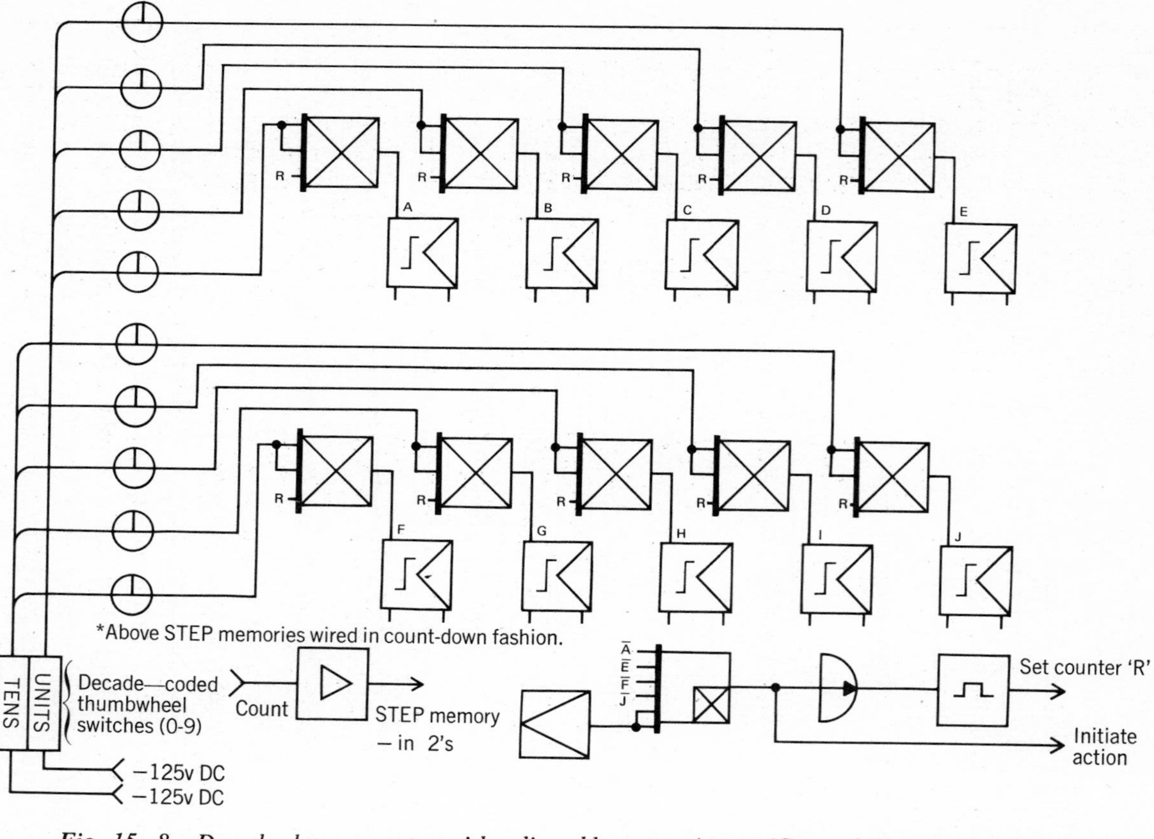

Fig. 15·8 Decade down counter with adjustable preset input. (General Electric Company)

Up-down counter. Figure 15·9 shows the connection for an up-down (add-subtract) counter. The *bi-step* MEMORY, having *two* independent steer-step inputs, is employed. The pin 5s (steer on) and pin 2s (step) are interconnected with the pin 8s in the standard manner for an up counter. The pin 6s (steer on) and pin 3s (step) are interconnected with the pin 8s in the standard manner for a down counter. The only exception is that in the *bi-step* MEMORY the steer-off inputs are the built-in NOTS of the steer-on inputs, and so it is not necessary to make any connections to pin 7 of the *step* MEMORYS.

The two *step* MEMORYS on the lower right side of Fig. 15·9 serve to determine whether the counter is counting up or counting down at the moment the second decade bank is to be stepped.

15 · 3 RING COUNTERS

The ring counter is not practical for pure counting but is a powerful tool when used for sequencing of events. As such it is the solid-state equivalent of a stepping switch and inherently has many additional advantages over conventional stepping switches. In addition to long life, the ring counter may be positively interlocked for foolproof sequencing (see Fig. 15·10), may return to home without retracing any steps, and may go forward or backward if desired.

The ring counter, shown in Fig. 15·11, may be made by chaining up *step* MEMORYS in conventional fashion (one *step* MEMORY for each step). Note that the standard output of the last *step* MEMORY is fed back into the steer-on input of the first *step* MEMORY and the NOT output of the last *step* MEMORY is fed back into the steer-off input of the first unit. Thus each *step* MEMORY is connected in the same manner (thus forming a ring).

With this connection, *one step* MEMORY must be turned on before any other *step* MEMORY will be steered on. The chosen *step* MEMORY may be turned on as shown in Fig. 15·11 or may be set on automatically by interchanging its pin 1 and pin 3 connections. Now with the first unit on, it is steering the second

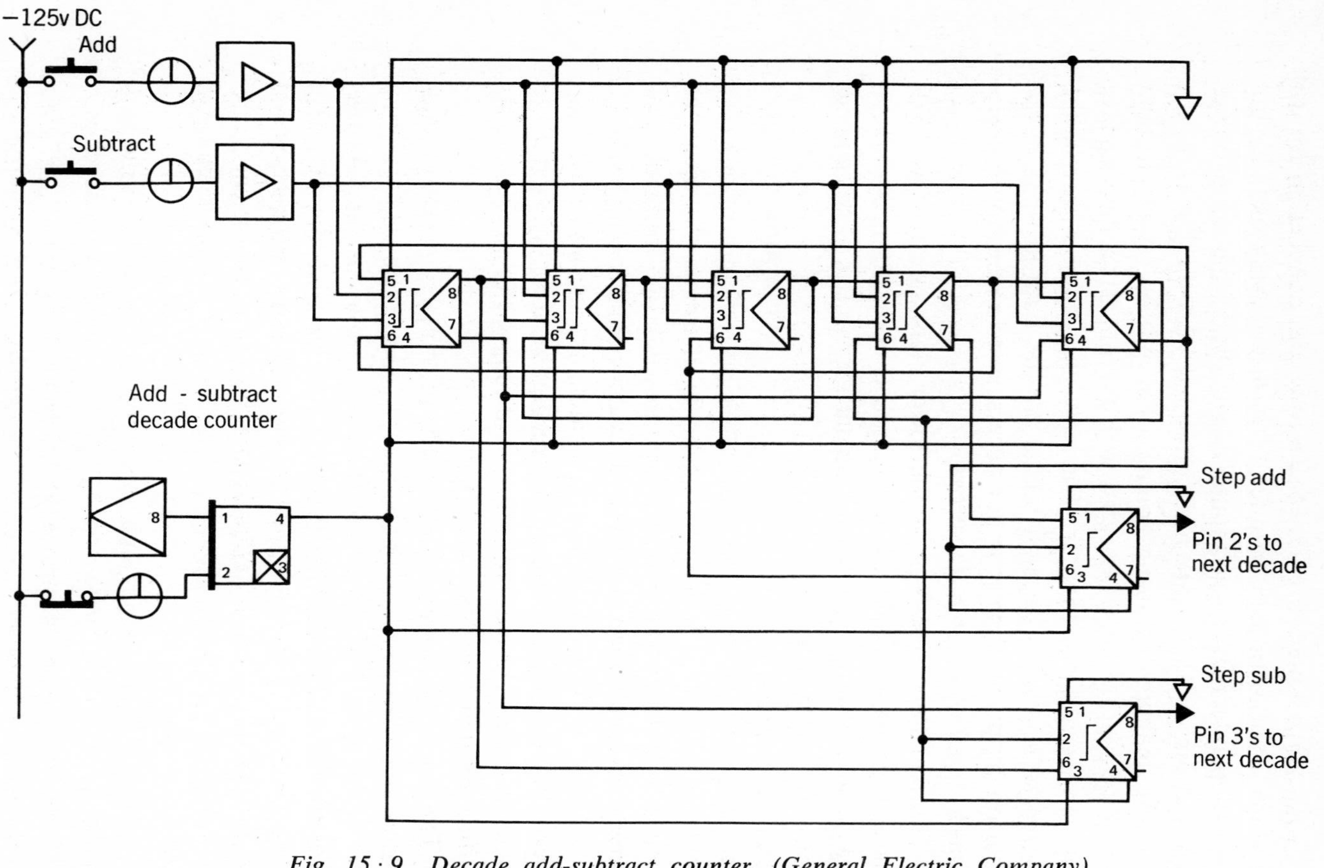

Fig. 15·9 Decade add-subtract counter. (General Electric Company)

unit on while the last unit is steering the first one off. When the step limit switch is closed, the second *step* MEMORY turns on, and the first one turns off. Now the third unit is being steered on, and the second one is being steered off. When the step limit closes again, the third one turns on, and the second one turns off. With each step, the output signal will move automatically to the next unit, and when it gets to the last unit, it will ring around to the first one and continue.

A common ring-counter application is one in which the ring counter automatically steps itself to the next operation upon completion of the current operation. Figure 15·10 shows this circuit. Positive operation of the system is guaranteed at all times by the two-input ANDs, which will not permit stepping unless the right operation is being performed and *that* operation is then completed.

15·4 BINARY COUNTERS

Understanding binary counters requires some understanding of binary arithmetic, and a study of this subject is highly recommended to those who wish to further their knowledge of complex numerical control and computer applications in industry. A brief explanation of some basic concepts of binary numbers here will suffice for an understanding of counter readout.

The decimal number system and the decade counter must utilize ten identifiable digits or conditions of the counter. This is true because there are ten numbers in the decimal system. The binary system of numbers has only two digits, 0 and 1. The decimal or decade counter must be able to indicate 10 conditions for each column or position and must have the ability to carry to the next column or position. The binary counter will only need to indicate two conditions for each column or position and have the ability to carry to the next column or position.

You may not be aware of it, but when you write a number in the decimal system such as 111, what you are actually saying is: one 100 plus one 10 plus one 1 equals 111. The positional

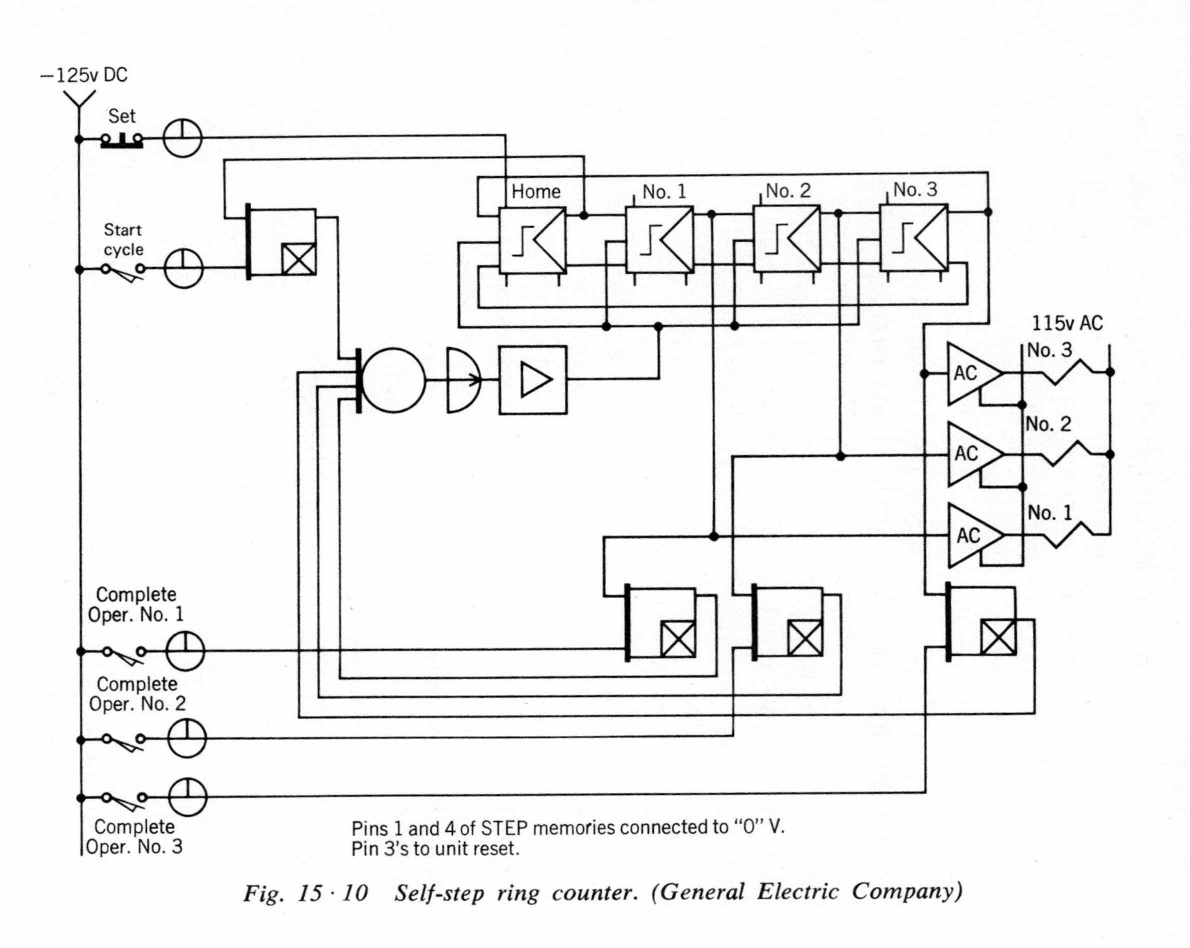

Fig. 15 · 10 Self-step ring counter. (General Electric Company)

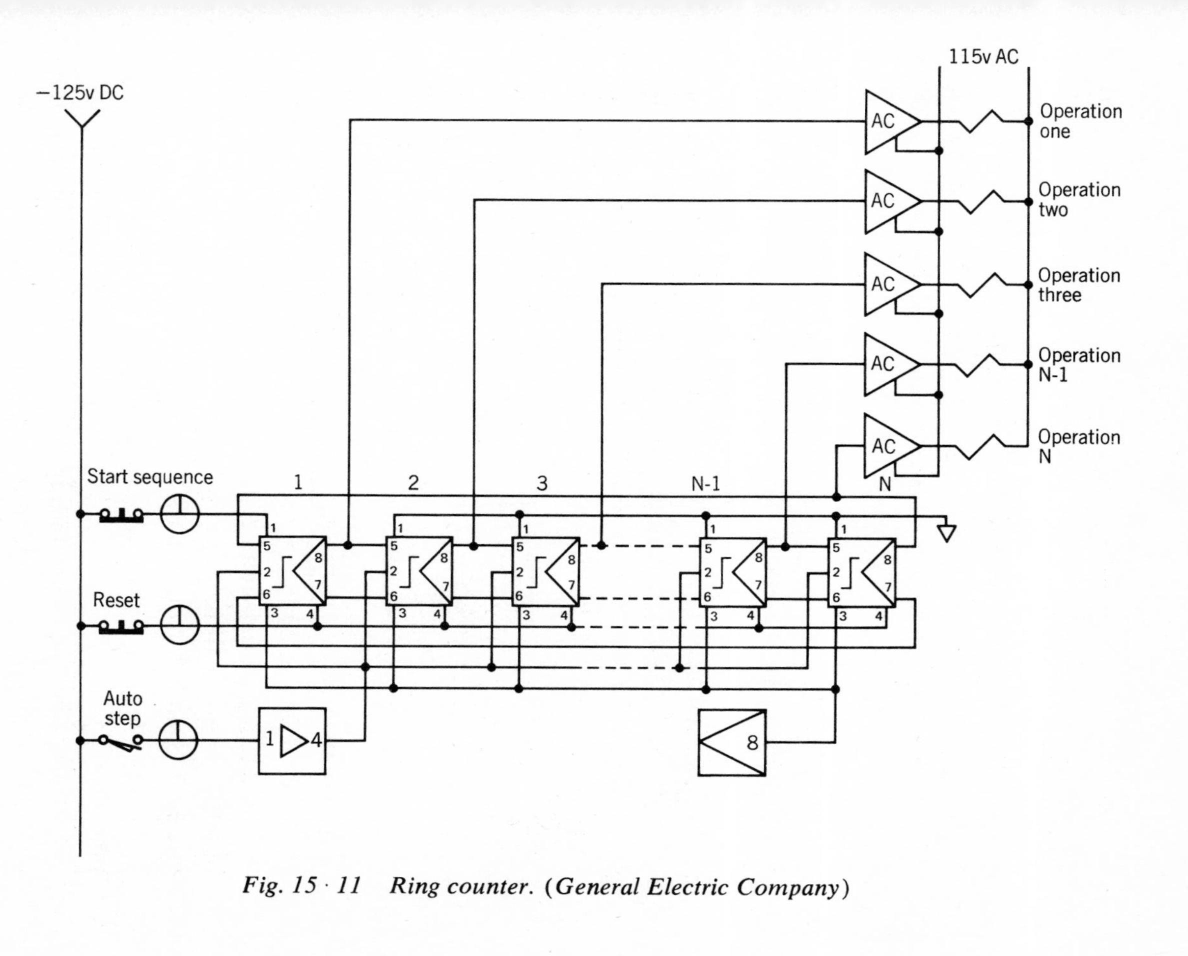

Fig. 15 · 11 Ring counter. (General Electric Company)

value of the units column is one. The positional value of the next column to the left is 10. The positional value of the next column to the left is 100, etc. Since there are nine digits possible in each column or position, the highest number possible with three positions is 999. This is actually $(9 \times 100) + (9 \times 10) + (9 \times 1)$.

The binary system provides only for the existence of a 1 or 0 in each position, as these are the only two digits in the system. This positional value of each column is $(2^4 2^3 2^2 2^1 2^0)$. The first position's value is 1, the second position's value is 2, and the third position's value is 4. Proceeding to the left, the positional values are 8, 16, 32, 64, 124, 248, etc. An easy way to remember this is to start with 1, and then each position to the left is the preceding position multiplied by two.

Binary numbers look a little strange until you get familiar with them. Consider the binary number 10101. This is not ten thousand one hundred one. Remember, each of the 1 digits indicates one times the positional value of its place in the number.

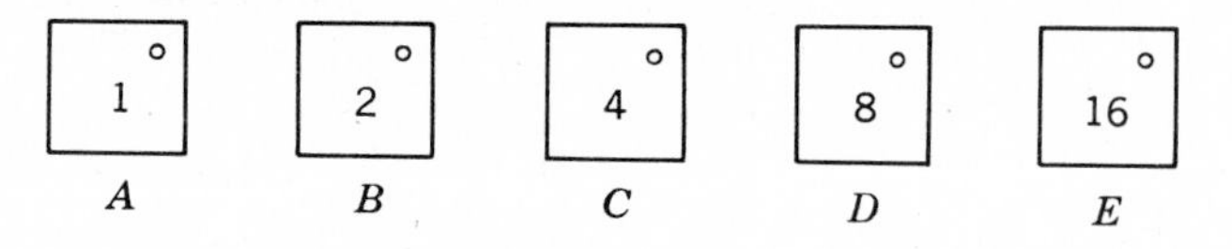

Fig. 15 · 12 Binary counter, block diagram.

The number 10101 actually says: $(1 \times 16) + (0 \times 8) + (1 \times 4) + (0 \times 2) + (1 \times 1)$ or $16 + 4 + 1$. This of course equals 21 in decimal numbers. A little experience working with binary addition will help you get this in mind. Remember, when you add 0 and 0, you get 0. When you add 0 and 1, you get 1. When you add 1 and 1, you get 1 plus a carry to the next position.

Consider the MEMORYS of Fig. 15·12. The small circle is the ON light so that we can identify the state of the MEMORY, ON or OFF. The numbers indicate the positional value of the MEMORY when it is ON. Pulses fed into this group of MEMORYS are to be counted by binary arithmetic. The first pulse turns on A to indicate a one count. The second pulse turns off A and turns on B to indicate a count of two. The third pulse must turn A on and leave B on for a three count. The fourth pulse must turn off A and B and turn on C. The fifth count

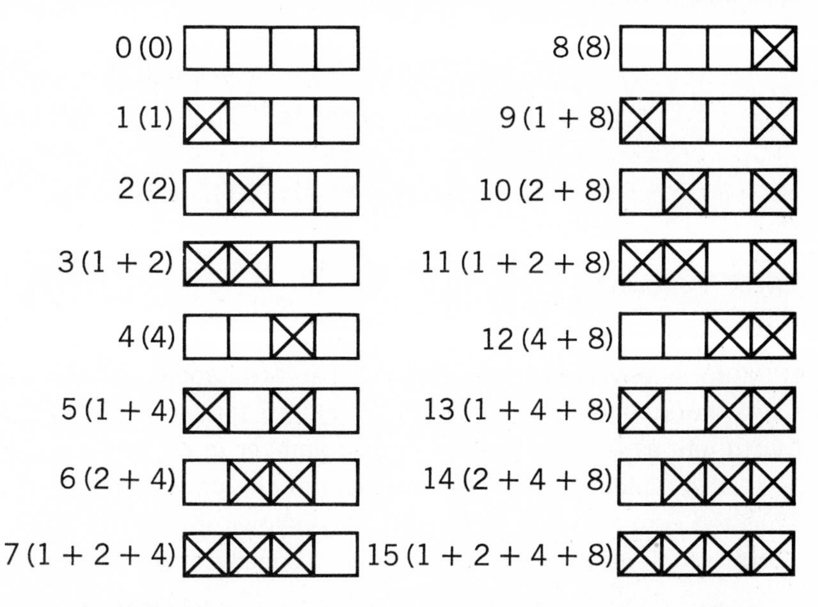

Fig. 15 · 13 Binary counter, coding. (General Electric Company)

will again turn on A to give a five indication. The above process continues until the maximum count of five MEMORYs, decimal number 31, is reached. The binary number 10101 or decimal 21 discussed previously would be displayed by A on, B off, C on, D off, E on. Numbers higher than 31 can be counted by adding MEMORY units. The addition of only one MEMORY would increase the total count to 63.

The greatest advantage to the binary counter is its economy in components. A decade counter of five elements will only count to 10, not to 31 as is possible with a binary counter.

The greatest disadvantage of the binary counter is the difficulty of readout, whether it be visual or electrical.

15 · 5 GENERAL ELECTRIC COMPANY BINARY COUNTER

In the binary counter, the number system used is one in which the decimal digits are represented by a binary code. This system of coding the numbers in the counter is called *binary-coded decimal* (BCD).

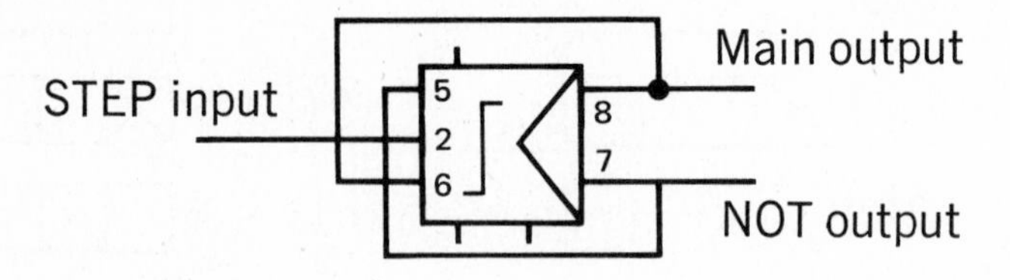

Fig. 15 · 14 Binary counter, connection for step MEMORY. (*General Electric Company*)

For this counter, we also use *step* MEMORYS. Each *step* MEMORY is assigned a binary code. The first unit is coded as 1 (2^0); the second unit as 2 (2^1); the third unit as 4 (2^2); the fourth unit as 8 (2^3); and so on. The number in the counter can be read by adding the coded numbers of the units that are ON. Thus the counter for 0 to 15 will be as shown in Fig. 15 · 13.

If we look at the first *step* MEMORY (coded 1), we see that it is ON every other count. Thus the unit must change state with every step input. To do this, we connect the *step* MEMORY as shown in Fig. 15 · 14.

With this connection, the *step* MEMORY is steering itself ON when the unit is OFF and conversely steering itself OFF when the unit is ON. Thus, with every step input the unit will change state.

If we look at the second *step* MEMORY (coded 2), we see that it changes state every time the first unit turns OFF. Thus by making the connections for the second unit as shown in Fig. 15 · 14 and using the NOT output of unit 1 as the step input of unit 2, we can make a binary counter to count to 3 and reset. By chaining up *step* MEMORYS in the same fashion, we can make a binary counter to count to 7 with three units, to 15 with four units, to 31 with five units, and so on. With 10 *step* MEMORYS, the counter can count to 1,023. Figure 15 · 15 shows the complete circuit for a binary counter which will count to 31.

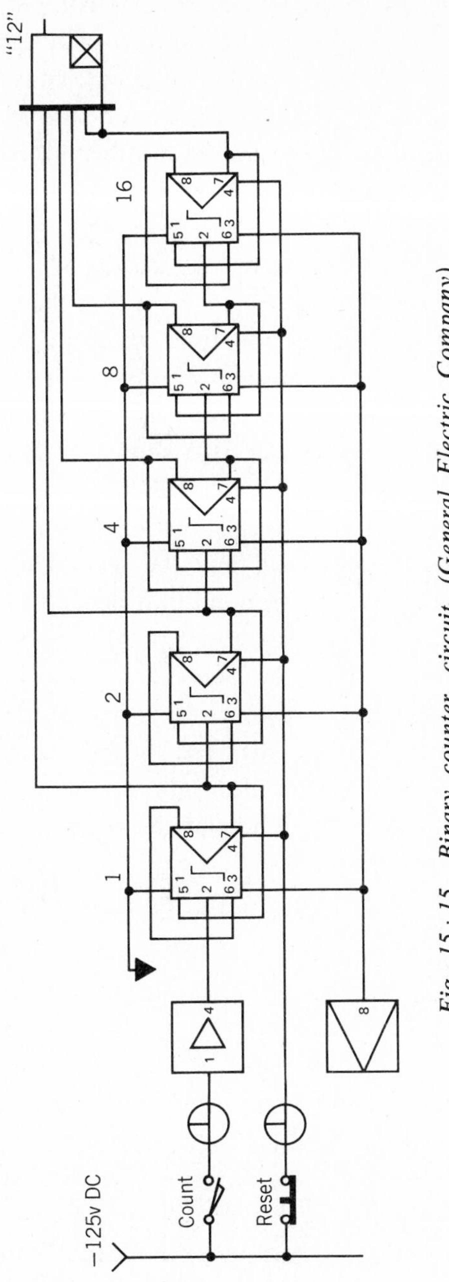

Fig. 15·15 Binary counter, circuit. (General Electric Company)

To remove a number from a binary counter, the decode circuit consists of an AND whose number of inputs equals the number of *step* MEMORYS in the counter. Thus to decode a number, you must look at the state of each *step* MEMORY in the counter. For example, to decode the number 12 from a 0 to 31 counter you would need a five-input AND with the connections as shown in Fig. 15·15.

A problem exists if the output of the decode AND feeds into a module with a flip-flop circuit in it, in that the AND may have a short output pulse during the time when the *step* MEMORYS are switching. This is because a *step* MEMORY must turn off before the next *step* MEMORY can respond to its step input and turn on. For example, the AND shown in Fig. 15·15 may produce a pulse output on the fourteenth pulse. The thirteenth pulse would have outputs at 1, 4, and 8. The next pulse would turn the first *step* MEMORY off and the second unit on. Before the second unit comes on, the condition needed to read out the number 12 exists, and a very short pulse output equal to the stepping response time of the *step* MEMORY will occur. This problem can be taken care of by making one of the inputs to the decode AND a signal from a NOT fed by the count limit switch. Now the readout will occur after all the *step* MEMORYS have switched. If it is desired to read out the counter on the leading edge of the count signal, a 100 μsec delay may be created as shown in Fig. 15·16. In this circuit the readout is a momentary signal (100 μsec).

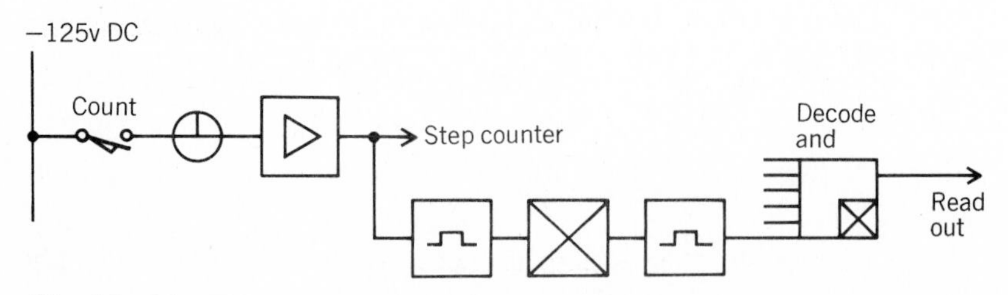

Fig. 15 · 16 Binary counter, readout circuit with delay. (General Electric Company)

15·6 SHIFT REGISTERS

Industrial processes often require that the control system be able to keep track of where some object is as it moves through the process. The system may require the control to identify and follow several different objects at the same time and process each one differently. The control circuit for doing this is referred to as a *shift register*.

Suppose the problem is to identify and follow assorted material on a conveyor. Each type of material is to be taken off at a different point. The conveyor must be equipped with lugs or bins so that each type or piece of material is in a separate bin. If there are 20 bins between the input and where material *A* is supposed to go, a counting process is necessary to keep track of where material *A* is and push it off the conveyor at the correct spot, i.e., after twenty bins have passed the input point.

This process requires a stepping-type shift register. Unless the shift register is coded, a separate channel would be required for each type of material. Coding can multiply the number of types of material which can use a single channel several times.

When the process calls for a random rate of input to conveyors without lugs or bins, the sequence type of shift register is required.

We will consider only a small sample of the many possibilities for shift registers. The application of this type of control is limited only by the demands of the process and the ingenuity of the design engineer.

15·7 GENERAL ELECTRIC COMPANY SHIFT REGISTERS

The student should not be misled by the brevity or simplicity of the material presented here. The General Electric Company control system is capable of the most complex shift-register installations.

A simultaneous-type shift register is shown in Fig. 15·17. The shift-register function is conveniently accomplished with *step* MEMORY units. The decade counter and the ring counter, described earlier, use the shift-register connection. Figure 15·17 shows the start of a three-channel simultaneous shift-register circuit and a readout bank which may or may not be the end of the shift register.

It is advantageous to code the input because of the simplicity of decoding the output. The binary code is suggested because it is simple and easy to learn. With the three channels it is possible to have eight different codes. The eighth is usually the code for all channels off. For this reason only seven push buttons are shown. The outputs of these push buttons are binary coded by connecting the outputs of the original inputs together and applying this to the inputs of the *step* MEMORYS. OR elements could be used to perform this logic.

The input bank uses *off-return* MEMORYS since they operate with direct logic and do not need to be stepped. A *step* MEMORY is not recommended as the unit in the first bank because of the requirement that steering connections be present while stepping. The first shift-register bank then consists of *step* MEMORYS. The *n*th bank is shown with a number 1 and a number 7 readout AND element. The unit reset is connected to all the MEMORY units. The signal amplifier is used to step the shift register and to turn off the *off-return* MEMORYS.

Pressing push button 7 will turn on MEMORYS 1, 2, and 3. With these elements on, the steering networks for *step* MEMORYS 4, 5, and 6 are set to turn them on when the step input is applied. Closing the step limit switch will apply this step input. Now units, 4, 5, and 6 are on and units 1, 2, and 3 are off. After the *n*th step pulse, *step* MEMORYS 7, 8, and 9 will be on. AND 11 will be on, giving us the desired output.

It is possible to store $n + 1$ signals in the shift register at one time. This information can be stepped at an approximate maximum rate of 1,800 steps per minute. The limitation is due to the time delay in the signal amplifier or driver. The maximum rate with a controlled step input is approximately 10 kilocycles per second. A controlled step input is one which originates from a logic, since it is impossible to operate contacts at this rate and to remove the bounce effect.

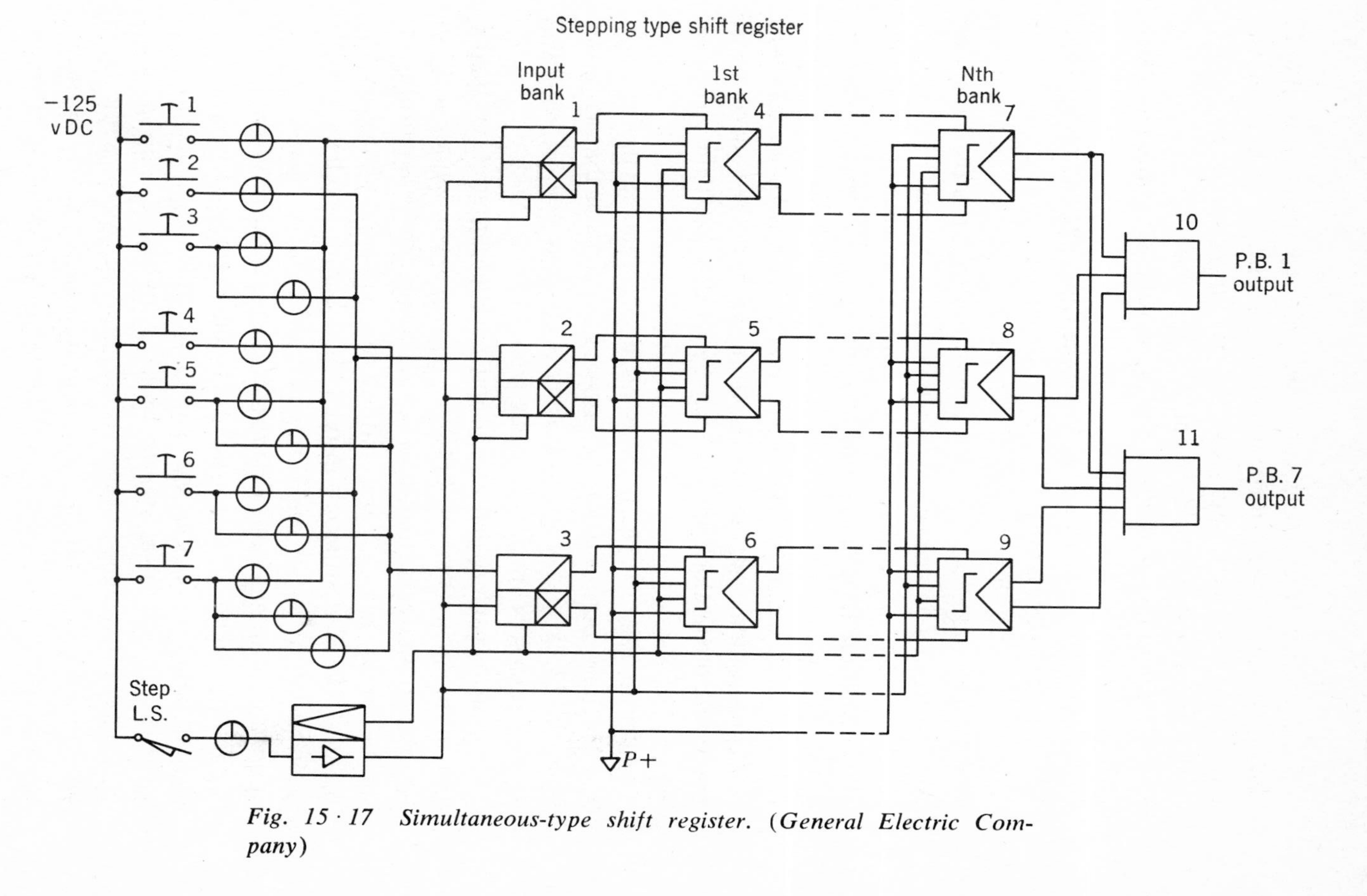

Fig. 15 · 17 Simultaneous-type shift register. (General Electric Company)

Sequence-type Shift Register. Figure 15·18 shows the connections to a two-channel shift register using *step* MEMORYS. Shown are banks 1, 2, and 3, last bank *N,* and second to last bank *N*-1. Intermediate banks would be connected in the same way as bank 3 or *N*-1. The codes of both channels being off cannot be used as a signal code since it is used to tell when a bank has no information, or has been reset after having passed information on to a subsequent bank.

The sequence shift register is designed to store random input information and present this on a first-in first-out basis. This type of shift register is needed for all types of conveyors that do not have lugs to position the material being handled.

Information is shifted from one bank to the next by means of a clock pulse obtained from a sine-to–square wave converter and a pair of single shots. Where faster shifting of information is required than by the signal from the sine-to–square wave converter, other clock means may be used to shift the information.

Refer to Fig. 15·18; there is no information in the register, and all steer connections to all MEMORYS are such that as pulses from the clock and NOT-clock signals are reaching all terminals 2s, the MEMORYS are not changing from their off condition.

In bank 1, each sealed-AND 1 and 2 has an input at terminal 1 and terminal 3. Pressing button 1, 2, or 3 will supply a third input, and the corresponding sealed AND turns on, putting information in bank 1 and putting the corresponding steer-on signal in the *step* MEMORY in bank 2. As limit switch 1 closes, a steer-on signal is put on *bi-step* MEMORY *A*. A clock signal is applied to terminal 2 of *bi-step* MEMORY *A;* it turns on, supplying the final input to AND 6, which turns on to put a steer-on signal to *step* MEMORY 7. On the NOT-clock pulse *step* MEMORY 7 turns on, which pulses *step* MEMORY 3 and/or 4. This puts the information that was in bank 1 into bank 2. At the same time, loss of signal at terminal 7 on *step* MEMORY 7 resets the sealed ANDs of bank 1. At the end of the NOT-clock pulse, *step* MEMORY 7 is reset by loss of pulse signal at terminal 4 (terminal 7 comes

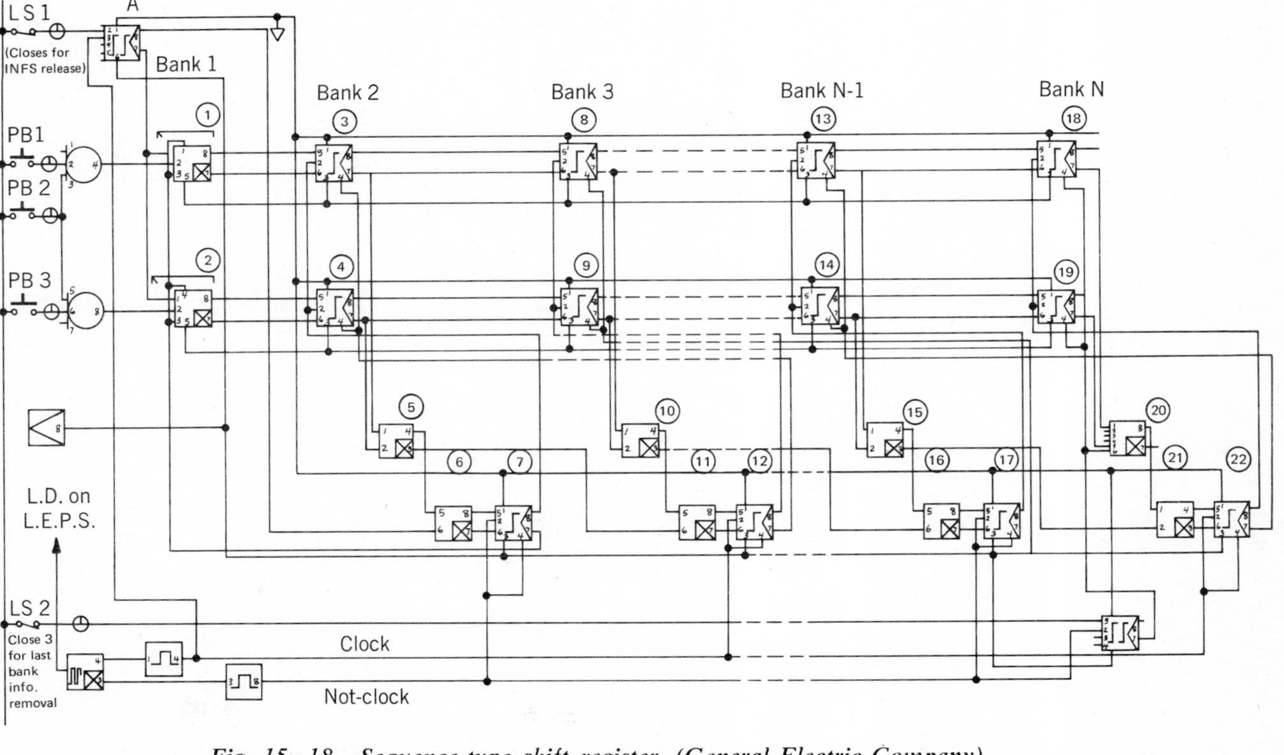

Fig. 15 · 18 Sequence-type shift register. (General Electric Company)

on), and the sealed ANDs 1 and 2 gain back their inputs to terminals 1 and 3 as soon as *LS*1 opens and the first clock pulse turns *bi-step* MEMORY *A* off (gains output on terminal 7).

Information is now in bank 2 AND, and 5 is off, which gives an input to AND 11 and, providing there is no information in bank 3, a second input is provided to AND 11 and the steer-on signal is applied to *step* MEMORY 12. The first clock pulse turns on *step* MEMORY 12, pulsing *step* MEMORYS 8 and 9 so that *step* MEMORY 8 and/or 9 turns on and at the same time *step* MEMORY 12 loses the output at terminal 7, which resets bank 2 MEMORYS.

Similarly, information is transferred automatically down to the last bank. As soon as a bank has passed on information, it is reset and can then accept new information from the previous bank on the next clock or NOT-clock signal, as the case may be.

Information in the last bank is different in that it must receive its reset signal from the function being controlled. With information in the last bank it may be decoded in the same way as information was decoded for the simultaneous-type shift registers. When the information has been used in the last bank, *LS*2 removes the information by closing. This puts a steer-on signal on *bi-step* MEMORY *B*, and the NOT-clock pulse turns MEMORY *B* on, removing the output from terminal 7, and resets bank *N*. Also an input from AND 20 is removed to prevent information from being shifted out of the register. As soon as *LS*2 opens, the first NOT-clock pulse turns *bi-step* MEMORY *B* off, returning the on signal to terminal 4 of *step* MEMORYS 18 and 19 and to AND 20 so that *step* MEMORY 22 has a steer-on signal. The first clock pulse then turns *step* MEMORY 22 on, which steps information into bank *N* and resets bank *N*-1.

15·8 OTHER APPLICATIONS OF STATIC CIRCUITS

The material on static control covered in this book has been limited to digital control using manufactured logic components. There are many custom designed and built static control systems, both digital and analog (feedback), which are finding wide usage in the mill-type industries.

Feedback systems sample the output in the form of voltage or current changes. These changes are then fed back into the input for a self-regulating action. This type of system is not an on-off action, such as logic control, but a modulated continuous output controlled by feedback. One of the chief applications of feedback control is speed control of d-c motors.

Analog systems utilize the solid-state devices—transistors, diodes, and SCRs—in almost every circuit. A good foundation in solid-state electronics is fast becoming a must for anyone engaged in control work, from design to service.

Summary

Counters and shift registers are a common and growing part of control systems. Many control manufacturers have developed basic components and circuits for the most common types of counters and registers.

The wide application of binary counters and other binary equipment indicates that the control man of tomorrow will be required to know binary arithmetic and codes.

The field of static control is changing so fast that challenging new circuits of today are often made obsolete before they are in full use. The serious student of static control will develop a good foundation of fundamental concepts and then maintain a program of continued study in solid-state devices and control developments.

Review Questions

1. How many MEMORYs would be required in a decade counter to count to 1,000?
2. How many MEMORYs would be required in a binary counter to count to 1,000?
3. List the positional values of the first six positions in the binary number system.
4. What is the decimal value of the following binary numbers: (*a*) 1101 (*b*) 10011 (*c*) 111011
5. The shift register which will keep track of particular units of material as they move along a conveyor is known as a ________ shift register.

INDEX